# International Scientific-Technical Conference on Innovative Engineering Technologies, Equipment and Materials (ISTC-IETEM-2019)



IOP Conference Series: Materials Science and Engineering
Volume 915

Kazan, Russia
4-6 December 2019

ISBN: 978-1-7138-4758-8
ISSN: 1757-8981

**Printed from e-media with permission by:**

Curran Associates, Inc.
57 Morehouse Lane
Red Hook, NY 12571

**Some format issues inherent in the e-media version may also appear in this print version.**



Institute of Physics
Dirac House, Temple Back
Bristol BS1 6BE UK

Phone: 44 1 17 929 7481
Fax: 44 1 17 920 0979

techtracking@iop.org

**Additional copies of this publication are available from:**

Curran Associates, Inc.
57 Morehouse Lane
Red Hook, NY 12571 USA
Phone: 845-758-0400
Fax: 845-758-2633
Email: curran@proceedings.com
Web: www.proceedings.com

# TABLE OF CONTENTS


Preface

Peer review declaration

Commissioning of a laboratory setup for studying the effects of ultraviolet and ozone on multifractional, unsorted production and consumption wastes ........ 1
*T M Abdullin, I R Gilmanshin*

Design of a laboratory installation for studying the effects of ultraviolet radiation and ozonation on organic compositions, unsorted municipal and medical waste ........ 5
*T M Abdullin, I R Gilmanshin*

Alternative model of project management at the stages of R&D ........ 11
*A V Antonov, A S Puryaev*

Developing a boss for hybrid electric motors ........ 16
*F F Bagdanurov, L N Shafigullin, F M Akhmetov*

Development of a parametric model for calculating cutting forces for external cylindrical turning of steel 20CrMNTi ........ 19
*I P Balabanov, O N Balabanova, V N Gilman*

Hardening of the working bodies of agricultural machines by electric spark and vibratory arc processing ........ 24
*A T Bayniyazova, M M Abzhaev, S N Sharifullin*

Research of parameters of hydrostatic transmission with axial flow divider for cases of multi-drive vehicle maneuvering ........ 27
*A M Belousov, D L Karelin, A V Boldyrev, S V Boldyrev*

Simulation of heat flows in a combined process of plasma spraying and coat hardening ........ 31
*A M Kadyrmetov, S N Sharifullin, E V Snyatkov, A A Plahotin, I A Mandrykin, V V Romanov*

Increase in efficiency of high-viscosity fluid transportation on the facilities of FEC ........ 36
*A N Dolgova, A V Akhmerov, A L Osipov, Yu V Snigireva, O G Dudarovskaya, E S Karataeva*

Comparative efficiency of pipeline heat-insulating materials ........ 40
*A N Dolgova, S A Yunak, A S Rudenko, A I Haibullina, E S Karataeva*

Study of the influence of hydroabrasive surface cleaning on a laser cutting ........ 44
*Az T Gabdrakhmanov, I H Israphilov, L N Shafigullin, T F Gabdrakhmanova*

Research of properties of polymer compositions based on aluminosilicate microspheres ........ 48
*Az T Gabdrakhmanov, AT Galiakbarov, I Y Karasev, T F Gabdrakhmanova*

Application of individual educational trajectories and adaptive content in engineering education ........ 52
*R D Galeev, I R Gilmanshin, N F Kashapov, S I Gilmanshina*

Experience in the application of nitriding in a glow discharge to increase the efficiency of gear processing by the method of grinding ........ 59
*V N Gilman, A I Fashutdinov, I P Balabanov*

Methodological features of using elements of distance learning in the training of competent specialists ........ 63
*I R Gilmanshin, S S Kosmodemyanskaya, A I Galeeva, K Saduakaskyzy*

Pedagogical conditions for training science teachers for sustainable development of society ........ 67
*S I Gilmanshina, R N Sagitova, I R Gilmanshin, A I Galeeva, I A Galeev*

Digital educational environment: training future teachers to design additional education ........ 72
*S I Gilmanshina, V A Minnakhmetova, I R Gilmanshin, A I Galeeva, G M Abyzbekova*

The digitalization of training of bachelors - the future science teachers ........ 76
*S I Gilmanshina, I R Gilmanshin, A I Galeeva, G M Abyzbekova*

How the radius of the screw insert core influences the flow behavior of the polymer solution ........ 81
*A I Kadyirov, J V Karaeva, V O Zdor, E M Khusnutdinova*

Formation of by-products during water chlorination as a result of corrosion of elements of the mechanical system. ........ 85
*O R Karataev, V F Novikov, R I Sayakhov, E S Karataeva*

Influence of various types of fertilizers and equipment on the physical and microbiological properties of soil ........ 89
*N F Kashapov, M M Nafikov, A R Nigmatzyanov, V V Medvedev, V A Chernov, M D Nigmatullin*

Features of robotization of agricultural processes ........ 94
*N F Kashapov, M M Nafikov, A R Nigmatzyanov, R F Sayfutdinov, R A Mingazov*

Using digital agricultural production based on navigation and information systems ........ 98
*N F Kashapov, M M Nafikov, M M Nafikov, A R Nigmatzyanov, O I Makarova, V V Medvedev, V A Chernov, M D Nigmatullin*

Modeling the processes of forming the organizational structure of management in itegrated formations ........ 101
*N F Kashapov, M M Nafikov, M X Gazetdinov, Sh M Gazetdinov, A R Nigmatzyanov*

Technological preparation of production for product quality management according to IATF 16949:16 requirements ........ 106
*S V Kasyanov*

The increasing of the electric vehicle maneuverability, excluding disconnecting of the survey from the contact network ........ 111
*A A Khafizov, R S Akhmetshin, R I Valiev*

Improving the quality of surfaces of products obtained by electroerosion treatment ........ 114
*I I Khafizov, I G Nurullin*

Application of robotic systems in production: new opportunities for development and risks ........ 119
*I I Khafizov, I G Nurullin, N F Kashapov*

Application of electric spark doping technology for strengthening the surface of the Part ........ 123
*I I Khafizov, I G Nurullin*

Engineering experiment as an underlying tool for training research engineers ........ 128
*G R Khamidullina, I A Konakhina, N F Kashapov, I I Khafizov, E M Khusnutdinova*

Estimation of precision parameters of milling machines ........ 131
*R M Khusainov, E R Ziangirova, G K Davletshina*

Comprehensive Test Procedure for Digital Instruments and Devices of Automated Versatile Systems ........ 135
*E M Khusnutdinova, P P Pavlov, V P Fandeyev, R N Khizbullin, A N Khusnutdinov, I V Cherepen'kin*

Designing the Fault-Detection and Troubleshooting Tests For the Troubleshooting Target Flowchart ........ 139
*E M Khusnutdinova, O A Filina, A N Khusnutdinov, A V Yashagina, G V Osetinskiy, N K Andreev*

Influence of nonlinear viscosity of waxy crude oils on the thermohydraulic characterization of their flow in helical channels: theory and experiment ........ 144
*E M Khusnutdinova, I A Konakhina*

Improving the performance of cylinder liners by creating thin-layer coatings in the interface ........ 147
*A T Kulakov, E P Baryshnikova, S V Snarsky, A A Gafiatullin, F L Nazarov, A F Malachowski*

Innovations in the management system of industrial enterprise to improve the efficiency of its activities ........ 151
*I A Kuzmina, I I Khafizov*

The use of mathematical models to improve the effective functioning of enterprises ........ 154
*I A Kuzmina*

Development of apparatuses based on kinetic calculation data of the process of removing liquid from solid materials by reducing pressure of vapor-gas medium ........ 157
*V A Lashkov, S G Kondrasheva, D A Khamidullina, E S Karataeva*

Quality management when operating the power equipment ........ 165
*A O Mamonova, E M Khusnutdinova, A N Khusnutdinov*

The current quality management system at Magnitogorsk iron and steel works ........ 168
*M N Moliboga, E M Khusnutdinova*

Development of cast dispersion-hardening ferrite-carbide steel ........ 171
*G F Mukhametzyanova, M S Kolesnikov, V I Astashchenko, I R Mukha-metzyanov*

Development of hardening technology for working surfaces of microsurgical instruments made of titanium alloy ........ 175
*F I Murataev, E R Galimov*

To the question of automation of the injection molding process ........ 186
*D L Pankratov, R V Gavariev*

About computer calculation error of vermicular graphite share in microstructure of CGI ........ 189
*A G Panov, I F Shaekhova, S A Ryzhkova, N G Degtyaryova, V A Ivanova*

Evaluation of the wear resistance and resource hardened by electrospark machining of the cutting surfaces of the fingers of the headers of combine harvesters ........ 192
*S A Velichko, S N Sharifullin, I S Kuznetsov, A V Kolomeichenko, R Y Solovyev, I N Kravchenko, N R Adigamov*

Energy saving in mechanical engineering through energy management ........ 196
*O A Plugatyreva, E M Khusnutdinova, A N Khusnutdinov*

Methods of Optimizing the Troubleshooting Parameters of Electric Power Facilities.................................... 199
*L S Sabitov, P P Pavlov, V P Fandeyev, V M Butakov, A N Khusnutdinov, A S Siyetinskaya*

Reliability study for traction electrical equipment of urban electric transport ................................................ 203
*L S Sabitov, A E Aukhadeyev, R G Idiyatullin, R S Litvinenko, L N Kisneeva, A N Khusnutdinov, O V Ilyushin*

Medical and Biological Problems in the Control of Therapeutic Exposure Parameters in Sports Medicine and Ways of Their Solution in New Medical Instruments and Systems.......................................... 208
*L S Sabitov, Rob N Khizbullin, Rad N Khizbullin, P P Pavlov, A N Khusnutdinov, A R Abdullina*

Digitalization of education: transformation of a natural science experiment .................................................. 211
*R N Sagitova, S I Gilmanshina, I R Gilmanshin, A I Galeeva, G M Abyzbekova, K Saduakaskyzy*

Recycling components made of polyurethane foam........................................................................................ 215
*L N Shafigullin, N V Romanova, A T Gabrakhmanov, O P Korchagin, I R Shamsutdinov, D A Agashkin*

Recommendations for making sound absorbing polyurethane materials ........................................................ 219
*L N Shafigullin, N V Romanova, S V Zonina, I E Valinova, I Yu Karasev*

Rationality and possibility of automation of precision calibration of gears .................................................... 223
*V G Shibakov, D L Pankratov, R S Nizamov, R A Hairullin*

Frame model of knowledge representation in the expert system of technological route selection.................. 231
*L A Simonova, G M Kapitonova*

Physical basis of thermal control of the operating mode of the refrigeration unit........................................... 235
*I F Timerbaev, I R Gilmanshin*

Industrial control system software patterns in manufacturing ......................................................................... 241
*R A Valiev, Sh Sh Khuzyatov*

Quantum chemical simulation for calculation of processes of sulfur composite materials ............................. 246
*A A Yusupova, A Y Akhmetova, L R Baraeva, R T Akhmetova, A I Khatsrinov, A A Bobyshev*

High-technology energy-efficient composite material made from waste of thermal power plants .................. 249
*A A Yusupova, G A Medvedeva, R O Akhmetova, A A Bobrishev*

The laser heat-treatment process control by the non-stationary convective heat exchange modeling.............. 253
*V V Zvezdin, R R Saubanov, R R Rakhimov, R M Khisamutdinov, A V Boldyrev, D L Karelin*

**Author Index**

IX International Scientific Conference "Innovative engineering technologies, equipment and materials - 2019" has been hosted by the exhibition center "Kazan Fair" within the framework of the 19th international specialized exhibition "Mechanical engineering. Metalworking. Kazan" and 13th specialized exhibition "TechnoWelding" on the 4th and 6th of December, 2019

The conference program was including 6 sections of work:

Section 1 - Innovative materials, technologies and equipment in mechanical engineering

Section 2 - Innovative developments and economics in engineering

Section 3 - Chemical Engineering

Section 4 - Mathematical and physical modeling of information, technical, technological and management systems and processes

Section 5 - Innovative Welding Technologies in Industry

Section 6 - Reliability, efficiency and safety of objects of the fuel and energy complex

Participants and guests of the conference discussed the scientific and technical achievements in the field of development, design and manufacture of engineering products. See the results of studies of the country's leading universities in the field of robotics, new materials (including on the basis of production waste), waste management and recycling, mechanical engineering, energy, fuel cells, manufacturing economy.

Presidium of the secsions headed by Director of the Institute of Engineering Prof. Nail Faikovich Kashapov and Director Engineering center «Energoprogress» Almaz Lyabisovich Garaev. The Presidium also included: Deputy. Director of the Institute for educational activities Ildar Hafizov Ilsurovich, Deputy Director of the Institute for Science Gilmanshin Iskander Rafailevich and director's advisor of Engineering center «Energoprogress» Linar Salihzanovich Sabitov.

Research and Innovation have been presented by the staff of the Engineering Institute of KFU, Institute of Naberezhnye Chelny, Institute of Chemistry named after A.M. Butlerov, Kazan State Power Engineering University and Kazan National Research Technical University named after A. N. Tupolev – KAI, Kazan State University of Architecture and Engineering. Representatives of Uzbekistan, Vietnam, Kazakhstan, Russian cities: Moscow, St. Petersburg, Voronezh, Yekaterinburg, Kazan, Nizhny Novgorod, Naberezhnye Chelny, Cheboksary, Lipetsk, Khimki, Tambov, etc., became participants of the IMTOM-2019. breakout sessions and round tables were made more than 200 reports, the total number of participants was 372 people.

Active participation in the conference were graduate students and students of the Engineering Institute. The experience of live communication with

IOP Conf. Series: Materials Science and Engineering **915** (2020) 011001 doi:10.1088/1757-899X/915/1/011001

scientists, practitioners, representatives of the real economy will allow buduyuschim experts to determine its scientific direction for a successful solution of important problems of import substitution and enhancing competitiveness assigned to the engineering industry of our country.

IOP Conf. Series: Materials Science and Engineering **915** (2020) 011002 doi:10.1088/1757-899X/915/1/011002

# Peer review declaration

All papers published in this volume of *IOP Conference Series: Materials Science and Engineering* have been peer reviewed through processes administered by the Editors. Reviews were conducted by expert referees to the professional and scientific standards expected of a proceedings journal published by IOP Publishing.

- **Type of peer review:** Single-blind

- **Conference submission management system:**

  for consideration are the articles discussed at the IX International Scientific Conference "Innovative engineering technologies, equipment and materials - 2019"

- **Number of submissions received:** 71

- **Number of submissions sent for review:** 71

- **Number of submissions accepted:** 65

- **Acceptance Rate (Number of Submissions Accepted / Number of Submissions Received X 100):** 91.5

- **Average number of reviews per paper:** 1

- **Total number of reviewers involved:** 4

- **Any additional info on review process:** ---

- **Contact person for queries:** Iskander Gilmanshin (is-er@yandex.ru)

IOP Conf. Series: Materials Science and Engineering **915** (2020) 012001 doi:10.1088/1757-899X/915/1/012001

# Commissioning of a laboratory setup for studying the effects of ultraviolet and ozone on multifractional, unsorted production and consumption wastes

**T M Abdullin[1], I R Gilmanshin[1,2]**

[1]Kazan Federal University, Kazan, Russia, 420008
[2] Kazan National Research Technical University named after A N Tupolev, Kazan, Russia

is-er@yandex.ru

**Abstract.** The article describes the process of commissioning carried out during the testing of the designed and manufactured laboratory installation for studying the effects of ultraviolet radiation and ozone on unsorted fractions of production and consumption waste, including medical waste. A detailed description of the installation is made, the characteristics of the elements of a power electric drive, metrological strapping and automation are given.

The study in the field of increasing the efficiency of thermal methods for the destruction of production and consumption wastes became especially relevant due to the sharp increase in the volume of municipal solid waste in large cities and megacities. The fractional composition of waste in the Russian Federation is shifting in favor of polymers and composites based on them. The recently actively developing sector of the separate collection of secondary use is still not able to provide 100% recycling. The main prerequisites are organic inclusions and low profitability of processing contaminated and composite fractions.

Installation developed by the joint scientific team of the Engineering Institute and the Chemical Institute. A.M. Butlerov Kazan University will allow to conduct targeted research on the interaction of ultraviolet radiation and ozone with model compositions of municipal and medical waste.

The key elements of the setup are a pair of ultraviolet lamps with differentiated emission peaks: a UV lamp with a peak wavelength of 185 nm and a UV lamp with a peak wavelength of 254 nm.

The choice of ranges was due to the need to provide a combined effect on the organic and polymer fractions of ozone wastes and the bactericidal effects of ultraviolet radiation. [15, 16]

Sources of ultraviolet radiation are traditionally used for the disinfection of various environments and the generation of ozone. The highest, 35-45%, efficiency of converting electric energy to ultraviolet radiation has low-pressure mercury lamps. The specific power of ultraviolet radiation ranges from 0.2 W / cm for mercury to 4 W / cm for amalgam lamps.

The radiation spectrum of a low-pressure mercury discharge is characterized by two resonant UV lines: 254 and 185 nm. Radiation with a wavelength of 254 nm has a pronounced bactericidal effect, while radiation with a wavelength of 185 nm serves as an active initiator of the formation of ozone. Radiation with a wavelength of 185 nm is actively absorbed by molecular oxygen and water vapor, followed by the generation of ozone and radicals ($OH^{\cdot}$).

IOP Conf. Series: Materials Science and Engineering **915** (2020) 012001 doi:10.1088/1757-899X/915/1/012001

In practice, a combination of 2 sources of ultraviolet radiation with wavelengths of 254 and 185 nm is used for the photochemical purification of various media from harmful substances and odors due to the generation of radicals and photochemical decomposition reactions. The paired effect is enhanced by the absorption of radiation at 254 nm by ozone with the formation of a highly active oxygen radical.

Figure 1 shows installation sketches with detailing of structural elements.

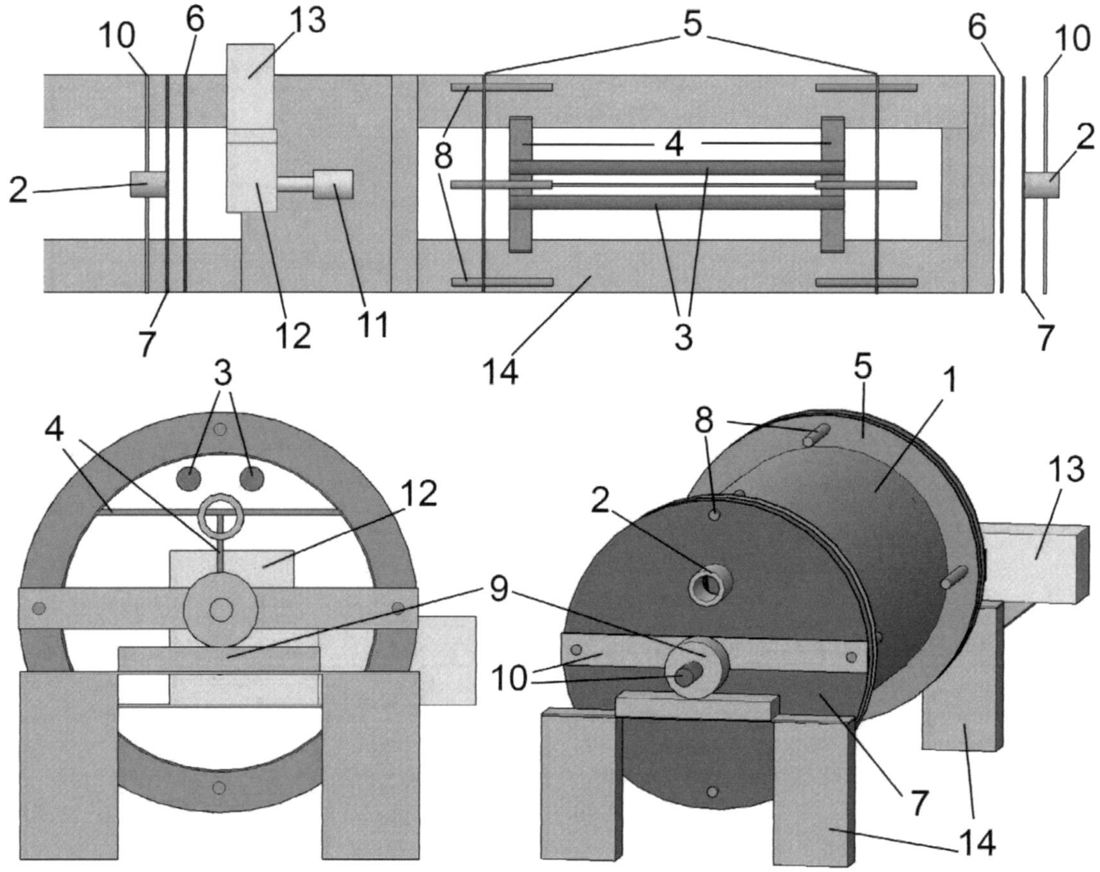


Fig. 1 of a laboratory setup for studying the effects of ultraviolet and ozone on multifractional, unsorted production and consumption waste.

1. Stainless steel working chamber, 2. Inlet and outlet gas valve, 3. Sources - UV radiation, 4. Flanged activator valves, 5. Flanges, 6. Fluoroplastic gaskets, 7. End caps, 8. Hardware, 9. Bearing nodes, 10. Drive shaft, 11. Protective coupling, 12. Lowering gear, 13. Power electric drive, 14. Installation bed.

In the process of experimental work, the initiation of ozone formation is provided by UV - Lama 1 (with a peak wavelength of 185 nm), UV - lamp 2 (with a peak wavelength of 254 nm) provides decontamination and reduction of the hazard class of the loaded volume of waste. Uniform irradiation and intensification of the chemical interaction provides a cyclic movement of the housing around its axis with an amplitude of 120 ° and a flag activator on a pendulum suspension.

Metrological piping of the installation provides monitoring of ozone concentration, temperature and humidity in the chamber.

Monitoring of ozone concentration is carried out by a pair of sensors of the MQ131 series (low concentration and high concentration) Fig. 2. The MQ131 is a semiconductor gas sensor consisting of a heater circuit and a sensor circuit. The MQ131 sensor requires a minimum preheat time of 48 hours before yielding consistent results (also called "burnout" times). At the same time, there is a calibration, it regulates the value of the base resistance and the time required to heat the sensor and obtain consistent readings (reading time). It is controlled by Arduino: heater, calibration, environmental control (temperature and humidity) to obtain more accurate data. Conclusion of values in ppm, ppb, mg / m$^3$, µg / m$^3$.

Humidity and temperature are monitored using HTU21D sensors. The calculations are made by the programmable controller Arduino Uno R3.

During commissioning, the operability and safety of all components of the structure were monitored, the operation of automatic control of the power electric drive and the operator's automated workplace was evaluated.

During the experiment, the operability of the developed design of the laboratory setup was confirmed. The performance of the metrological harness of the installation of an automatic power drive automatic control system was confirmed.

A low concentration sensor (connected to the workstation for centralized monitoring and process control) monitors the safety of structures during the experiment (tightness control of welded and detachable joints is carried out).

A high concentration sensor (connected to the automated workstation for centralized monitoring and process control) monitors the concentration of ozone in the working chamber of the installation during the experiment (values are displayed in ppm, ppb, mg / m3, µg / m3, depending on the settings of the calculator).

A combined temperature and humidity sensor (connected to the workstation for centralized monitoring and process control) monitors the state of the gas environment in the installation chamber.

The next step is a series of experiments to study the effects of ultraviolet and ozone on unsorted waste:

- The quantitative and qualitative composition of the exhaust gases after the primary waste treatment (by means of gas mass spectrometry);
- Change in sample weight before and after in primary waste treatment;
- Dependence of the elemental composition of the outgoing gas on the fractional and elemental composition of the waste.

**References**

[1] Shashkov I.V. Utilization and processing of municipal solid waste. Publishing house of FSBEI HPE "TSTU", 2015.

[2] Abdullin, T.M. Educational laboratorial facility for researching the technology of plasma gasification of production and consumption wastes including medical wastes / T.M. Abdullin, I.R. Gilmanshin, N.F. Kashapov, S.I. Gilmanshina, A.I. Galeeva, D.R. Krainova // IOP Conference Series: Materials Science and Engineering Volume 412, Issue 1, 23 October 2018, #012001

[3] Federal Law of December 30, 2009 No. 384-ФЗ "Technical Regulations on the Safety of Buildings and Structures"

[4] Ershov A. G., Shubnikov V. L. Thermal disposal of waste: theory and practice, myths and legends. MSW Journal. 2014. No. 5. P. 47 - 52.

[5] Federal Law of January 10, 2002 No. 7-FZ "On Environmental Protection"

[6] Bilitiewski B. Waste incineration: German experience. Municipal solid waste. 2007. No. 1. From 47 - 49.

[7] Federal Law of March 30, 1999 No. 52-FZ "On the Sanitary and Epidemiological Well-Being of the Population"

[8] Federal Law of November 23, 2009 No. 261-ФЗ "On Energy Saving and on Improving Energy Efficiency and on Amending Certain Legislative Acts of the Russian Federation"

[9] Gilmanshin I.R. Energy utilization of landfill gas as a way to form a new model of waste management / I.R. Gilmanshin, N.F. Kashapov, S.I. Gilmanshina // Problems and prospects of innovative development of the economy: Materials of the scientific forum (XXI international scientific and practical conference), Alushta, September 19-24, 2016 Simferopol: Scientific and Technical Union of Crimea; Moscow: EkOOnis Publishing House - 2016. - 264 p.
[10] Kozhinov V.F., Kozhinov I.V. Ozonation of water. M., Stroyizdat, 1974.
[11] Abdullin T.M. A technique for researching the technology of plasma gasification of medical waste and production and consumption waste / Gilmanshin I. R., Kashapov N. F., Gilmanshina S. I., Krainova D. R. // Materials of the IX International Scientific and Technical Conference "Innovative Engineering Technologies, equipment and materials - 2018 "(ISTC" IMTOM-2018 "). Part 1. - Kazan, 2018 .-- 237-241 p.
[12] Bernadiner M. N., Bernadiner I. M. High-temperature processing and disposal of liquid, paste-like and solid industrial and medical wastes. Ecology and industry of Russia. 2011. April. From 19 - 21.
[13] Gilmanshin I.R. Study of projects for the rehabilitation of landfills for production and consumption waste in terms of the organization of energy utilization of landfill gas / I.R. Gilmanshin, N.F. Kashapov, S.I. Gilmanshina, A.I. Galeeva, L.S. Sabitov // Materials of the VIII International Scientific and Technical Conference "Innovative Engineering Technologies, Equipment and Materials - 2017" (ISTC "IMTOM-2017"). Part 1. - Kazan: Foliant Publishing House - 2017. - P. 220-224.
[14] Abdullin T.M., Gilmanshin I.R., Kashapov N.F., Sabirzyanov R.G., Gilmanshina S.I., Galeeva A.I. and Gadirova E.M. 2019 About waste disposal problem in Russian Federation. IOP Conf. Series: Materials Science and Engineering570 (2019) 012001
[15] Vasilyak L.M. The use of flash lamps for disinfection // Electronic processing of materials. - 2009. - No. 1. - P. 30-40
[16] Voronov A. New generation of low pressure mercury lamps for producing ozone // Proceedings of World congress on ozone and UV technologies. - LA, California, USA. - 2007 .-- P. 2166 -2172.

IOP Conf. Series: Materials Science and Engineering **915** (2020) 012002 doi:10.1088/1757-899X/915/1/012002

# Design of a laboratory installation for studying the effects of ultraviolet radiation and ozonation on organic compositions, unsorted municipal and medical waste

**T M Abdullin[1], I R Gilmanshin[1,2]**

[1]Kazan Federal University, Kazan, Russia, 420008
[2] Kazan National Research Technical University named after A N Tupolev, Kazan, Russia

is-er@yandex.ru

**Abstract.** The article presents the design process of a laboratory setup for studying the ultraviolet effect and ozonation on multifractional unsorted production and consumption wastes, including medical ones. The tasks, structural and functional diagram of the installation are formulated. The requirements for the constituent elements are defined. The process of designing and selecting components is described..

The existing system of solid municipal waste management in the Russian Federation is not capable of providing access to a model with conditionally zero burial by 2025. However, it is necessary to note a significantly increased attention from the federal and regional authorities to this problem and systemic positive changes. At the same time, as of today, the collection and disposal of MSW without separation still accounts for about 80% and leads to the withdrawal of land from sustainable circulation (over 50 thousand hectares are occupied by landfills).

The key guidelines for sustainable development of the waste management industry should be identified:

- bringing the total share of recycled and purposefully processed waste to 100%;
- elimination of the negative impact of waste and its products on the ecosystem through 100% replacement of technologies and materials that do not support natural circulation and biodegradation.

The current state of the problem clearly contrasts with the proposed guidelines for the sustainable development of the industry, which determines the relevance of the technologies for thermal destruction (burning) of unsorted fractions of municipal waste, as technologies for the transition process. This approach allows reducing the volume of waste disposal by an order of magnitude and eliminates the possibility of biological pollution of ecosystems. At the same time, protection of the atmosphere from pollution by waste products requires increased attention. The modern morphology of wastes and the imperfection of the widespread technological solutions of thermal degradation predetermine the need for research aimed at solving the following problems:

- search for solutions to improve energy efficiency and environmental safety of technological solutions for thermal destruction of waste;
- development of technologies for non-thermal methods of intensifying the destruction of composite materials and polymers;

IOP Conf. Series: Materials Science and Engineering **915** (2020) 012002 doi:10.1088/1757-899X/915/1/012002

- Search for effective solutions for the decontamination and reduction of the hazard class of transported waste (especially medical waste).

To increase the efficiency of the process, reduce the mass of unsorted waste and the cost of disposal, it is proposed to introduce the stage of preliminary processing and preparation for subsequent disposal by thermal methods. The integration of the pre-treatment stage makes it possible to reduce waste volumes, reduce the hazard class and increase the energy efficiency of subsequent thermal waste destruction.

During the analysis of literary sources, a choice was made in favor of ultraviolet radiation and ozonation technologies as the most promising in solving the tasks.

Ultraviolet exposure to the far spectrum penetrates the cell walls and destroys the nucleic acids of DNA and RNA. Ultraviolet disinfection consists in the absorption of radiation by nucleic acids, as a result of which cells lose their ability to divide. As a result, the selected wavelength spectrum of ultraviolet radiation in the range of 160 - 260 nm.

The ozonation process is triggered by far UV radiation in the selected range due to the use of UV lamps with quartz glass. Ozonation contributes to the destruction of hazardous chemicals, tobacco smoke, organic substances, detergents and cleaning products, products of combustion and mold, fungi and bacteria. All chemicals that are in the air, reacting with ozone, decompose into harmless compounds: carbon dioxide, water, oxygen and saturated oxides.

The technology itself plays a key role in the environmentally friendly and safe disposal of waste. In order to make it possible to study the mechanism of the effect of ultraviolet radiation and ozonation on unsorted waste fractions, we formulated the basic requirements for the designed laboratory installation:

1. Sealed enclosure with a volume of 14 liters.
2. An interned pair of UV sources of different spectra.
3. Monitoring ozone concentration in the chamber.
4. Inlet and outlet gas channel with emergency pressure relief system.
5. Temperature monitoring in the range of 25-55 ° C.
6. Automation of supervisory control of process parameters in the chamber.

In accordance with the specified requirements, a structural-functional scheme has been developed:

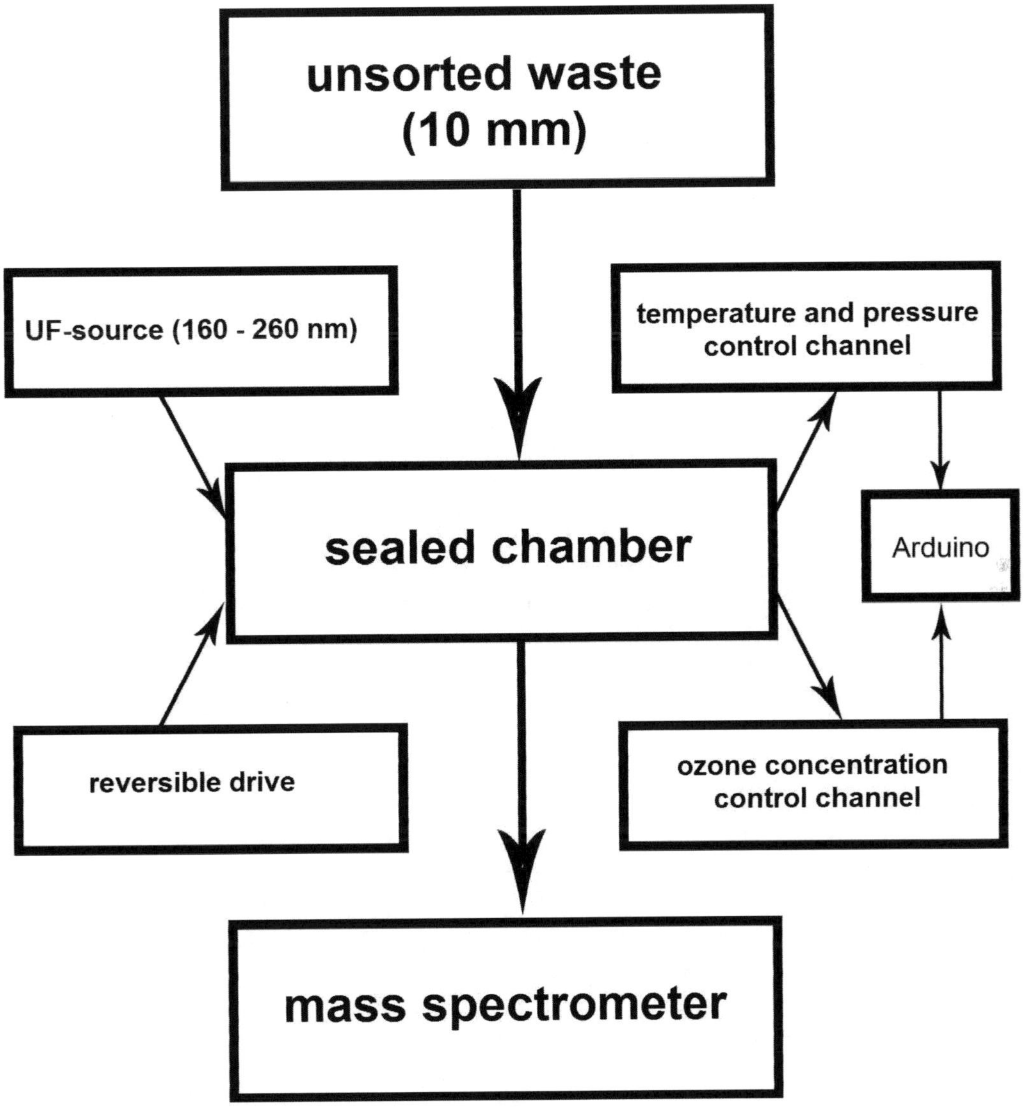


Fig. 1 Structural and functional diagram of the installation

In the study of the interaction of ultraviolet radiation and ozone with waste, it is necessary to ensure the maximum interaction area, which is impossible with the samples stationary. Thus, the installation functionality should be supplemented by the possibility of mixing samples. The solution to this problem is provided by the following technical solution: automated reversible drive, forced waste mixing system.

The efficiency of the mixer is determined by the geometric characteristics of the fractional composition of the waste. Based on an analysis of the experience of the Dresden Technical University of Germany in terms of studying the physicochemical properties of waste, we proposed a fraction dimension of 10 mm.

Preliminary study of the installation model was carried out in the Autodesk AutoCAD software environment, which allowed us to optimize a number of design decisions at the design stage and accelerate the creation of an experimental sample.

IOP Conf. Series: Materials Science and Engineering **915** (2020) 012002 doi:10.1088/1757-899X/915/1/012002

The danger of hard ultraviolet radiation to humans, along with the aggressiveness and toxicity of ozone, have determined high demands on the materials of the installation casing. During the analysis of the properties of structural materials, we selected as the main ones: stainless steel and fluoroplastic.

To achieve the tightness of the structure, fluoroplastic gaskets are used, since fluoroplast has unique technical characteristics:

- Resistance to chemical attack.
- Low coefficient of friction.
- Resistance to adhesion to other surfaces.
- Heat resistance - the flexibility and elasticity of the material are maintained at temperatures in the range from -70 ° to + 270 ° C.
- Minimum surface tension.
- Ftoroplast perfectly lends itself to mechanical processing (drilling, grinding, milling).

Stainless steel is inert to these substances and does not corrode upon contact with acetone, toluene, alcohols, ammonia, flammable solvents, nitrocellulose varnishes, perchlorethylene, trichlorethylene, potassium and sodium alkali, phosphoric acid, sulfuric acid, etc.

For uniform ultraviolet irradiation, the housing makes a cyclic rotational movement around the longitudinal axis with an amplitude of 120 degrees by means of an electric drive paired with a reduction gear. The drive is controlled using standard Arduino solutions. Power supply is organized through a standard 5A driver. The intensification of the ozonation process is provided by a flag type initiator freely suspended in the chamber.

The tightness of the housing is ensured by fluoroplastic gaskets. Tubular type UV lamps are fixed with stainless steel clamps.

The installation is mounted on a welded steel frame 250 × 850 × 370 mm from a rectangular profile 1 mm thick.

IOP Conf. Series: Materials Science and Engineering **915** (2020) 012002 doi:10.1088/1757-899X/915/1/012002

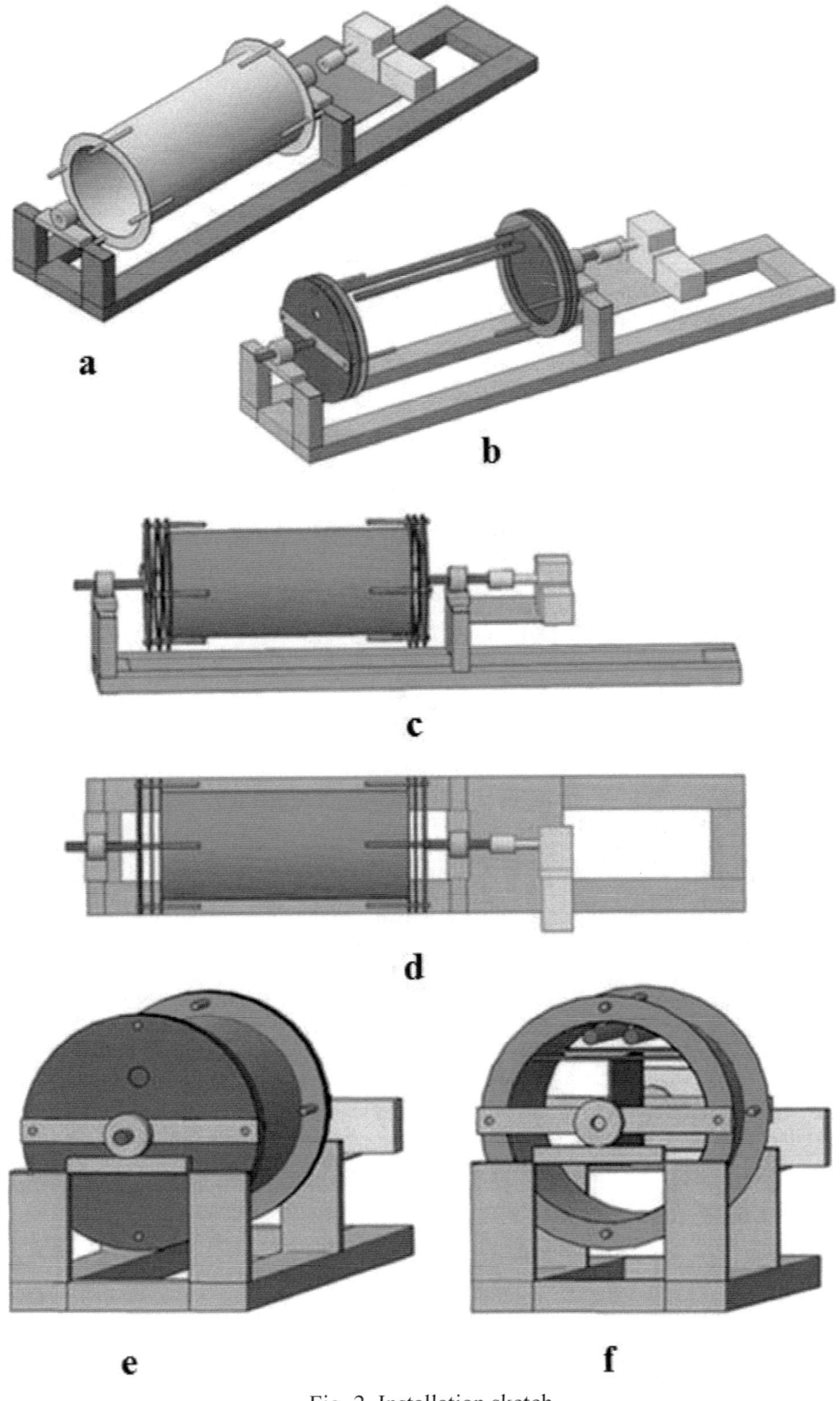


Fig. 2. Installation sketch

The developed and designed installation will allow obtaining experimental data on the morphological and elemental composition of the exhaust gases, optimal from the point of view of the development of auto-destruction, the morphological and elemental composition of the samples of unsorted waste, as well as the change in the mass of the samples after processing.

The next step is to conduct experiments to study the effects of ultraviolet and ozone on unsorted waste:

- The amount of harmful chemicals in the exhaust gas after the primary treatment of waste
- The amount of harmful chemicals in the mass of samples before - after the primary treatment of waste
- Change in sample weight before - after in primary waste treatment
- Changes in the elemental composition of the exhaust gas after the addition of various fractions after the primary treatment of waste.

**References**

[1] Shashkov I.V. Utilization and processing of municipal solid waste. Publishing house of FSBEI HPE "TSTU", 2015.

[2] Abdullin, T.M. Educational laboratorial facility for researching the technology of plasma gasification of production and consumption wastes including medical wastes / T.M. Abdullin, I.R. Gilmanshin, N.F. Kashapov, S.I. Gilmanshina, A.I. Galeeva, D.R. Krainova // IOP Conference Series: Materials Science and Engineering Volume 412, Issue 1, 23 October 2018, #012001

[3] Federal Law of January 10, 2002 No. 7-FZ "On Environmental Protection"

[4] Bilitiewski B. Waste incineration: German experience. Municipal solid waste. 2007. No. 1. From 47 - 49.

[5] Federal Law of March 30, 1999 No. 52-FZ "On the Sanitary and Epidemiological Well-Being of the Population"

[6] Kozhinov V.F., Kozhinov I.V. Ozonation of water. M., Stroyizdat, 1974.

[7] Abdullin T.M. A technique for researching the technology of plasma gasification of medical waste and production and consumption waste / Gilmanshin I. R., Kashapov N. F., Gilmanshina S. I., Krainova D. R. // Materials of the IX International Scientific and Technical Conference "Innovative Engineering Technologies, equipment and materials - 2018 "(ISTC" IMTOM-2018 "). Part 1. - Kazan, 2018 .-- 237-241 p.

[8] Bernadiner M. N., Bernadiner I. M. High-temperature processing and disposal of liquid, paste-like and solid industrial and medical wastes. Ecology and industry of Russia. 2011. April. From 19 - 21.

[9] Abdullin T.M., Gilmanshin I.R., Kashapov N.F., Sabirzyanov R.G., Gilmanshina S.I., Galeeva A.I. and Gadirova E.M. 2019 About waste disposal problem in Russian Federation. IOP Conf. Series: Materials Science and Engineering570 (2019) 012001

IOP Conf. Series: Materials Science and Engineering **915** (2020) 012003 doi:10.1088/1757-899X/915/1/012003

# Alternative model of project management at the stages of R&D

**A V Antonov[1], A S Puryaev[1]**

Kazan Federal University, [1]Naberezhnye Chelny Institute, 18 Kremlyovskaya street, Kazan, 420008, Russia Federation

aidarp@mail.ru

**Abstract.** The purpose of the work is to cover the practical application of innovative project management mechanisms using the principles of Target costing in the field of engineering. The approaches and requirements for the application of tools in practice are identified. Mechanisms for applying approaches based on the Target Costing concept in the development and updating of the technical and economic model are proposed.

**1. Introduction** Nowadays the development of an economy based on innovation requires organizations to own the entire spectrum of management tools, taking into account an understanding of their role and degree of influence on the economic result. Of course, project innovation has always been a high-risk area of investment, considering the multifactorial nature of forecasting and a high degree of uncertainty about the attainability of the set targets. We previously studied various methods and approaches in comparison analyzing project management and evaluating their effectiveness [1-4].

The most objective assessment of a new product before making a decision on investing should be given by a technical and economic model that justifies the economic feasibility of development and market competitiveness. But often, an excellent technical and economic model is destroyed already at the stage of an idea and even a concept in the process of forming design stages for reasons of technical unrealizability at the appropriate time with a planned volume of investments and a deviation in cost, after the start of production. Economic analysis when designing a new product gives positive and negative feedback regarding the presence of deviations from the target cost levels set at the concept stage. Thus, controllability and completeness of data is created for making an informed decision, both for a specific change and for the product as a whole.

## 2. Aim of Research

Developing and proposing an alternative model for project management at the stages of R&D.

## 3. Research Results and Discussion

To implement an assessment of the impact of design and technological changes during product development, the authors propose an algorithm for updating the initial data on the feasibility study, stated at the concept stage. Elements of the algorithm are applied [5].

This algorithm (Figure 1) generally reflects the recommendations of the methodology based on the principles of Target Costing at the top level in terms of updating the financial model of the product, taking into account the factors of the investment component during the development and implementation of a new product.

IOP Conf. Series: Materials Science and Engineering **915** (2020) 012003 doi:10.1088/1757-899X/915/1/012003

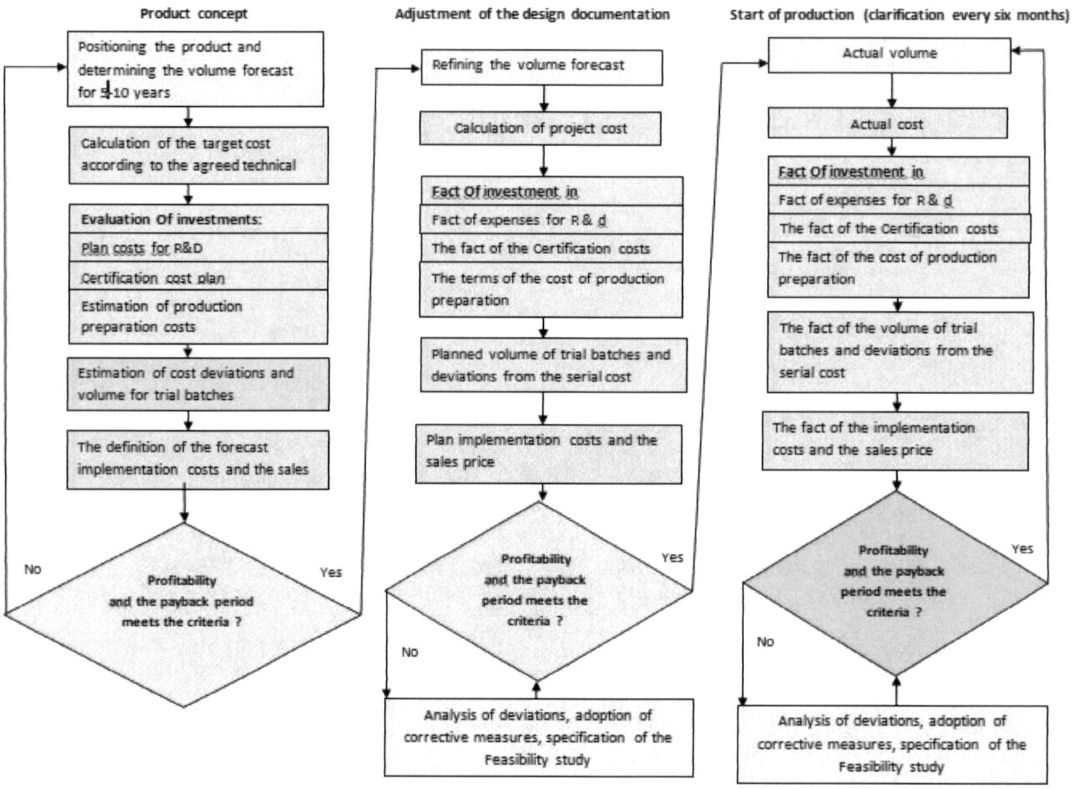


*Figure 1 – The algorithm of updating the technical and economic model*

This algorithm involves three iterations of the formation and updating of the technical and economic model, taking into account the current revaluation of the economic parameters of the product itself, as well as investment parameters based on actual or updated forecast data. This model for evaluating the effectiveness can be considered as a lever for monitoring and innovation management for the project manager or the organization's leadership in product or portfolio investment.

The proposed model covers the key indicators of investment and product efficiency: price positioning, sales volume, cost of production, investment assessment at all stages of the project (R & D costs, certification costs, costs of project's mechanic training, selling costs). Also in the model are elements for predicting permissible deviations from the target serial cost for reasons of rising prices for the period before development, taking into account serial technologies (increased cost of pilot batches of mastered components manufactured by bypass technologies).

An example of filling out a feasibility model is presented in Figure 2.

**Economic justification of product sales prices** | **96537**

**alternative design changes for the base product** | **96635**

| | | | Stage of the development Task | The stage of issuance of design documentation | | The stage of serial production | |
|---|---|---|---|---|---|---|---|
| № | Name | Unit of measure | The new model (estimated FCA) | Project cost | Deviations | Standard cost price | Deviations |
| 1 | Sales volume (approximate) total | pieces | 100 | 95 | -5 | 142 | 42 |
| 2 | Total cost of the chassis | rub | 3 237 552 | 3 239 552 | -2 000 | 3 593 683 | -356 131 |
| 3 | Direct chassis costs | rub | 2 504 899 | 2 504 999 | -100 | 2 805 487 | -300 588 |
| 4 | Conditionally variable expenses | % | 0,0% | 0,0% | | 0,0% | |
| 5 | Cost of the add-on (including installation and transport costs) | rub | 671 297 | 687 398 | | 687 398 | |
| 6 | Estimated price (EXW) | rub | 3 683 634 | 3 683 634 | | 3 683 634 | |
| 7 | Delivery costs (under the terms of the contract) | rub | 262 240 | 262 240 | | 262 240 | |
| 8 | Contract price without VAT (calculated) minimum | rub | 4 617 171 | 4 633 272 | | 4 633 272 | |
| 9 | Contract price without VAT (offer) | rub | 4 617 171 | 4 633 272 | | 4 633 272 | |
| 10 | The profit margin on the units chassis | rub | 1 178 735 | 1 178 635 | | 878 147 | |
| 11 | Marginal profitability of the chassis | % | 32% | 32% | | 24% | |
| 12 | Profit (oss) | rub | 446 082 | 444 082 | | 89 951 | |
| 13 | Profitability of sales of the chassis | % | 12% | 12% | | 2% | |
| 14 | R & d expenses excluding VAT | rub | 5 493 783 | 5 467 128 | | 814 400 | |
| 15 | The cost of certification | rub | 679 200 | 0 | | 0 | |
| 16 | Costs of technological preparation of production | rub | 1 050 000 | 1 050 000 | | 250 000 | |
| 17 | Total investment | rub | 7 222 983 | 6 517 128 | | 1 064 400 | |
| 18 | The profit margin of the contract with the cost of R & d | rub | 112 379 717 | 106 503 197 | | 123 882 474 | 11 502 757 |

*Figure 2 - The example of technical and economic model being updated*

Almost all the elements of the technical and economic model at the concept stage are formed on the basis of functional-cost analysis data, rough expert estimates on market expectations, and if the assessment is given by an experienced expert, the indicators include promotion risks taking into account changes in the market situation. The same uncertainty factor in forecasting can be attributed to cost parameters and investment cost estimates. These initial overestimations can be considered as elements of risky suboptimization of subdivision indicators to ensure their attainability. In this way, at the first stage the technical and economic model is distorted by the excessive probability of deviations from normal expectations. Carrying out the current updating of the model, it is possible to track unrealized embedded deviations that improve performance indicators. If some of the elements worsen the model, then it is necessary to determine the cause, this may be a forecast error due to poor study of the forecasting area, an event with a low probability of expectation, force majeure, etc.

Using the economic analysis tool (Cash-Flow) and sample input from different stages of the model give an opportunity to provide an updated payback forecast considering current deviations from the baseline scenario (Figure 3).

| The stage of serial production | | | | | | | | | | | |
|---|---|---|---|---|---|---|---|---|---|---|---|
| Indicator | 0 year | 1 year | 2 year | 3 year | 4 year | 5 year | 6 year | 7 year | 8 year | 9 year | 10 year |
| **CASH FLOW** | | | | | | | | | | | |
| **Investments, thousand rubles** | **1 064,40** | | | | | | | | | | |
| The volumes of all pieces. | | 10 | 30 | 50 | 10 | 7 | 7 | 7 | 7 | 7 | 7 |
| - the same cumulative total, PCs. | 0 | 10 | 40 | 90 | 100 | 107 | 114 | 121 | 128 | 135 | 142 |
| | | | | | | | | | | | |
| The stage of serial production | **1 064,40** | 10 | 30 | 50 | 10 | 7 | 7 | 7 | 7 | 7 | 7 |
| **Total adjusted profit, thousand rubles** | | **899,51** | **2 698,53** | **4 497,55** | **899,51** | **629,66** | **629,66** | **629,66** | **629,66** | **629,66** | **629,66** |
| Total profit, thousand rubles | | 899,51 | 2 698,53 | 4 497,55 | 899,51 | 629,66 | 629,66 | 629,66 | 629,66 | 629,66 | 629,66 |
| *including per unit* | | | | | | | | | | | |
| The stage of serial production | **89,95** | 90 | 90 | 90 | 90 | 90 | 90 | 90 | 90 | 90 | 90 |
| Net cash flow, thousand rubles | -1 064,40 | 899,51 | 2 698,53 | 4 497,55 | 899,51 | 629,66 | 629,66 | 629,66 | 629,66 | 629,66 | 629,66 |
| - the same cumulative total, thousand rubles. | -1 064,40 | -164,89 | 2 533,64 | 7 031,18 | 7 930,69 | 8 560,35 | 9 190,00 | 9 819,66 | 10 449,32 | 11 078,97 | 11 708,63 |
| Discount rate (comparison rate), % | 12,00 | 12,00 | 12,00 | 12,00 | 12,00 | 12,00 | 12,00 | 12,00 | 12,00 | 12,00 | 12,00 |
| Discounted cash flow, thousand rubles | -1 064,40 | 803,13 | 2 151,25 | 3 201,26 | 571,65 | 357,28 | 319,00 | 284,82 | 254,31 | 227,06 | 202,73 |
| - the same cumulative total, thousand rubles. | -1 064,40 | -261,27 | 1 889,98 | 5 091,25 | 5 662,90 | 6 020,18 | 6 339,19 | 6 624,01 | 6 878,32 | 7 105,38 | 7 308,11 |
| **PAYBACK** | | | | | | | | | | | |
| **Simple payback period, years** | **1,06** | | | | | | | | | | |
| **Discounted payback period, years** | **1,12** | | | | | | | | | | |
| **NPV, thousand rubles.** | **7 308** | | | | | | | | | | |
| **return on investment ROI** | **6,866** | | | | | | | | | | |
| **IRR, %** | **155,7%** | | | | | | | | | | |

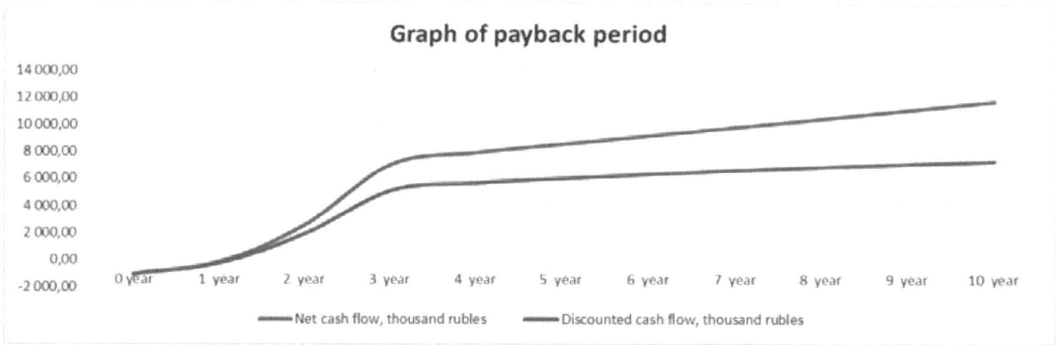


*Figure 3 – Assessment of efficiency of technical and economic model being updated*

For managing effectively the project progress, the data entered into the technical and economic model at the early stage of the concept are updated twice - for the first time at the stage of completion of the design documentation with an estimate of the project's mechanic training volumes and for the second time based on the results of the project's mechanic training and the generation of evidence. For the goal-setting of indicators in the implementation of all types of work within each stage, it is necessary to structure the indicators into components and evaluate their attainability at the lower level at the time of making decisions on the investment components of costs or the design and technological model of the cost of the product.

## 4. Conclusion

Thus, the offered algorithm will allow providing timely and adequate assessment of risks of decisions being taken and give a clear forecast of the impact of deviations on the technical and economic model.

The organization need to have a certain level of development in following fields to put into practice offered actualized technical and economic model:

1. It is required to possess methodology of calculation of design cost of the product at different stages of development with an increase of the reliability level as the structure, composition and updating of data on the technology of manufacturing individual bundles, the cost of purchased components.

2. It is necessary to have elements of project data management for investment and financial flows, forecasting the timing of exploitation and the start of production.

3. It is required to organize storage aggregation information and data processing systems for technical and economic model.

4. It is necessary to have competent personnel who are able to adequately predict and evaluate risks, motivated to escalate them in a timely manner before making decisions, digitize these risks and clearly follow certain criteria of appropriateness.

Execution of requirements for the organization of work with using the proposed algorithm in innovative projects seriously reduces the risk of exploration the start of production of an unprofitable or non-competitive product. The using of the proposed algorithm for updating the technical and economic model will allow creating additional analytical tools for prompt management of the project at the R&D stages, getting up-to-date forecast on the reachability of primary estimates and a clear understanding of the reasons for deviations The updated model allows to accumulate and analyze the change in input data at different stages of the project and build a digital model of the innovative business process to find ways to optimize.

## References

[1] Anon 2014 *A Guide to the Project Management Body of Knowledge (PMBOK Guide)* (Moscow: Olimp-Biznes)
[2] Gol'dshtejn G Y 2000 *Strategic Aspects of R&D Management* (Taganrog: Izdatel'stvo TRTU)
[3] Spasennyh M Y 2010 *Innovative Business: Corporate R&D Management* (Moscow: Izdatel'stvo «Delo»)
[4] Vertiy B D 2009 Determining R&D performance indicators *Economics* **8(57)** 222–7
[5] Puzynya T A 2012 Target costing and kaizen costing - innovative concepts of domestic accounting *Mod. Res. Innov.* **7**

IOP Conf. Series: Materials Science and Engineering **915** (2020) 012004 doi:10.1088/1757-899X/915/1/012004

# Developing a boss for hybrid electric motors

**F F Bagdanurov[1], L N Shafigullin[1] and F M Akhmetov[1]**

[1]Naberezhnye Chelny Institute (branch), Kazan Federal University
Sjujumbike street 10a, Naberezhnye Chelny, 423800 Tatarstan, Russia

misharin_82@mail.ru

**Abstract**. The existing types of electric motors were analyzed. The paper describes the concept of a hybrid electric motor. The paper describes the process of developing a boss for hybrid electric motor. The paper describes the technology of making particulate filler from electrical steel.

**1. Introduction** Today, electric motors are widely used in various industries: machinery, processing industry, etc. Conventional asynchronous motors are used in many industries, their advantage is high reliability. The disadvantage of these motors is low efficiency. These machines were invented over one hundred years ago. There are more advanced stepper motors which require expensive controllers, with typical slew range. The advantage of motors is a high torque, chopper control to adjust a number of revolutions per minute precisely.

**2. Body text** The hybrid electric motor under development (see Figure 1) is based on the operating principles of stepper and asynchronous motors and uses the advantages of both of them. The hybrid electric motor consists of various types of composite materials (85% of its content). The key motor structure components are as follows: shaft, housing cover, housing, rotor disc, coil with winding, stator ring, rotor disc.

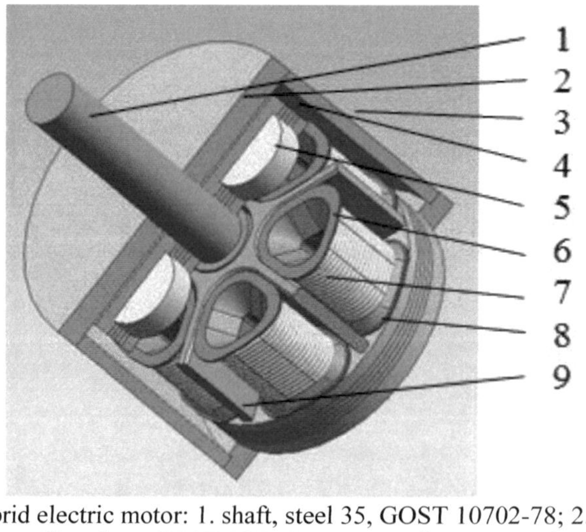


Figure 1. Section of hybrid electric motor: 1. shaft, steel 35, GOST 10702-78; 2. cover "1", fiberglass, GOST 8865-93; 3. housing, fiberglass, GOST 8865-93; 4. rotor disc "1", section; 5. permanent magnet, GOST 21559-76; 6. coil with winding; 6. stator ring; 7. rotor disc "2"; 8. cover "2", fiberglass, GOST 8865-93

The paper studies the process of developing a boss with its main function being transfer of electromagnetic fields from a winding to a rotor (Figure 2)

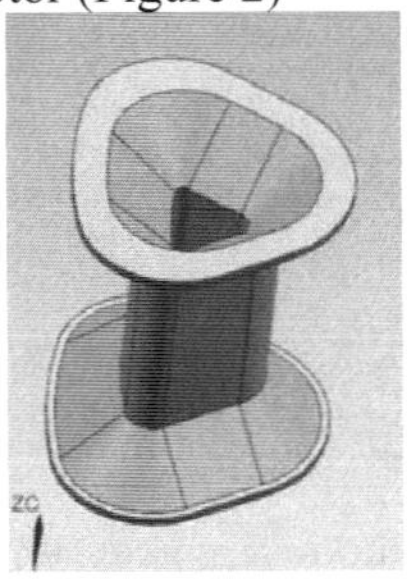


Figure 2. Boss, general view

Due to a complex part geometry, a new material needs to be developed in order to get a complex product geometry and maintain electromagnetic values.

The analysis of the literature showed the absence of particulate-filled composite materials with the required electromagnetic values. In view of this, it was required to make a new composition which would consist of a polymer matrix and particulate filler. The exploratory research was carried out. The most suitable binder was Norsodyne H 44173 polyester resin [1]. It has the required physical mechanical and processing properties [2]. It was suggested to use a powder made of electrical steel as a filler, which has the necessary parameters of "hysteresis line". Today, the Russian industry does not produce powders with required characteristics. So, the process was developed to produce a powder from electrical steel. A grinding machine was developed to get required powder fraction and characteristics, elongated shape (Figure 3).

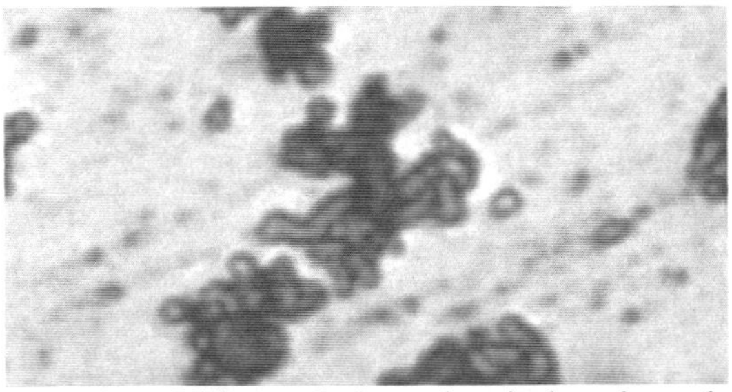

Figure 3. Particles of filler made from electrical steel powder (x 200)

Resulting particles have a length of 0.1 mm, thickness of about 0.02 mm. The advantage of resulting powder shape is improved transfer of magnetic field. As magnetic fields tend to form parallel lines the elongated particle shapes are most suitable for this implementation. When exposed to a magnetic field, magnetic fields are concentrated at the ends of filler particles. A unique forming process was developed to prevent filler particles from sublimation and damage when added to matrix material. It involves the use of a special instrument with specified shape and size, which has optimized conditions for mixing composite systems.

A mixing process demanded the development of requirements for process of manufacturing composite materials [3]:

- a mixing tool should have a cylindrical shape with the rounded end;
- powder agglomerates should not contact the tool [4];
- tool speed should be within 20-30 mm/s.

These requirements make it possible to provide a uniform particle distribution in the binder and prevent their damage.

Plastic molds and permanent magnets are used to ensure unidirectionality of particles inside a mold and their contact with each other during molding [5-6]. The direction of permanent magnet fields

coincides with the axis of a product, which makes it possible to guide filler particles along the axis of a part. And every particle becomes magnetized and starts gravitating towards other particles.
Polymer compound is made by vacuum forming to ensure minimum number of defects in materials. The resulting boss has unique electromagnetic properties due to different directions of magnetic fields inside a composite material.

## 3. Conclusions
An innovative design of a boss for electrical motor ensures high heat transfer, concentration of magnetic fields in some restricted space, prevents sticking effect for permanent magnet over the coil.

## References
[1] Karikh F G and Shafigullin L N Automation of process of manufacturing products from a mixture of fluids fast polymerized in air using the jet centrifugal method *IOP Conf. Series: Materials Science and Engineering* 570 (2019) 012038 IOP Publishing doi: 10.1088/1757-899X/570/1/012038
[2] Zharin D E, Butorin I V and Shafigullin L N 2008 Optimized compositions of matrix particulate-filled and fiber polymer composite materials with high processing properties *Collection of papers and abstracts of VII Congress of production engineers in the automotive indutry* pp. 94 - 5
[3] Galimov E R, Fedyaev V L and Morenko I V 2014 About the basic laws of non-isothermal cross-flow past of a circular cylinder with beads mix *Journal of Physics: Conference Series* vol 567 number 012012
[4] Galimov E R, Bobryshev A N, Lakhno A V, Voronov P V, Galimova N Ya and Sharafutdinov R F 2015 New approaches for evaluating rheological models in composites *IOP Conf. Series: Materials Science and Engineering* vol 86 number 012001
[5] Kashapov N F, Nafikov M M, Gazetdinov M X, Nafikova M M and Nigmatzyanov A R 2016 Innovative production technology ethanol from sweet sorghum *IOP Conference Series: Materials Science and Engineering* vol 134 issue 1 8 number 012012
[6] Sokolova Yu A, Zharin D E and Shafigullin L N 2009 Developing compositions of special polymer composite materials *Academia. Architecture and Construction* vol 2 pp. 104-7

IOP Conf. Series: Materials Science and Engineering **915** (2020) 012005 doi:10.1088/1757-899X/915/1/012005

# Development of a parametric model for calculating cutting forces for external cylindrical turning of steel 20CrMNTi

**I P Balabanov[1, 2], O N Balabanova[1] and V N Gilman[1]**

[1] Kazan Federal University, 18 Kremlyovskaya street, Kazan, 420008, Russian Federation

[2] Kazan National Research Technical University named after A. N. Tupolev –KAI, 10 K. Marks street, Kazan, 420111, Russian Federation

balabanovip@mail.ru, olgachelny75@mail.ru, tim_vg@mail.ru

**Abstract.** The paper presents a mathematical model for calculating the cutting force when processing 20crmnti steel with the Sandvik CNMG 120408 16P25T tool. The stand based on the 16d25 machine is presented. It allows measuring the actual values of the spindle speed, longitudinal feed, cutting depth, and cutting force. The results are transmitted to the computer via the LTR-EU-8 workstation with galvanically isolated LTE modules and a synchronization interface. According to the test results, the obtained theoretical model has a deviation from the actual measurements of no more than 0.12%. It is shown that the calculation results provided by the leading tool manufacturers are expected to be overstated. The difference in cutting forces can be more than 1.5 times.

## 1. Introduction

The cutting force is a key parameter which is necessary for predicting the required cutting power, the temperature in the cutting zone and the cutting tool, and the quality of the formed surface [1]. At the same time, the models for predicting cutting forces themselves have recently been significantly simplified [2] or tool manufacturers are carefully bypassing this concept by replacing them with their derivatives, for example, power consumption or tool life [3]. For example, the well-known tool manufacturer Walter (https:// www.walter-tools.com) recommends calculating the cutting forces by means of the formula (1) [4]

$$F_c = A \times k_c \times h^{-mc} \ (N) \quad (1)$$

where A – chip cross section (mm2);
h – chip thickness (mm);
kc – specific cutting force (H/mm2);
mc – correction facto.

But the cutting force depends on many additional parameters, for example, on the hardness of the material or the presence of a crust on the surface of the workpiece. The coefficients kc and mc themselves are approximate, obtained on some "similar" material and instrument. It can be concluded that the presented model is most likely inaccurate and can only be used for superficial estimates of the probability value of the parameter.

In recent years, we have become more and more interested in modeling [5]. Modeling allows you to analyze existing solutions without significant costs. But approximate models give a very conditional result, which is unacceptable in the modern development of computer-aided design systems [6,7,8,9]. Therefore, there is an urgent need for adequate models to predict cutting forces.

The most complete and simple enough to implement model for calculating the cutting force is the empirical Taylor equations (2) [1], [2] allowing you to take into account any processing conditions. At the same time any special processing conditions can be included in the K coefficients.

$$P = 10 \times Cp \times t^{x} \times s^{y} \times V^{n} \times K \qquad (2)$$

where t, S, V – the parameters of the cutting conditions;
K – correction factor that allows you to take into account cutting conditions;
Cp, x, y, n – empirical coefficients and degree indicators.

Based on the above, the aim of the work is to develop an empirical cutting model based on the Taylor equation for processing 20crmnti steel with a specific cutting tool. The final goal of the work can be stated as the development of a method for quickly obtaining a model for determining the cutting force for a specific material and cutting tool [9,10,11,12,13]. In the future, these models can be used to accurately simulate computer processing.

## 2. Methods

The standard universal lathe 16D20 was used as a stand for conducting experiments. To fix the actual cutting speed, a non-contact induction speed meter was installed directly on the spindle. To account for real movements, displacement meters were installed with brackets up to 1000 mm for longitudinal movement, and up to 100 mm for transverse movement relative to the spindle rotation axis. A three-dimensional force meter was used with a DCLNR 2525 M12 holder. The force meter has been calibrated. Sandvik CNMG 120408 16P25T cutting insert is new. The devices and the sensor are connected to the LTR-EU-8 workstation with galvanically isolated LTR modules and a synchronization interface. The force meter is connected to the TR212M with an 24-bit / 7.6 kHz ADC designed to connect up to eight load cells with resistance from 100 Ohms to 1 kOhm. The remaining sensors were connected to LTR24 (4 strictly parallel ADC channels 24 bit / 117 kHz) and LTR11 supporting multi-channel data acquisition mode: up to 32 single-phase signal acquisition channels (ADC 14 bit/ 400 kHz).

To obtain the coefficients of the empirical model of cutting, it is proposed to use the search for the optimal solution by exhaustive search method taking into account the limitations [14,15,16,17,18]. This method is quite accurate [5] and is quickly solved by specialized software. In our case, we will use the "solution search" function of the MS EXCEL application.

To calculate the coefficients of the empirical cutting model, the cutting forces were measured at various cutting speed, longitudinal feed and the size of the removed allowance. The range of cutting speeds was 150-370 m / min. The average increment of the speed is 50 m / min. The machine has a stepped, mechanical system for changing the spindle speed. The cutting speed depends on the diameter being machined (2). During the experiment, the actual spindle speeds were taken into account.

$$V = \frac{\pi \times d \times n}{1000} \text{ (m/min)} \qquad (3)$$

where, d – machined diameter (mm); n – rotations (rpm).

The size of the removed allowance varied from 0.5-2.5 mm with a measurement step of about 0.5. The measurement step depends on the gearbox. In order to eliminate the error of the longitudinal feed due to the error in the manufacture of the machine and wear during the operation period, the actual feed was measured based on the displacement meter. The feed values were 0.11-0.13-0.22-0.25 mm / rev. Workpieces were processed from one batch. The hardness of each preform was previously measured, it was at 156 HB. Processing without coolant.

## 3. Results

IOP Conf. Series: Materials Science and Engineering **915** (2020) 012005 doi:10.1088/1757-899X/915/1/012005

Table 1 presents sample V-s-t-P dependency measurement data. The software interface for calculating the coefficients of the empirical cutting model based on the search for the optimal solution by enumeration taking into account the limitations [19] by the built-in MS EXCEL tools is shown in Figure 1.

**Table 1.** Selected experiment results.

| | V (mm/min) | S ($mm^{-1}$) | T (mm) | P (H) |
|---|---|---|---|---|
| 1 | 306 | 0,11 | 1,53 | 263,7 |
| 2 | 178 | 0,22 | 1,54 | 451,5 |
| 3 | 369 | 0,22 | 0,5 | 121,4 |
| 4 | 367 | 0,13 | 2,5 | 433,1 |
| 5 | 375 | 0,13 | 3 | 512 |
| 6 | 217 | 0,25 | 0,5 | 154 |

Figure 1 in cells C12-C17 presents data obtained during the experiment. In cells B5-F5, selected by the "search for a solution" function in the MS EXCEL application based on restrictions (2). In cells H12-H17, values calculated based on the Taylor empirical model (1). As you can see, the discrepancies in the data obtained during the experiment and the values of the cutting force calculated on the basis of the selected coefficients are insignificant and amount to no more than 0.12% (0.54 H at a value of 450 H).

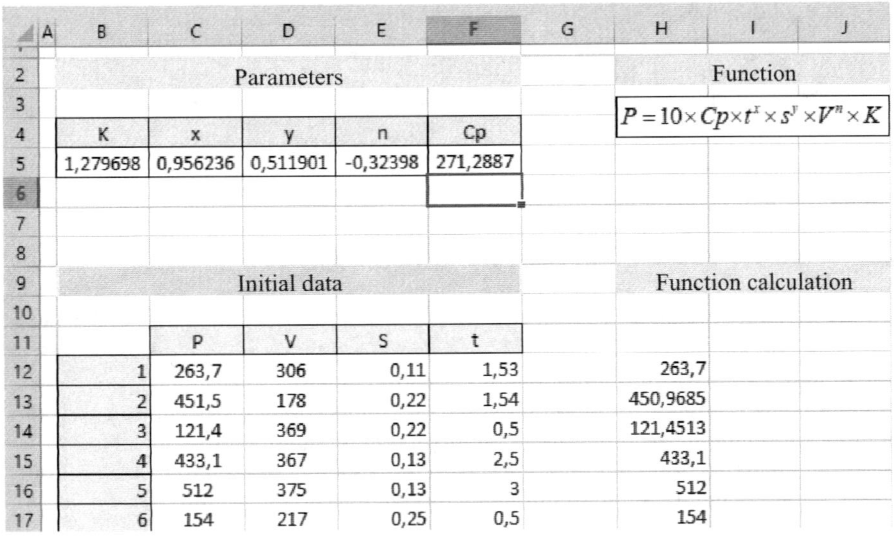

Parameters

| K | x | y | n | Cp |
|---|---|---|---|---|
| 1,279698 | 0,956236 | 0,511901 | -0,32398 | 271,2887 |

Function

$P = 10 \times Cp \times t^{x} \times s^{y} \times V^{n} \times K$

Initial data / Function calculation

| | P | V | S | t | Function calculation |
|---|---|---|---|---|---|
| 1 | 263,7 | 306 | 0,11 | 1,53 | 263,7 |
| 2 | 451,5 | 178 | 0,22 | 1,54 | 450,9685 |
| 3 | 121,4 | 369 | 0,22 | 0,5 | 121,4513 |
| 4 | 433,1 | 367 | 0,13 | 2,5 | 433,1 |
| 5 | 512 | 375 | 0,13 | 3 | 512 |
| 6 | 154 | 217 | 0,25 | 0,5 | 154 |

Fifure 1 – The coefficient selection interface for the construction of the cutting model

## 4. Conclusions

As a result of the experiment of processing the 20CrMnTi material with the Sandvik CNMG 120408 16P25T tool, a model for calculating the cutting force was obtained (4):

$$P = 10 \times 271.2887 \times t^{0.956236} \times s^{0.5119} \times V^{-0.32398} \times 1.2797 \quad (4)$$

Where t, S, V – parameters of cutting conditions.

As calculations show, the deviation of the theoretical model was not more than 0.12% of the results obtained at the test bench. According to the results of preliminary modeling of the cutting forces of a similar material with a tool with a similar geometry and cutting conditions s = 0.11 mm / rev, t = 1.53 mm, V = 306 m / min, the following results were obtained:

- according to the source https://www.walter-tools.com, the cutting force will be 494.15Fc | N, with a specific material removal of 49.73cm³ / min and a power of 2.58 Pmot | KW;
- according to the source https://www.sandvik.coromant.com, the specific volume of material removal will be 48.00 cm³ / min and cutting power 2.32 Pmot | KW; according to indirect calculations, it can be assumed that the cutting force will be about 460 H;
- due to the use of new materials and increased cutting speeds, the calculation of cutting forces according to the methods of Baranovsky and Kosilova is not relevant.

As you can see, the results presented by online services are expected to be overstated. Their task is to guarantee the declared durability. At the same time, the results of the actual test show cutting forces that are overstated by more than 1.5 times.

**References**

[1] Baranovsky Yu.V. Modes of cutting metals / Ed. 4th. rewrited and add. M .: Research Institute Avtoprom, 1995 .-- 456 p.

[2] Metal cutting and cutting tools: Textbook / V.G. Solonenko, A.A. Ryzhkin. - M .: INFRA-M, 2011 .-- 416 p. 60x90 1/16. - (Higher education). ISBN 978-5-16-004719-5 - Access mode http://znanium.com/catalog/product/258644

[3] SANDVIK COROMANT Manual 2019 Manual Machining (https://www.sandvik.coromant.com) Retrieved: 31.10.2019

[4] Turning/ lathe machining ISO (https:// www.walter-tools.com/ SiteCollectionDocuments /wmc /index-bakup.html#content) Retrieved:: 31.10.2019

[5] Computer modeling: textbook / V. M. Gradov, G. V. Ovechkin, P. V. Ovechkin, I. V. Rudakov- M.: COURSE: INFRA-M, 2018. - 264 p..

[6] A.G. Kondrashov, D.T. Safarov, A.I. Faskhutdinov, G.K. Davletshina Single-Turn Worm Mills for Conical Round-Tooth Gears / Russian Engineering Research, 2017, Vol. 37, No. 9, pp. 812–813.

[7] Weiss Denis, The Effect of Cutting Edge Rounding Geometry on Tool Wear, Machine tool Park № 12,2016.

[8] Kasyanov S.V., Kondrashov A.G., Safarov D.T. Research of characteristics of wearproof coating for cutting tools / INTERFINISH-SERIA 2014: International Conference on Surface Engineering for Research and Industrial Applications. 2014. P. 62.

[9] Faskhutdinov A. I., Emelyanov D. V., Blinova A. S. Influence of the size of the cut layer on the productivity of mechanical processing / Proceedings of scientific articles of the 2nd international youth scientific and practical conference "Progressive technologies and processes" September 24-25, 2015 in 3 volumes. Kursk.: ZAO Universitetskaya kniga. - 2015. Pp. 111-115.

[10] Chatterjee-Fisher, R, Eisell, F. et al., Nitriding and carbonitriding, TRANS./ edited by Supov A.V. M.: metallurgy, 1990. 280s.

[11] Khusainov R. M., Golovko A. N., Petrov S. M., Yurasov S. Yu., Balabanov I. P., grechishnikov V. A., Romanov V. B., Pivkin P. M. / / Determination of tool parameters in technological cutting systems / STIN. 2016. No. 10. Pp. 17-20.

[12] S. V. Kasyanov, A. G. Kondrashov, D. T. Safarov Rapid Assessment of Wear-Resistant Tool Coatings / Russian Engineering Research, 2017, Vol. 37, No. 11, pp. 969–973.

[13] Khusainov R.M., Khaziev R.R. Modeling of forming technological errors in processing by gear shaping machine // IOP Conference Series: Materials Science and Engineering. - 2017. - Vol.240, Is.1. - Art. № 012045.

[14] Leushin, I.O. Technologies for Use in the Formation of a Differentiated Structure in Iron Billets Used in Glass Molds / I.O. Leushin, D.G. Chystyakov // METAL SCIENCE AND HEAT

TREATMENT. - 2016. - №5-6. - T. 58. - p. 299-302.

[15] Leushin, I.O. Creep of the investment mold material / I.O. Leushin, L.I. Leushina, O.S. Koshelev // Russian Metallurgy. - 2017. - №13 (DEC. 2017). - C. 1092-1095.

[16] A.V. Shaparev and I.A. Savin, Influence of the State of the Contact Surfaces on the Formation of the Joint of Steel and Brass during Cold Cladding. Solid State Phenomena, Vol. 284, pp. 319-325, 2018 DOI: https://doi.org/10.4028/www.scientific.net/SSP.284.319

[17] R.V. Gavariev and I.A. Savin, Research of the Mechanism of Destruction of Compression Molds for Casting under Pressure of Color Alloys. Solid State Phenomena, Vol. 284, pp. 326-331, 2018 DOI: https://doi.org/10.4028/www.scientific.net/SSP.284.326

[18] Altintas Y, Engin S. Generalized modeling of mechanics and dynamics of milling cutters. Cirp Annals-Manufacturing Technology. 2001;50:25-30.

[19] Dong JL, Yu TB, Chen H, Li BC. An improved calculation method for cutting contact point and tool orientation analysis according to the CC points. Precision Engineering-Journal of the International Societies for Precision Engineering and Nanotechnology. 2020;61:1-13.

[20] Gaitonde VN, Karnik SR, Figueira L, Davim JP. Analysis of Machinability During Hard Turning of Cold Work Tool Steel (Type: AISI D2). Materials and Manufacturing Processes. 2009;24:1373-82.

[21] Mabrouki T, Rigal JF. A contribution to a qualitative understanding of thermo-mechanical effects during chip formation in hard turning. Journal of Materials Processing Technology. 2006;176:214-21.

[22] Reddy NSK, Rao PV. Selection of an optimal parametric combination for achieving a better surface finish in dry milling using genetic algorithms. International Journal of Advanced Manufacturing Technology. 2006;28:463-73.

[23] Saha P, Singha A, Pal SK. Soft computing models based prediction of cutting speed and surface roughness in wire electro-discharge machining of tungsten carbide cobalt composite. International Journal of Advanced Manufacturing Technology. 2008;39:74-84.

[24] Shrot A, Baker M. Determination of Johnson-Cook parameters from machining simulations. Computational Materials Science. 2012;52:298-304.

[25] Stephenson DA, Agapiou JS. CALCULATION OF MAIN CUTTING EDGE FORCES AND TORQUE FOR DRILLS WITH ARBITRARY POINT GEOMETRIES. International Journal of Machine Tools & Manufacture. 1992;32:521-38.

[26] Suresh R, Basavarajappa S, Gaitonde VN, Samuel GL. Machinability investigations on hardened AISI 4340 steel using coated carbide insert. International Journal of Refractory Metals & Hard Materials. 2012;33:75-86.

[27] Tournier C, Duc E. Iso-scallop tool path generation in 5-axis milling. International Journal of Advanced Manufacturing Technology. 2005;25:867-75.

[28] Yallese MA, Chaoui K, Zehib N, Boulanouar L, Rigal JF. Hard machining of hardened bearing steel using cubic boron nitride tool. Journal of Materials Processing Technology. 2009;209:1092-104.

IOP Conf. Series: Materials Science and Engineering **915** (2020) 012006 doi:10.1088/1757-899X/915/1/012006

# Hardening of the working bodies of agricultural machines by electric spark and vibratory arc processing

**A T Bayniyazova[1], M M Abzhaev[1], S N Sharifullin[2]**

[1]Kyzylorda State University named after Korkyt at, 120014 City of Kyzylorda, 29A Aiteke bie str. Kazakhstan
[2]Kazan Federal University, 18 Kremlyovskaya Str., Kazan 420008 Russia

akmabay@mail.ru, marat_78@mail.ru, Saidchist@mail.ru

**Abstract:** The paper presents the results of studies of some physical and mechanical properties of metal samples made of 65G steel, hardened by electric spark and vibratory arc methods. Research is related to solving the problems of increasing the working resource of the working bodies of agricultural machines. The studied elemental composition, phase composition, and microstructure of the surface of the samples.

## 1. Introduction

Among plasma methods of hardening the surfaces of products of electric spark and vibro arc methods have certain advantages. They do not overheat the base of the product, but only allow you to change the physico-mechanical and physico-chemical properties of the surface layer of the material. In this case, it is possible to harden the surface layer with alloying elements of carbide materials. As a result of plasma-chemical reactions, a new structure of the surface layer of the material of the part is formed, not only with alloying of the electrode elements, but also with the appearance of various compounds. Moreover, not only the elements of the electrode and the material of the part are involved in the process. The plasma-chemical process occurs in the atmosphere of the environment, which is usually not only air. It can contain dust particles of products of one or another production, and various gases. In addition, any environment has a certain humidity. It should also be taken into account that the plasma-chemical process with the vibration method is intermittent. In the interval between discharges, the process of structural changes in the material continues, the formation of new compounds and, possibly, new chemical elements appear. The next category already interacts with this "ensemble" of chemical elements and various compounds. Then this process is repeated again. Therefore, in recent years, scientists have been paying increasing attention to the problems of studying plasma technologies using vibration methods [1 - 5]. This paper presents some results of studies of the physicomechanical properties of samples made of 65G steel treated with electric spark and vibration arc discharges.

## 2. Materials and methods

The studies were carried out on samples of 65G steel measuring 30x30 mm with a thickness of 2 mm. Processing was carried out at technological installations of electric spark and vibration arc hardening developed by GOSNITI (Moscow). The tungsten-cobalt VK6 rod with a diameter of 4 mm and a copper-graphite rod with a diameter of 8 mm served as an electrode. The morphology of the surface and elemental composition of the samples were studied using a Zeiss EVO 50 XVP scanning electron microscope (SEM) with INCA Energy-350 and INCA Wave-500 elemental analyzers. The crystal structure and phase composition were studied on a Dron-7 X-ray diffractometer with a CuK$\alpha$ X-ray source (0.154178 nm). The surface layer of the following samples was investigated: sample No. 1 – initial, sample No. 2 with electric spark treatment, sample No. 3 with vibratory arc treatment.

## 3. Results and discussion

IOP Conf. Series: Materials Science and Engineering **915** (2020) 012006 doi:10.1088/1757-899X/915/1/012006

On the figure 1 shows microphotographs of the surface of the samples before and after processing. It is seen that after processing a developed surface is observed. The unevenness of the treated surface is visible. Such processing is due to the fact that the electrode moves on the surface in manual mode. Therefore, part of the surface areas did not directly contact the electrode. But the main part turned out to be under the influence of electrospark processing, there are also local processing sites with a molten zone. Figure 2 presents spectrum of elemental composition of the surface layer of samples before and after processing by vibroplasma. An analysis of the spectra showed that the plasma treatment of products allows them to modify their surface layer and form a hard and wear-resistant coating.

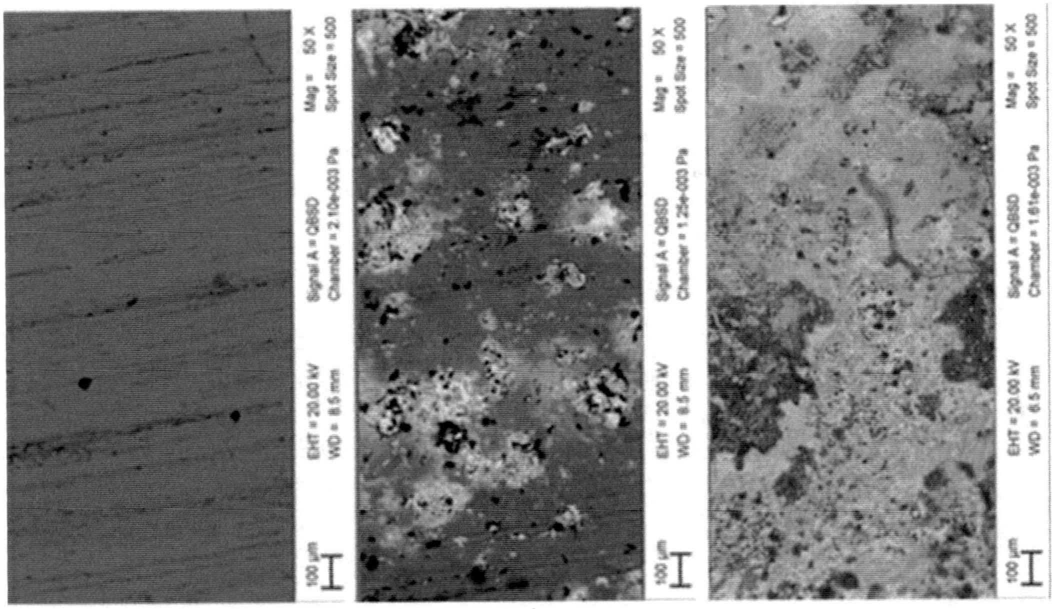

a b c

**Figure 1.** Microphotographs of samples No. 1 - initial (a), No. 2 - after electrospark processing (b) and No. 3 - after vibration-arc processing (c)

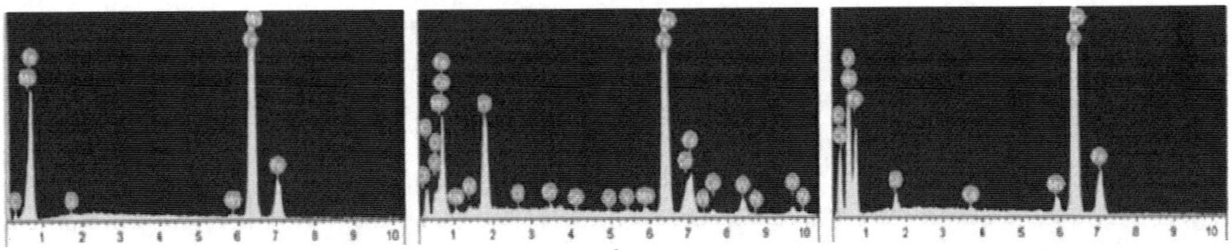

a b c

**Figure 2.** The spectrum of the elemental composition of the surface layer of the samples before and after processing by vibroplasma: a) - the initial sample; b) - sample after electrospark processing; c) - sample after vibratory arc treatment

## 4. Conclusions

The surface layer formed by the vibroplasma represents a new composite structure. The upper thin layer consists of anode material modified by elements of the cathode material and interelectrode medium. Under the top layer is a modified hardened layer consisting of a mixture of anode and cathode materials. Underneath is a zone of thermal influence formed as a result of pulsed heat. Its structure differs from the structure of the material of the workpiece by the structure and grain size. The hardening of the base material in the depth of the part during electrospark machining occurs by 0.5 - 1.0 mm, during vibration-arc machining - up to 3-4 mm.

**References:**

[1] Denisov V.A. Recovery of basic parts of diesel engines with emergency defects coatings based on composite materials: dissertation ... Doctors of technical sciences: 05.20.03 / Denisov Vyacheslav Aleksandrovich. - Saransk, 2015 .- 383 p.

[2] Velichko S.A. Development of high-performance repair technologies for tractor mounted hydraulic systems using the electric spark treatment method: dissertation ... Doctor of Technical Sciences: 05.20.03 /Velichko Sergey Anatolevich. - Saransk, 2017 .- 480 s.

[3] Kolomeichenko A.V. Investigation of nanometallokeramic composite coatings obtained by vibro-arc surfacing /A.V. Kolomeichenko, N.V. Titov, Yu.A. Kuznetsov, L.V. Kalashnikova, O.O. Bagrintsev, S.N. Sharifullin //IOP Conf. Series: Materials Science and Engineering 412 (2018) 012049 doi: 10.1088 / 1757-899X / 412/1/012049.

[4] Kadyrmetov, A.M. Efficiency of surface cleaning by a glow discharge for plasma spraying coating /A.M. Kadyrmetov, N.F. Kashapov, S.N. Sharifullin, A.I. Saifutdinov and S.A. Fadeev //IOP Conference Series: Materials Science and Engineering 134 (2016) 012010 doi: 10.1088 / 1757-899X/134/1/012010.

[5] Kashapov, L.N., Kashapov, N.F., Kashapov, R.N. Influence of plasma-electrolyte discharge to the glass surface (2014) Journal of Physics: Conference Series, 567 (1), art. № 012024.

[6] Kashapov, R.N., Kashapov, L.N., Kashapov, N.F. Research of plasma-electrolyte discharge in the processes of obtaining metallic powders (2017) Journal of Physics: Conference Series, 927 (1), art. № 012086.

[7] Kashapov, L.N., Kashapov, N.F., Kashapov, R.N. Research of cobalt chromium alloy surface morphology after sandblasting (2014) IOP Conference Series: Materials Science and Engineering, 69 (1) art. № 012017.

[8] Martynov A.V. Improving the technology of repair of hydraulic distributors by restoration and hardening of parts by the method of electric spark processing: the dissertation ... candidate of technical sciences: 05.20.03 /Martynov Aleksey Vladimirovich. - Saransk, 2012 .- 254 p.

[9] Chumakov, Pavel Vasilievich. Technology of repair of power hydraulic cylinders of agricultural machinery by the electrospark method: the dissertation ... candidate of technical sciences: 05.20.03 / Chumakov Pavel Vasilyevich. - Saransk, 2013 .- 238 p.

IOP Conf. Series: Materials Science and Engineering **915** (2020) 012007 doi:10.1088/1757-899X/915/1/012007

# Research of parameters of hydrostatic transmission with axial flow divider for cases of multi-drive vehicle maneuvering

**A M Belousov, D L Karelin, A V Boldyrev and S V Boldyrev**

Kazan Federal University, Naberezhnye Chelny Institute, Naberezhnye Chelny, Republic of Tatarstan, Russian Federation

belousoff.alexey@yandex.ru

**Abstract**. The article considers a mathematical model of hydrostatic transmission with an axial flow divider for a multi-drive vehicle during its maneuvering. The new method for controlling the energy parameters of the wheel drive by adjusting the angles of swash plates for each section of the axial flow divider depending on the radius of rotation of the multi-drive vehicle is presented and static characteristics are constructed.

**1. Introduction** The use of hydrostatic transmission (HST) for multi-drive vehicles has several significant advantages relative to transmissions with electrodynamic or mechanical drives. Therefore, the works of many authors are devoted to the study of new designs of HST and its control systems to increase its energy efficiency. In particular, the works [1-4] consider mathematical models of the HST with special emphasis on the swash-plate mechanism of adjustable pump [1], the HST using a hydraulic accumulator for energy recovery [2] and the HST of off-road multi-axle wheeled vehicles [3, 4].

One of the functions of transmission is the distribution of moments and frequencies of rotation of the wheels when the vehicle is moving along a curved path. When using hydrostatic transmissions, the wheels speed is controlled by changing the fluid flow through the hydraulic motor-wheels.

## 2. Materials and methods

In this paper, the use of an axial flow divider [5] is considered for independent control of angular velocity of each hydraulic motor in case of curvilinear movement of a vehicle. For the plotting of static characteristics of HST there was used the theoretical method of numerical research using the mathematical equations presented in the article.

Calculation of static characteristics was carried out using MathCad v.14 program under the following assumptions: the roughness of the roadway, the leakage and compressibility of the working fluid was not taken into account.

To calculate the kinematics of the vehicle's turning there was used the scheme shown in Fig.1.

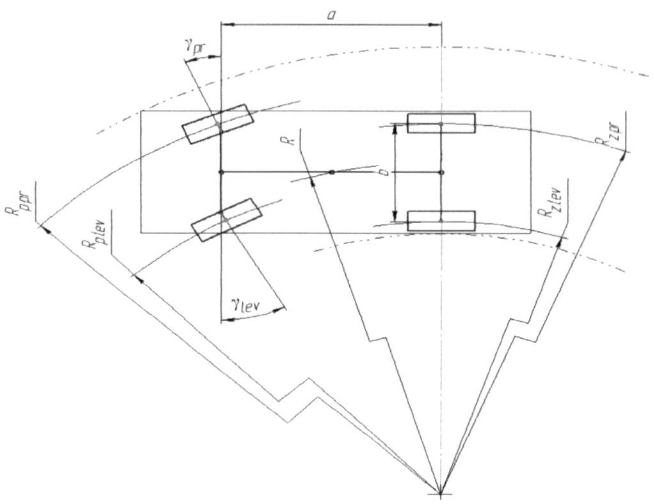


Fig.1. Scheme of the turning kinematics of two-axial vehicle

The turning radii of the rear left, rear right, front left and front right wheels are determined from equations (1) - (4), respectively.

$$R_{z.lev} = R - b/2 \quad , \tag{1}$$

$$R_{z.pr} = R + b/2 = R_{z.lev} + b \quad , \tag{2}$$

$$R_{p.lev} = ((R + b_p/2)^2 + a^2)^{0,5} - b_k \quad , \tag{3}$$

$$R_{p.pr} = ((R - b_p/2)^2 + a^2)^{0,5} + b_k \quad , \tag{4}$$

here $R$ – turning radius of the center of the rear axle, m; $a$ – wheelbase of a vehicle, m; $b$ – distance between centers of rear left and rear right wheels, m; $b_p$ – distance between front axle joints, m; $b_k$ – distance from joint to wheel center, m.
The turning angles of kinematically connected right and left front wheels are determined from equations (5), (6):

$$\gamma_{pr} = arctg(a/(R + b/2)) \quad , \tag{5}$$

$$\gamma_{lev} = arctg(a/(R - b/2)) \quad . \tag{6}$$

Transforming equations (5), (6), we obtain expressions for determining the radius of the center of rear axis:

$$R = a/tg(\gamma_{pr}) - b_p/2 = a/tg(\gamma_{lev}) + b_p/2 \quad . \tag{7}$$

Angular velocity of the vehicle when turning

$$\omega_a = \upsilon / R \quad , \tag{8}$$

here $\upsilon$ – linear velocity of the vehicle, m/s.
Rotation frequencies of each wheel

$$n_{z.lev} = 2\omega_a R_{z.lev} / (\pi^2 R_k) \quad , \tag{9}$$

$$n_{z.pr} = 2\omega_a R_{z.pr} / (\pi^2 R_k) \quad , \tag{10}$$

$$n_{p.lev} = 2\omega_a R_{p.lev} / (\pi^2 R_k) \quad , \tag{11}$$

$$n_{p.pr} = 2\omega_a R_{p.pr} / (\pi^2 R_k), \tag{12}$$

here $R_k$ – the wheel radius, m.

For the control system it is necessary to determine the total rotation frequency of all hydraulic motor-wheels $n_{max}$ according to the equation:

$$n_{\max} = n_{z.lev} + n_{z.pr} + n_{p.lev} + n_{p.pr}. \tag{13}$$

and inclination rates of swash plates of divider's sections

$$k_{z.lev} = n_{z.lev} / n_{\max}, \tag{14}$$

$$k_{z.pr} = n_{z.pr} / n_{\max}, \tag{15}$$

$$k_{p.lev} = n_{p.lev} / n_{\max}, \tag{16}$$

$$k_{p.pr} = n_{p.pr} / n_{\max}. \tag{17}$$

Then the rotation frequency of the axial divider shaft

$$n_d = \frac{q_m}{q_{\max.s}} \cdot \frac{n_{z.lev} + n_{z.pr} + n_{p.lev} + n_{p.pr}}{k_{z.lev} + k_{z.pr} + k_{p.lev} + k_{p.pr}}, \tag{18}$$

here $q_m$ – displacement of hydraulic motor, cm$^3$; $q_{max.s}$ – maximum displacement of divider section , cm$^3$.
Fluid flow rates through hydraulic motors without leakage

$$Q_{z.lev} = n_{z.lev} \cdot q_m, \tag{19}$$

$$Q_{z.pr} = n_{z.pr} \cdot q_m, \tag{20}$$

$$Q_{p.lev} = n_{p.lev} \cdot q_m, \tag{21}$$

$$Q_{p.pr} = n_{p.pr} \cdot q_m. \tag{22}$$

Total fluid flow rate supplied by the pump group:

$$Q_{pump} = Q_{z.lev} + Q_{z.pr} + Q_{p.lev} + Q_{p.pr}. \tag{23}$$

According to the presented equations, the calculation of the static characteristics of HST for curvilinear motion of a vehicle was performed using the KAMAZ-43501 as an example. As input data for calculations the actual parameters of the vehicle were used: $R_k$=570 mm; $q_m$=$q_{max.s}$= $8{\cdot}10^{-5}$ m$^3$; $a$=3670 mm; $b$=2105 mm; $b_p$=1710 mm; $b_k$=197,5 mm; $\upsilon$=50 km/h.

## 3. Results and Discussion

Figure 2 presents the results of calculations obtained using the mathematical model. It can be noticed that in case of one front driven axle, the rotation frequencies of hydraulic motor-wheels on the right side in the motion direction of the vehicle turning to the left increase with increasing turning angle (fig.2a). At the same time, the rear left motor-wheel slows down, and the front left motor-wheel first slows down in the range of rotation angles 5°…25°, and then it smoothly accelerates. This is explained by different values of the radii $R_{z.lev}$, $R_{z.pr}$, $R_{p.lev}$, $R_{p.pr}$ relatively to instantaneous turning center of O (fig. 1). The flow rate of the group of hydraulic pumps increases with increasing turning angle of the vehicle (fig. 2b).

IOP Conf. Series: Materials Science and Engineering **915** (2020) 012007 doi:10.1088/1757-899X/915/1/012007

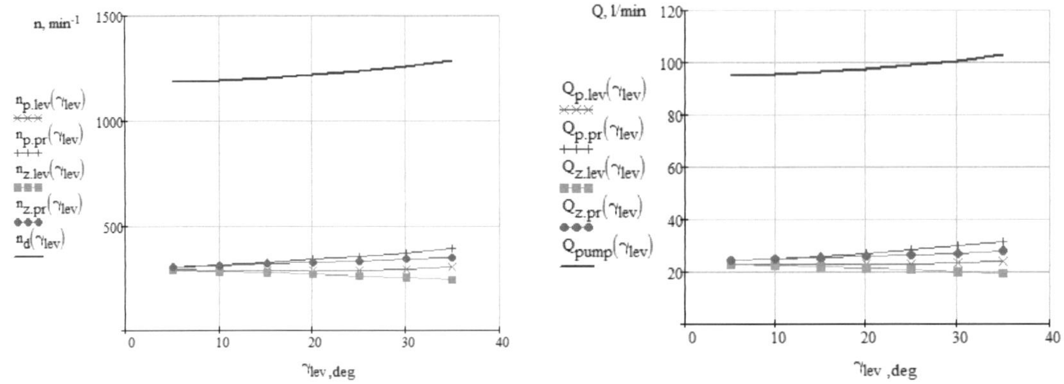


a) b)

Fig.2. Fluid flows rates of hydraulic motors and required pump flow rate (a) and rotation frequencies of hydraulic motor-wheels and the divider's shaft (b) depending on the turning angle of the front left wheel

## 4. Conclusion

The presented static characteristics show the possibility of implementing independent volumetric regulation of rotation frequency of several undjustable hydraulic motor-wheels (for example, high-torque radial-piston hydraulic motors) having a common drive from one or a group of adjustable pumps. In this case, the electronic control system for the drive of considered HST, based on the developed mathematical model, will provide a qualitative ratio between rotation frequencies of the hydraulic motor-wheel shafts by adjusting the inclination of swash-plates of corresponding sections of the axial flow divider.

## References


[1] Kugi A, Schlacher K, Aitzemuller H, Hirmann G 2000 Modeling and simulation of a hydrostatic transmission with variable-displacement pump *Mathematics and Computers in Simulation* Vol. 53 pp.409-414

[2] Ho T H, Ahn K K 2010 Modeling and simulation of hydrostatic transmission system with energy regeneration using hydraulic accumulator *Journal of Mechanical Science and Technology* Vol. 24 pp.1163-1175.

[3] Comellas M, Pijuan J, Potau X, Nogues M, Roca J 2012 Analysis of a hydrostatic transmission driveline for its use in off-road multiple axle vehicles *Journal of Terramechanics* Vol. 49 pp.245-254.

[4] Comellas M, Pijuan J, Potau X, Nogues M, Roca J 2013 Efficiency sensitivity analysis of a hydrostatic transmission for an off-road multiple axle vehicle *International Journal of Automotive Technology* Vol. 14 No. 1 pp.151-161.

[5] Mulyukin V L, Boldyrev A V, Karelin D L, Belousov A M 2017 Simulation of parameters of hydralic drive with volumetric type controller *IOP Conf. Series: Materials Science and Engineering*, vol. 240

# Simulation of heat flows in a combined process of plasma spraying and coat hardening

**A M Kadyrmetov[1], S N Sharifullin[2], E V Snyatkov[1], A A Plahotin[1], I A Mandrykin[1], V V Romanov[1]**

[1] Voronezh State Forestry Engineering University named after G.F. Morozov, Russia,
[2] Kazan (Volga) Federal University, Russia

kadyrmetov.a@mail.ru, Saidchist@mail.ru

**Abstract:**The simulation technique of heat distribution in a detail for applying plasma coating in combination with electro-mechanical treatment and cooling by water jet is worked out. It is shown that water cooling substantially decreases the workpiece surface temperature and provides the formation of more rugged coat by reducing of internal stresses in it.

## 1. Introduction

One of the effective method of detail surfaces restoring is the plasma coating with concomitant electromechanical breaking in by the roller (Fig. 1) [1, 2]. However, in such process details can get burned and the coats can accumulate excess internal stresses.

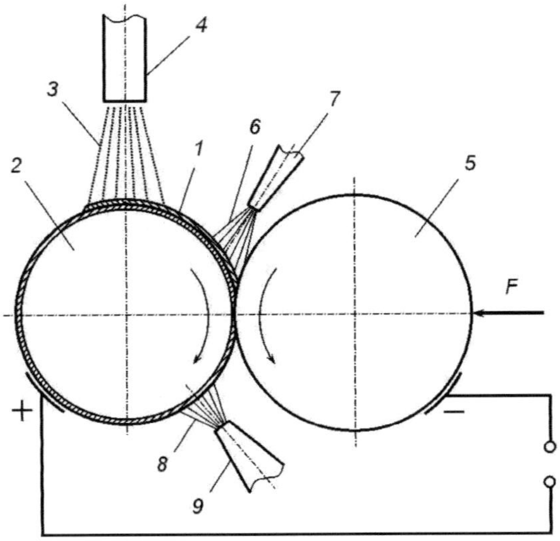


**Figure 1.**. Diagram of combined plasma spraying coating. *1* – coat, *2* – detail, *3* – plasma jets, *4* –plasmatron, *5* – roller, *6, 8* – cooling jet, *7, 9* – injectors, *F* – roller loading force

Modeling of heat extension in the combined method of plasma spraying coating for the actual shapes of details, several ambients and heat dynamics is exceedingly difficult task [3, 4]. In this paper we developed a model, which is based on the equations of classical thermodynamics. In this case, the task complexity for accounting of all external conditions overcomes by using the disretization of space, the algorithmization by the numerical methods of calculation on the base of the grid finite-difference methods and programming [4-5].

## 2. Materials and methods

In the three-dimensional case, heat propagation is described by the heat equation:

$$\frac{\partial}{\partial t}T(\vec{r},t) = \left(\nabla, \chi(\vec{r},t)\nabla T(\vec{r},t)\right) + Q(\vec{r},t), \quad (1)$$

wherein $T(\vec{r},t)$ – the temperature distribution in the space and its variation during some time; $\vec{r}$ – radius vector of the investigated point in space; $t$ – is the time; $\nabla = \frac{\partial}{\partial x}\vec{\imath} + \frac{\partial}{\partial y}\vec{\jmath} + \frac{\partial}{\partial z}\vec{k}$ – differential operator nabla; $x$, $y$, $z$ – Cartesian coordinates of the investigated point in space; $\vec{\imath}$, $\vec{\jmath}$, $\vec{k}$ – unit vectors in Cartesian space; ( , ) – scalar product; $\chi(\vec{r},t)$ – substance temperature conductivity coefficient(generally depends on the position in space and time); $Q(\vec{r},t)$ – the thermal field changing over time from the sources of heating and cooling in this coating scheme. Temperature conductivity coefficient can be expressed through the coefficient of thermal conductivity $\kappa$, heating capacity $c$ and substance density ρ by the dependence $\chi = \kappa / (c\rho)$ [4].

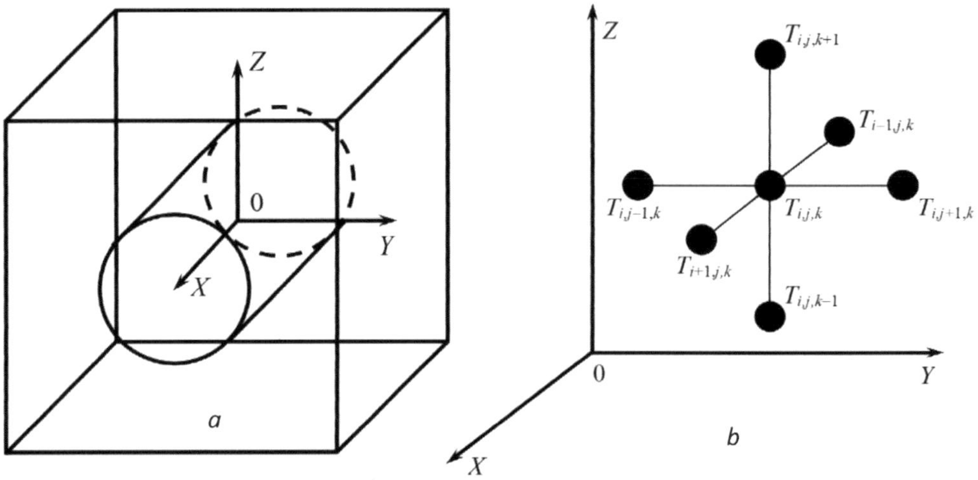


**Figure 2.** Representation of the geometry of the shaft in the model (*a*) and the crucial scheme of accounting for adjacent nodes with the grid option of the thermal conductivity equation (*b*).

For sufficiently adequate repeating of the detail form in the model the space in which the modeling is performing, is discretised in the rectangular grid with a step $h$ = 1 mm (Fig. 2, a). The length of the space in each of the three spatial directions $X$, $Y$, $Z$ is 100 grid cells (corresponding to a size of 100 mm). In this case, the total number of cells is $100^3$ = 1 000 000.

The grid for the solution of the thermal conductivity equation has the form, represented in Fig. 2, *b*. Each grid node has six adjacent nodes from which reception or transmission is possible. In the finite-difference statement of the problem, the equation (1) is transformed as follows. For each node ($i,j,k$) on each step of integration in time temperature $T_{i,j,k}$ depends on the temperature of adjacent nodes as follows.

$$\frac{\Delta T_{i,j,k}}{\Delta t} = \chi_{i,j,k}\left(\frac{\Delta T_{i,j,k}}{(\Delta x)^2} + \frac{\Delta T_{i,j,k}}{(\Delta y)^2} + \frac{\Delta T_{i,j,k}}{(\Delta z)^2}\right) + Q_{i,j,k} \quad (2)$$

where $\Delta t$ is the time discretization step; $\Delta x = \Delta y = \Delta z = h$ is the discretization step of space; $\chi$ is the temperature conductivity coefficient; $Q_{i,j,k}$ – heat input from the external environment to a given cell or heat removal from it. If we spell out details of finite differences in (2) we obtain the following final formula for thermal calculating [5]:

$$\frac{T^{\tau+1}_{i,j,k}-T^{\tau}_{i,j,k}}{\Delta t} = \frac{\chi_{i,j,k}}{h}\left(T^{\tau}_{i+1,j,k} + T^{\tau}_{i-1,j,k} + T^{\tau}_{i,j+1,k} + T^{\tau}_{i,j-1,k} + T^{\tau}_{i,j,k+1} + T^{\tau}_{i,j,k-1} + 6T^{\tau}_{i,j,k}\right) + Q_{i,j,k}. \quad (3)$$

This formula enables the calculating of each cell ($i$, $j$, $k$) temperature $T^{\tau+1}_{i,j,k}$ for the next step of the integration time $\tau + 1$ based on the actual temperature $T^{\tau}_{i,j,k}$ of actual integration step $\tau$.

Heat propagation problem is solved for media of two types: metal of the shaft ($\chi = 0.25$W / (m·K)) and gas surrounding the shaft ($\chi = 0.026$ W/(m·K)). At the initial moment of time, all grid nodes have the same room temperature ($T = 20$ °C). Unlike from the actual coating process, a detail in the model is stationary and the heating and cooling spots representing the system of coating, move around the detail in the plane $x = 0$.

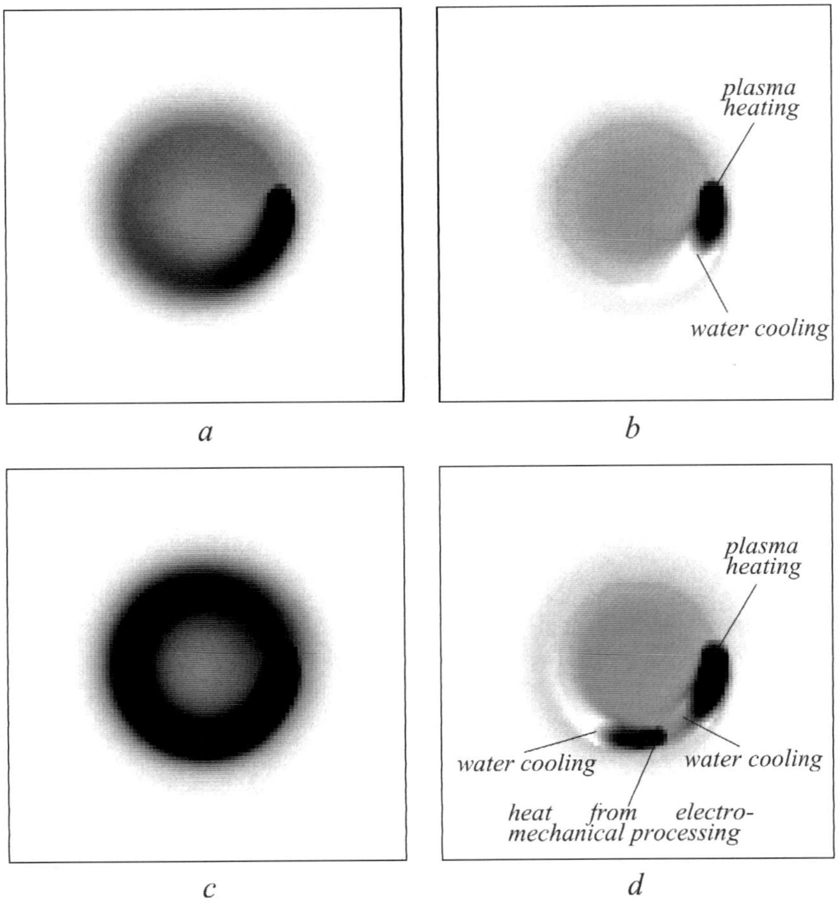


**Figure 3**. Illustration of improved surface cooling of the detail: temperature distribution maps in the cross section of the neck of the crankshaft for plasma spraying
*a, c* - without water cooling; *b, d* - with cooling of the shaft by one or two water jets; a, b - without electromechanical processing; *c, d* - with electromechanical processing

Thermal process simulation program is developed in the Object Pascal programming in the integrated programming environment Borland Delphi 7. The essential program technical limitations: a step of integrating for differential equations is not more than 0.01 s; the maximum size of the cube space discretization is not more than 1 mm. The parameters of the computer model were close to the following real parameters of the coating process: substrate - steel; Ni55%+Ti45% powder; the diameter of the powder particle (model element) is 10 microns; effective roller radius is 10 mm; typical thickness of a single coat layer is 50 microns; the speed of movement of the plasmatron and the roller relative to the surface is 1 sm/s; powder consumption is 10 mg / s; characteristic roller pressure is 100 MPa.

## 3. Results and discussions

Sing the developed model, five computer experiments were conducted to simulate the main options for the operation of a plasma spraying system (Fig. 3). On the temperature distribution maps dimming level is proportional to temperature. The temperature of the detail surface is reduced from 600 °C (Fig. 3, *a*, *c*) to 200 °C with cooling by one water jet subsequently the spraying spot (Fig. 3, *b*) and up to 150 °C at cooling by two water jets (Fig. 3, *c*). Fig. 4 represents the calculation example of temperature profile.

Analyzing the cases of coating without additional electromechanical treatment (lanes 1 and 2), it can be concluded that the cooling by water jet reduces the average temperature of the detail surface along the line of movement of the plasmatron to 50-200 °C. In the case of additional electromechanical treatment without the water cooling the average surface temperature is about 200 °C higher than in the case of its absence. And using two water cooling jets allows to maintain the average temperature of the detail at an acceptable level (about 150 °C). Analyzing the profiles of the curves we can conclude that the every single water jet gives the reducing of the average temperature along the line of movement of the plasmatron to about 150-200 °C.

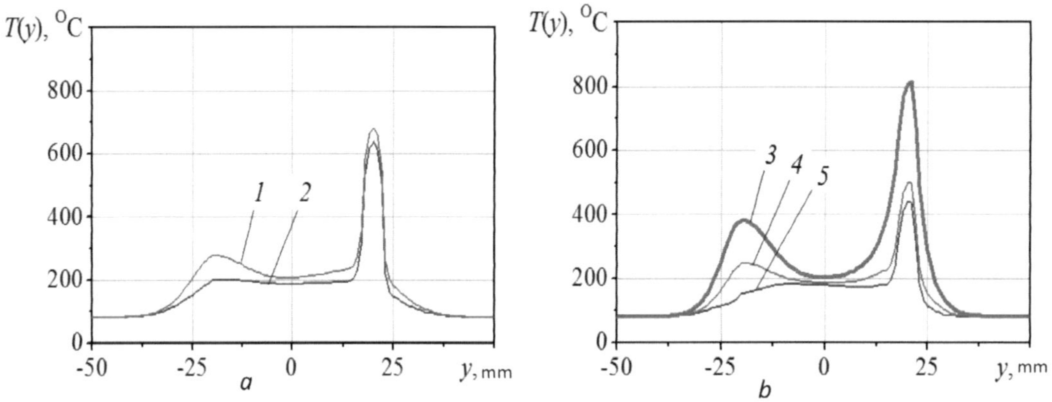


**Figure 4**. The temperature distribution in the cross section of the shaft (in the "plasma heating spot") during plasma coating without (*a*) and additional electromechanical treatment (*b*): *1-5* - number of computer experiment

## 4. Conclusions

Water cooling substantially reduces the temperature of the detail surface what contributes the formation of a more durable coating by reducing the internal stresses in it. In the case of the additional electromechanical surface treatment it is appropriate to use two cooling water jets: before and after roller pass.

## References

[1]. Kadyrmetov, A. M., Pat. Russia №2480533, IPC S23S 4/18, V24V 39/06, 9/00 V23N Method of combined hardening of parts /Kadyrmetov A. M., Posmetyev V. I., Posmetyev V. V., Nikonov V. O., Suhochev G. A., Maltsev A. F. – No. 2011140996/02; appl. 10/11/2011; publ. April 2, 2013, b Jul. No. 12. – 8 p.

[2]. Suhotchev, G. Computer generated spray coating modeling with simulta-neous mechanical and electromechanical processing /G. Suhotchev; A. Kadyrmetov; E. Pamfilov //2015

International Conference on Mechanical Engineering, Automation and Control Systems (MEACS), IEEE Conference Publications, 1-4 Dec. 2015, Tomsk, Russia, DOI: 10.1109/MEACS.2015.7414923. – 2015. – pp. 1-4.

[3]. Sovetov, B. Ya. Modeling of systems: a training manual / B. Ya. Sovetov, S. A. Yakovlev. – M.: Higher School, 1998. – 319 p.

[4]. Polyanin, A. D., Linear problems of heat- and mass transfer: Basic formulas and results / A. D. Polyanin // Theoretical Foundations of Chemical Engineering. –2000. – T. 34. – No. 6. – P. 563-574.

[5]. Grinchik, N. N. To the problem of non-isothermal mass transfer in porous media /N. N. Grinchik, P. V. Akulich, P. S. Kuts, N. V. Pavlyukevich, V. I . Terehov //Engineering physical journal, 2003. – T. 76. – No. 6. – P. 129-142.

IOP Conf. Series: Materials Science and Engineering **915** (2020) 012009 doi:10.1088/1757-899X/915/1/012009

# Increase in efficiency of high-viscosity fluid transportation on the facilities of FEC

**A N Dolgova[1,2], A V Akhmerov[2], A L Osipov[2], Yu V Snigireva[2], O G Dudarovskaya [2] and E S Karataeva[3]**

[1] Rostov State University of Railway Engineering, Rostov-on-Don city, square of Rostovskiy strelkoviy polk narodnogo opolcheniya, bld. 2, Russia
[2] Kazan State Energy University, Kazan city, Krasnoselskaya str., bld. 51, Russia
[3] Kazan Federal University, Engineering Institute, 420021, Kazan, S.Saydashev st., bld. 12

e.s_karataeva@mail.ru

**Annotation.** The thermal method of increasing the efficiency of high-viscosity fluid transportation is considered, using the example of M100 fuel oil. The graphical dependences of the dynamic fluid viscosity on the temperature, the kinematic viscosity on the consumption of a heating steam and on the power of heating pads provided that high-viscosity fluid is heated, as well as the Prandtl number on the heat output of a heater, are obtained.

The high-viscosity fluid is a fluid in which the kinematic viscosity at a temperature of the medium surrounding the pipeline is so great that pumping of such a substance is associated with very large pressure losses. To increase the efficiency of high-viscosity fluid transportation, thermal heating is widely used. Let us consider this method as an example of the high-viscosity fluid - residual fuel oil of mark M100.

Nowadays mazout is widely used on the facilities of FEC (Fuel and Energy Complex) as a fuel for steam boilers and industrial furnaces, and certain types of residual fuel oils are used to lubricate rough machinery. The methods and duration of drain and bulk operations, the fuel transportation conditions are determined by viscosity.

The viscosity is a main technical property of residual fuel oils and it determines their marking. The viscosity of residual fuel oils is determined by a composition of the oil from which they are obtained, by a nature of the technical process and a depth of selection of the oil and lighter fractions. It strongly depends on temperature and can double when the temperature decreases by 10 °C. Under such the conditions, a flow of the residual fuel oil does not qualitatively differ from a flow of some ordinary homogeneous fluid and corresponds to the Newton equation, however, it requires increased energy consumptions, since the viscosity of residual fuel oil is much higher than the viscosity of light oil products [1-3].

The dynamic viscosity of the residual fuel oil of mark M100 can be calculated according to the Walter formula, which is valid in the temperature range 40-100°C [1]:

$$\mu = \left[\exp_{10}\left(\exp_{10}\left[9{,}855 - 3{,}751 \cdot \lg(t + 273)\right]\right) - 0{,}8\right] \cdot 10^{-6} \cdot \rho$$

where $\mu$ *is* a coefficient of dynamic viscosity of fuel oil, Pa·s; $t$ is a temperature of residual fuel oil, $\rho$ is a density of residual fuel oil, kg/m³.

Fig. 1 shows a dependence of the dynamic viscosity of fuel oil on temperature. Verification of the adequacy of the expression was carried out by comparing with the experimental data of Zdor V.O. The experimental data correlate well with the calculated data. The discrepancy was less than 10 %.

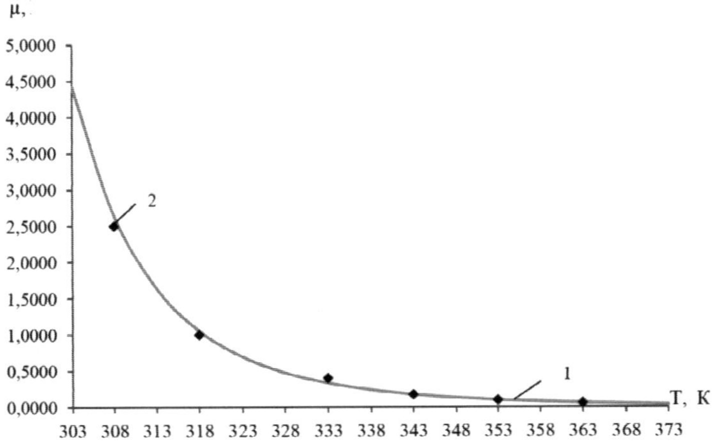


Fig. 1. Dependence of dynamic viscosity on temperature: 1 – calculation according to the Walter equation; 2 – experimental data of Zdor V.O. [2-4]

Thermal heating of the high-viscosity fluid in FEC is carried out in pipelines and special heaters. For heating in pipelines, heaters are used in the form of steam satellites, which are located in one heat-insulating casing with a pipeline being heated. Electric heating systems are also used: of direct (current is passed directly through the pipeline being heated) and indirect heating (with the use of a satellite heater which is parallel or spirally wound onto the pipeline and in the form of heating cables and tapes). Heat-exchange apparatuses of various designs are used as special heaters.

The amount of heat, obtained by the high-viscosity fluid, is determined by the expression:

$$Q = Gc_p\rho(t_2 - t_1).$$

where $G$ is a volume flow of the high-viscosity fluid, m$^3$/h; $c_p$, $\rho$ are a density (kg/m$^3$) and a specific thermal capacity of the high-viscosity fluid (J/(kg·K)); $t_2$, $t_1$ are a final and initial temperature of the high-viscosity fluid at the outlet and inlet of the preheater, K.

The Prandtl number for the high-viscosity fluid can be determined by the formula:

$$\mathrm{Pr} = \frac{\nu \rho c_p}{\lambda},$$

where $\nu$ is a kinematic viscosity, m$^2$/s; $\lambda$ is a thermal conductivity, W/(m·K).

Using these expressions, the dependence of the Prandtl number on a thermal performance of the heater is determined (Fig. 2). The calculation is made at a flow rate of the residual fuel oil 0.36 m$^3$/h and a temperature range 30-100 °C. The Prandtl number characterizes physical properties of the fluid; it is a measure of the similarity of temperature and velocity fields in a heat carrier flow. With an increase in the thermal performance of the heater, the Prandtl number decreases [7-9].

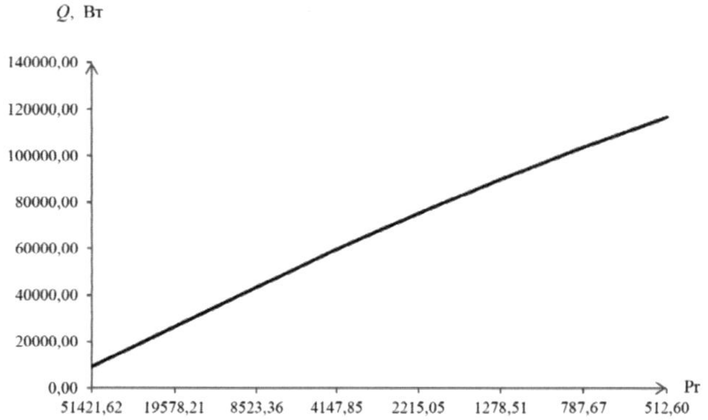


Fig. 2. Dependence of the Prandtl number on thermal performance of the heater

Fig. 3 and 4 show the dependences of the kinematic viscosity changes on thermal heating by a heating steam and electric pads.

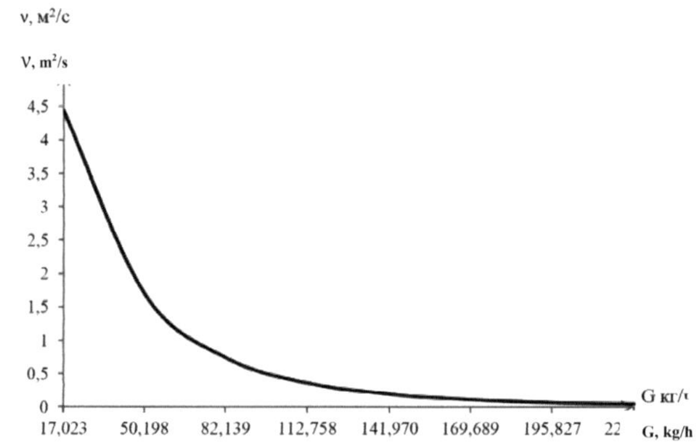


Fig. 3. Dependence of the kinematic viscosity on the heating steam consumption

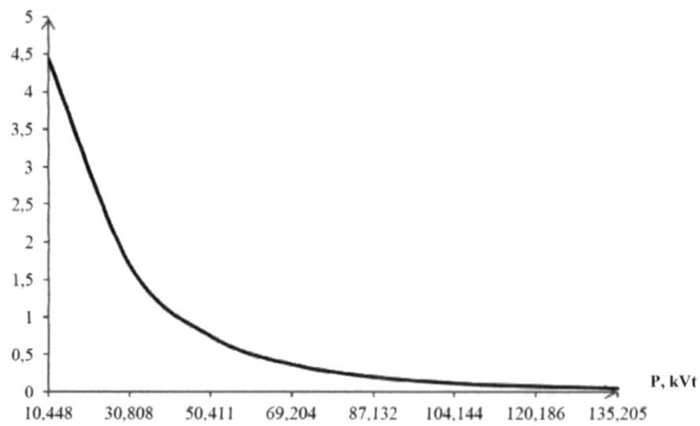


Fig. 4. Dependence of the kinematic viscosity on the electric pad power capacity

According to the obtained dependences, it is clear that during the temperature increase, the viscosity of the residual fuel oil M100 decreases. The optimal dynamic viscosity parameters for transportation of the considered high-viscosity fluid through pipelines and for a normal operation of residual fuel oil pumps are 0,17-0,32 Pa·s, which corresponds to 60-70°C, with a further increase in the temperature of the substance, the viscosity reduction is insignificant. At a consumption of the high-viscosity fluid 3,6 $m^3/h$, the required amount of the heat energy transferred to reduce the viscosity is 112-147 kg/h of heating steam, and the required power of electric pads at 1 hour of operation is about 70 000 W.

**References**

[1] Nazmeev Yu G Residual fuel oil facilities of TPS. M.: Publishing House MPEI, 2002. 612 p.
[2] Zdor V O, Vachargina E K, Kamalov R F and Karaeva Yu V Rheological properties of residual fuel oil of mark M100 *Bulletin of the Technological University*, 2018. Vol.21. No. 9. pp.66-68.
[3] Yarullina A A Improving the rheological properties of high-viscosity fluids: bac. of WRC. Kazan, 2019. 70 p.
[4] Gil'Manshin, I.R., Kashapov, N.F. Energy service contracts in regional engineering center for small and medium businesses IOP Conference Series: Materials Science and Engineering, 69 (1), статья № 012010.
[5] Karataev O.R, Karataeva E.S., Mechanical filtration, based on elective concentration of particles, as an innovative method of water treatment//IOP Conference Series: Materials Science and Engineering. - 2018. - Vol.412, Is.1. - Art. No. 012030.
[6] Karataev O.R, Novikov V.F, Karataeva E.S, Use of modified molecular sieves in mechanical filters//IOP Conference Series: Materials Science and Engineering. - 2019. - Vol.570, Is.1. - Art. No. 012035.
[7] Haibullina A.I., Ilyin V.K., Sabitov L.S., Hayrullin A.R., Dolgova A.N., Zinnatullin N.H. Thermal and hydraulic efficiency of the staggered tube bundle in pulsating flow// IOP Conference Series: Materials Science and Engineering. - 2018. - Vol.412, Is.1. - Art. No. 012027.
[8] Akhmerov A.V., Osipov A.L., Dolgova A.N., Faizullina G.F., Sabitov L.S. Study of process of intensification of dehydration of secondary activated sludge for its utilization as a secondary energy source//
[9] Zinnatullin N.H., Zinnatullina G.N., Haibullina A.I., Sabitov L.S., Kashapov N.F. Minimum density of irrigation of the rotor of a film centrifugal heat and mass-exchange apparatus// IOP Conference Series: Materials Science and Engineering. - 2019. - Vol.570, Is.1. - Art. No. 012006.

IOP Conf. Series: Materials Science and Engineering **915** (2020) 012010 doi:10.1088/1757-899X/915/1/012010

# Comparative efficiency of pipeline heat-insulating materials

**A N Dolgova[1,2], S A Yunak[1], A S Rudenko[1], A I Haibullina[2] and E S Karataeva[3]**

[1] Rostov State University of Railway Engineering, Rostov-on-Don city, square of Rostovskiy strelkoviy polk narodnogo opolcheniya, bld. 2, Russia
[2] Kazan State Energy University, Kazan city, Krasnoselskaya str., bld. 51, Russia
[3]Kazan (Privolzhsky) Federal University, engineering institute, Kazan city, Saidashev street, bld.12, Russia

e.s_karataeva@mail.ru

**Annotation.** The paper considers the problem of reducing heat losses in the heat supply systems of the FEC facilities (fuel and energy complex) due to the the use of heat insulation of the pipelines. A comparative analysis of heat insulation of the heat lines: PUF (polyurethane foam), mineral wool, foam rubber, ultrathin heat insulation of the mark "Corund". The graphical dependences of heat losses and the efficiency factor of heat insulation on a diameter the pipelines are obtained.

Insulation of the pipelines is a very important measure that protects them from various external influences, but also increases the efficiency and also the operational life of the main pipeline. The structure of heat insulation consists of a main heat-insulating layer, an external protective coating and fastening. The main heat-insulating layer provides protection of the insulated surface from heat loss, the outer protective coating protects the main heat-insulating layer from mechanical damage, moisture, exposure to aggressive environments, etc. To isolate the heating networks of the FEC facilities, the task of which is to deliver heat from boiler houses to consumers, different insulating materials are used (tabl. 1). First of all, such materials are faced with the task of reducing heat losses.

When choosing heat-insulating materials and coating layers, the resistance of elements of the heat-insulating structure to aggressive environmental factors should be taken into account, including possible effects of the substances contained in the insulated object, and also these materials must meet hygiene requirements and fire safety ones [1, 2,10].

Mineral wool, which is a fibrous material obtained from slags of metals or molten rocks, is widely used as a heat insulator.

Mineral wool, which is a fibrous material obtained from metal slag or molten rock, is widely used as a heat insulator.

Foam concrete insulation is a light, environmentally friendly material, it is not susceptible to corrosion and over the time it only gains strength.

**Table 1.** Characteristics of insulating materials

| Name of heat-insulating material | Heat conduction coefficient, W/m$^2$ K | Heat resistance, °C | Combustibility |
|---|---|---|---|
| Polyurethane foam (PUF) | 0,036 | -100°C...+150°C | C (combustible) 2-C4 |
| Foam concrete | 0,06 | -150°C...+660°C | NC (non-combustible), LC (low-combustible) |

| | | | |
|---|---|---|---|
| Mineral wool | 0,070 | -180°C...+700°C | NC, C1 |
| Superthin heat insulation «Corund» | 0,0012 | -60°C...+260°C | NC, C1 |
| Foam rubber | 0,03 | -200°C...+5°C | C1-C3 |

PUF insulation is a synthetic substance of a cellular structure from the group of gas-filled plastics, it is environmentally friendly, durable and easy to use. [1].

A liquid ceramic heat-insulating coating «Corund» is a liquid composition based on water. It consists of closed-porous microgranules of foamglass, as well as pigmenting, inhibitive, antipyrene and rheological additives [2,5].

A foamed synthetic rubber has minimal hygroscopicity and vapor permeability, reliably protecting the pipeline from moisture which is condensed with sharp variations in temperature, it does not decay, it is also not subject to burning.

According to the methodology in [3,4, 7-9], a comparative analysis of the following heat-insulating coatings of a heat pipe during above-ground routing was carried out: mineral wool with a coating layer of galvanized steel; polyurethane foam (PUF) with a coating layer of galvanized steel; foam concrete with a coating layer of galvanized steel; ultrathin heat insulation of a mark «Corund»; foam rubber.

The heat calculation of the pipeline was carried out with the following parameters: a temperature of the heat carrier was 100°C; an ambient temperature was -10°C; a wind speed was 5,1 m/s; a diameter of the pipeline was considered in the range of 15-500 mm; a thickness of the heat-insulating material depending on the diameter was 34-54 mm, thin heat-insulating materials were 0,21-25 mm.

Graphical dependences of the efficiency and heat loss coefficients of various heat-insulating materials on a diameter of the pipeline are obtained (Fig. 1, 2).

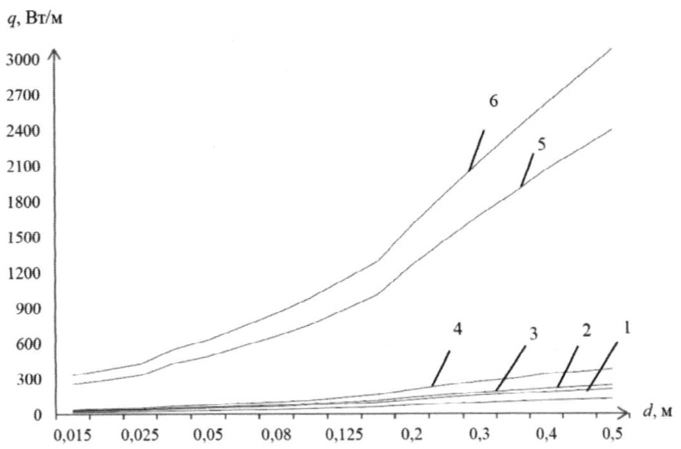


Fig. 1. Dependence of heat losses on a diameter of the pipeline:
1 – PUF; 2 – foam concrete; 3 – mineral wool; 4 – ultra-thin insulation «Corund»; 5 – foam rubber; 6 – pipeline without insulation

Having considered various heat-insulating materials when laying the pipelines above ground, it is advisable to give preference in favor of PUF insulation. The figures 1 and 2 show that it has a high coefficient of heat insulation efficiency and insignificant heat losses compared to other insulating materials.

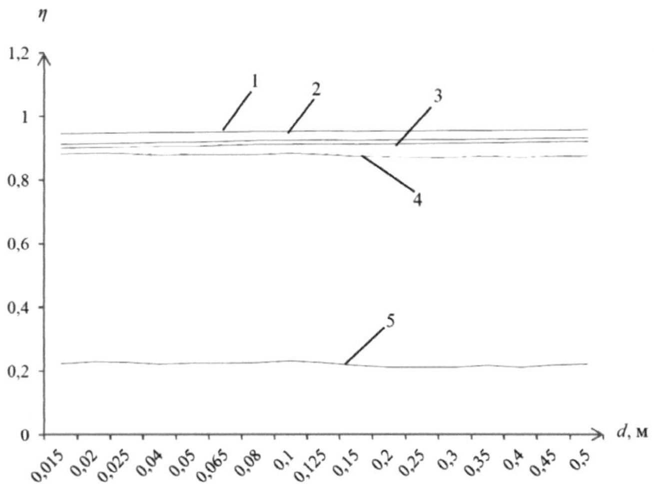


Fig. 2. Dependences of the efficiency coefficient on a diameter of the pipeline: 1 – PUF; 2 – foam concrete; 3 – mineral wool; 4 – ultra-thin insulation «Corund»; 5 – foam rubber

Based on the data obtained, in order to increase the efficiency of this heat-insulating material, 2 options for installing a coating layer are proposed: 1st option - the use of PUF insulation with a coating layer of foam rubber; 2nd option – the use of PUF insulation with a coating layer of galvanized steel and ultra-thin heat insulation «Corund».

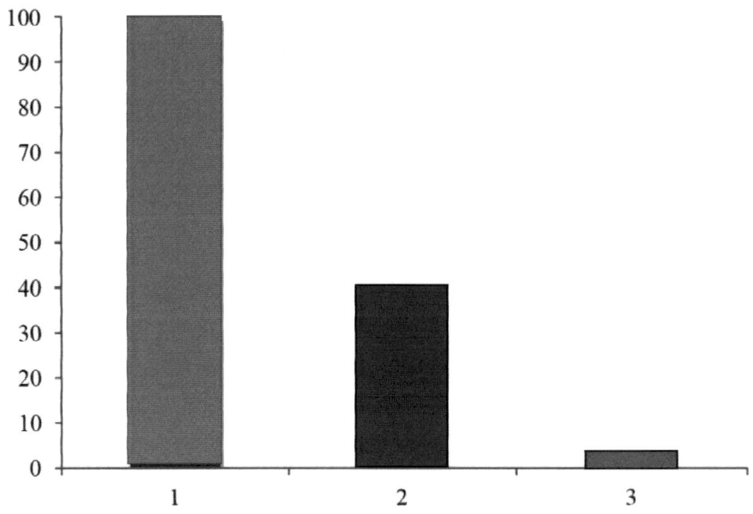


Fig.3. Percentage diagram of the average values of heat losses in three ways of installing a coating layer of PUF insulation: 1 – a coating layer of foamed rubber; 2 – a coating layer of galvanized steel and ultra-thin heat insulation «Corund»

With the expansion of construction activities, reliability of the operation of pipelines and cost effectiveness of laying them become increasingly important, which requires the development of the most rational solutions, confirmed by a technical and economic feasibility. The figure 3 shows that the most effective way of insulating the pipelines during its laying is the use of polyurethane foam insulation with a coating layer of galvanized steel and ultra-thin heat insulation «Corund». This method of insulation has the lowest percentage of heat losses compared with the other proposed. In the proposed methods of insulation, an attempt is made to generalize the existing experience in the design, construction and operation of pipelines and improve this knowledge by introducing new means of insulating pipelines during above-ground routing.

**References**

[1] GOST 30732-2006. Steel pipes and shaped objects with heat insulation from polyurethane foam with a protective sheath. Moscow, 2007. 3 p.
[2] R-NP SRO SSK - 04 -2014. Recommendations on the use of ultrafine liquid insulation «Corund». Chelyabinsk, 2016. 8 p.
[3] Kashapov, L.N., Kashapov, N.F., Kashapov, R.N. Research of the impact acidity of electrolytic cathode on the course of the plasma-electrolytic process (2013) Journal of Physics: Conference Series, 479 (1), № 012011
[4] Moiseev B.V., Zemenkov Yu.D., Nalobin N.V., Zemenkova M.Yu., Dudin S.M. Methods of heat calculation of pipelines for various purposes. Tyumen: TIU, 2016. 19 p.
[5] Safarov D T, Kondrashov A G, Khafizov I I. Development of the design of a laboratory vibro-grinding machine for preparing samples for metallographic research. IOP Conf. Series: Materials Science and Engineering 412 (2018) 012064
[6] Dolgova A.N., Bondar D.S. Aapplication of gas-piston unit in the scheme of energy installation of generation of energy carriers//Proceedings of the Rostov State University of Railways. 2018. No. 4. P. 47-52.
[7] Dautov, G., Kashapov, N., Fayrushin, I., Dautov, I. Characteristics of potential well for nondegenerate electronic gas in a equilibrium dusty plasma (2013) Journal of Physics: Conference Series, 479 (1), № 012014.
[8] Taimarov M.A., Ilyin V.K., Osipov A.L., Dolgova A.N., Akhmerov A.V. Heat and pump complex for disposal of secondary energy resources of petrochemical plants //News of higher educational institutions. Energy issues. 2019.Vol. 21. No. 3-4. S. 8-16.
[9] Gilmanshin, I., Kashapov, N., Gilmanshina, S., Galeeva, A. Landfill energy complex based on the renewable energy installations// IOP Conference Series: Materials Science and Engineering, 134.
[10] Sultanov S.M., Kim C.V., Karataev O.R. Analysis of experimental data and development of methods for calculating the work of screw conveyors//Actual problems of the humanities and natural sciences. 2015. No. 4-4. S. 27-32.

# Study of the influence of hydroabrasive surface cleaning on a laser cutting

**Az T Gabdrakhmanov, I H Israphilov, L N Shafigullin, T F Gabdrakhmanova**

Kazan Federal University, 423800, 68/19 Mira avenue, Naberezhnye Chelny, Russia

ATGabdrahmanov@kpfu.ru, veyron000@mail.ru

Abstract. The example of laser cutting of 1.0038 steel shows the features of waterjet (EPS) cleaning. The influence of laser cutting parameters on ensuring the necessary quality and properties of steel is considered.

## 1. Introduction

The EPS process is an etching process, which does not require the use of acids or harsh chemicals. The process is environmentally friendly, since it does not require the use of hazardous chemicals, like conventional pickling. The finished product has a cleaner and rust-resistant, dross-free surface. Unlike acid etching, EPS is a mechanical process which uses steel corner shots and water circulating in the system for reuse. Since EPS-treated materials are corrosion resistant, they, unlike acid-etched materials, do not require an oil coating to protect against rust. This means that solvents didn't require to remove the oil before painting. In addition, EPS technology actually creates a new surface because it removes secondary scale. On this new surface, there are no such undesirable effects as shells, rust or silicon strips and etchings, which are often found on etched and oiled materials.

The results of a study of the EPS effect cleaning on the quality of laser cutting presents in this paper [1-2].

## 2. Experimental studies

The object of the study was steel 3SP without and with EPS cleaning 4 mm thick. EPS - Eco Pickled Surface. As a working medium, a mixture of water and steel angle shot is used, which is fed by means of turbines with rotating impellers to the surface of the material.

Laser cutting of the samples was carried out on the installation of FL-CUT-3015 (production of LLC NTO IRE-Polyus). The laser cutting unit is designed for high-quality cutting of metal up to 30 mm thick. The laser used was an LS-2 fiber ytterbium laser [3-7].

Samples after laser cutting are shown in Figure 1.

IOP Conf. Series: Materials Science and Engineering **915** (2020) 012011 doi:10.1088/1757-899X/915/1/012011

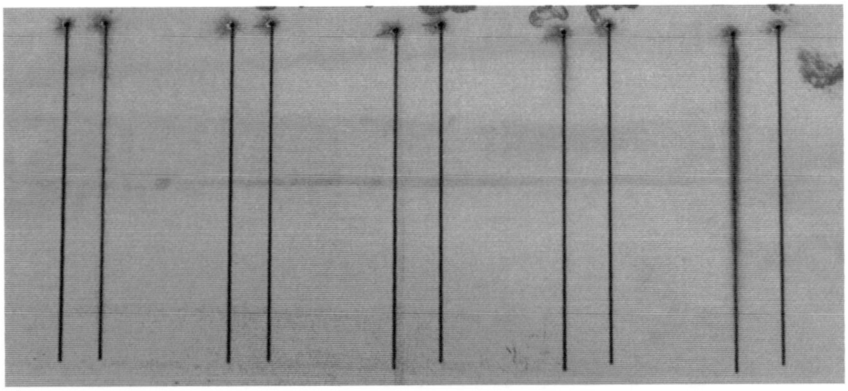

a)

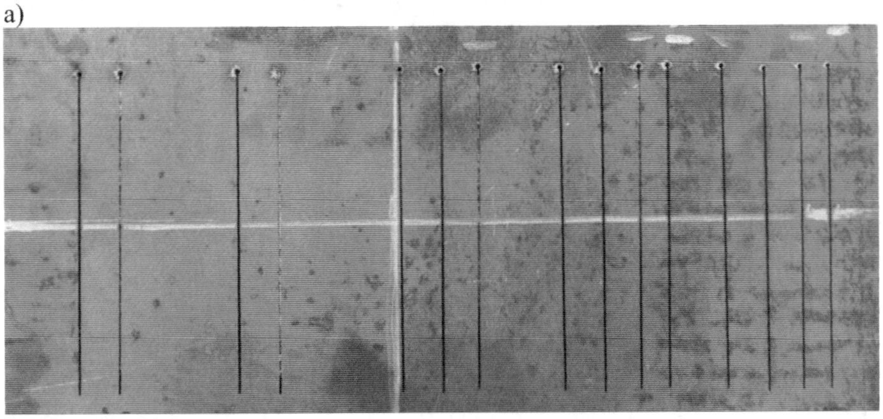

b)

**Figure 1.** Samples after laser cutting; a) with EPS cleaning; b) control without processing

Laser cutting modes are presented in table 1. In all cases, oxygen was used as a working gas at a pressure of 15 atm. The nozzle diameter is 1.2mm.

Figures 2-3 show samples after laser cutting.

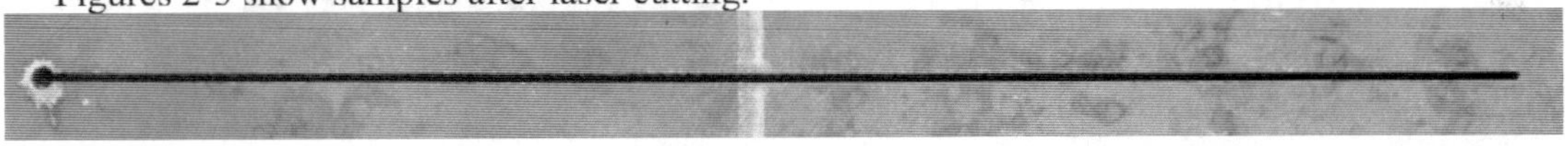

a)

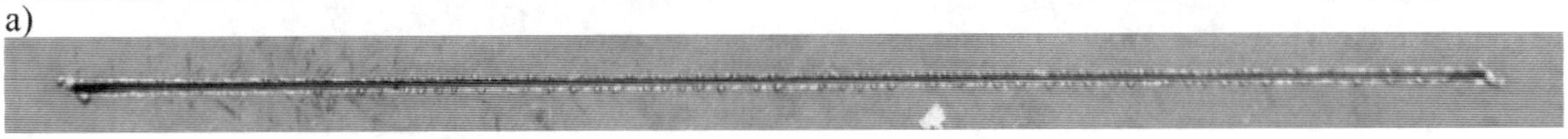

b)

**Figure 2 .** Sample 1 after laser cutting; a) top view; b) bottom view

a)

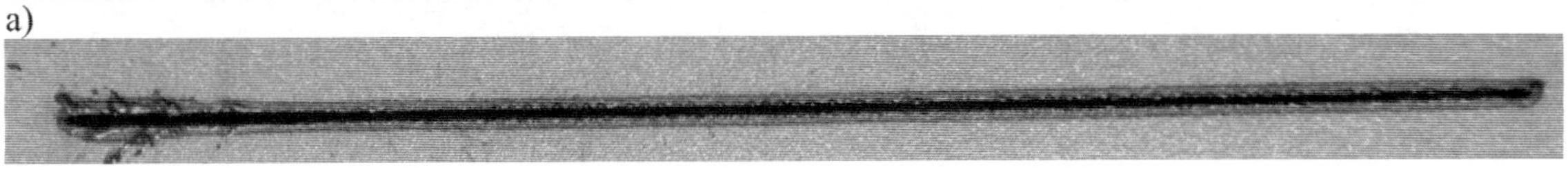

b)

**Figure 3.** Sample 9 after laser cutting; a) top view; b) bottom view

**Table 1.** Laser Cutting Modes

| Sample No. | Radiation power, kW | Laser head moving speed, mm / min |
|---|---|---|
| control | | |
| 1 (optimal) | 2 | 3000 |
| 2 | 2 | 3200 |
| 3 | 2 | 3300 |
| 4 | 2 | 3400 |
| 5 | 2 | 3500 |
| 6 | 2 | 5200 |
| 7 | 1,6 | 3000 |
| 8 | 1,4 | 3000 |
| with EPS cleaning | | |
| 9 (optimal) | 2 | 3000 |
| 10 | 2 | 3100 |
| 11 | 2 | 3200 |
| 12 | 2 | 3400 |
| 13 | 2 | 2800 |
| 14 | 1,8 | 3000 |
| 15 | 1,6 | 3000 |
| 16 | 1,4 | 3000 |
| 17 | 1,2 | 3000 |
| 18 | 1 | 3000 |

The microstructure of the samples was condacted in cross section on a Thixomet motorized metallographic complex with software. Particle microhardness was measured in accordance with GOST R ISO 6507-1-2007 on a Durimet microhardness tester [8-10]. The research results are shown in Figure 4.

×100 ×100

a) b)

**Figure 4.** Microstructures and transverse sections of the samples: a) sample 1, b) sample 9

## 3. The conclusion

Based on visual inspection of the samples after laser cutting according to the modes (table 1), metallographic studies and the results of measuring the microhardness of the base metal and the cutting zone revealed:

1. Not all samples were cut to the entire depth due to a lack of laser power density in some modes. In some cases, grata is formed, especially when the laser radiation power is reduced; in the case of samples with EPS cleaning, the grata value was less.

2. Metallographic studies of the base metal and the metal of the cutting zone revealed an identical ferrite-pearlite structure. There are no microcracks in the heat affected zone.

3. The microhardness of the metal of the heat-affected zone is in the range of 281-488 HV, the base metal is 112-147 HV.

4. The size of the heat-affected zone does not exceed 0.323 mm, and the average cut width is 0.55 mm. In samples with EPS purification, these values were lower.

Comprehensive studies of the quality of laser cutting found that preference is given to samples obtained by EPS cleaning.

**References**

[1] Gabdrakhmanov Az T, Israphilov I H, Shafigullin L N, Gabdrakhmanova T F 2019 Study of laser cutting of copper and copper alloys *IOP Conference Series: Materials Science and Engineering*. Volume 570, Issue 1, , Article number 012020.

[2] Grigoryants A G, Shiganov I N, Misyurov A I 2006 Processes of laser processing *Moscow publishing house MGTU im. N. Uh. Bauman*, 664 p.

[3] Kashapov N F, Sharifullin S N 2015 Hardening of the surface plasma jet high-frequency induction discharge of low pressure *IOP Conference Series: Materials Science and Engineering*, 86 (1), Article number 012021.

[4] Galimov E R, Samoylov V M, Danilov E A, Fedyaev V L, Galimova N Ya and Orlov M A 2017 Production of thermally conductive carbon foams and their application in automobile transport // *IOP Conf. Series: Materials Science and Engineering* vol 240 issue 1 p 1-4.

[5] Kashapov R N, Kashapov L N, Kashapov N F 2017 Formation of cracks in the selective laser melting of objects from powdered stainless steel 17-4 *PH IOP Conference Series: Materials Science and Engineering*, 240 (1), статья № 012074.

[6] Kashapov L N, Kashapov N F, Kashapov R N, Semushin V D 2016 Changes of thermal balance in a plasma-electrolyte system according to the shape of the applied voltage *Journal of Physics: Conference Series*, 669 (1), статья № 012028.

[7] Gabdrakhmanov A T, Galiakbarov A T, Samigullin A D, Galiakbarov R T 2016 The calculation of a thermal field in the surface of a processed partunder the influence of a low-temperature plasma *IOP Conference Series: Materials Science and Engineering* Volume 134, Issue 1, Article number 012040. DOI: 10.1088/1757-899X/134/1/012040.

[8] Gabdrakhmanov A T, Israphilov I H, Galiakbarov A T 2014 The study the erosion of the electrodes under the influence moving electric arc *Journal of Physics: Conference Series* Volume 567, Issue 1, Article number 012013. DOI: 10.1088/1742-6596/567/1/012013.

[9] Gabdrakhmanov A T, Galiakbarov A T, Samigullin A D, Galiakbarov R T 2016 The calculation of a thermal field in the surface of a processed partunder the influence of a low-temperature plasma *IOP Conf. Ser.: Mater. Sci. Eng.*, Volume 134, Issue 1, Article number 012040.

[10] Kashapov L N, Kashapov N F, Kashapov R N, Denisov D G 2016 Plasma electrolytic treatment of products after selective laser melting *Journal of Physics: Conference Series*, 669 (1), статья № 012029, .

IOP Conf. Series: Materials Science and Engineering **915** (2020) 012012 doi:10.1088/1757-899X/915/1/012012

# Research of properties of polymer compositions based on aluminosilicate microspheres

**Az T Gabdrakhmanov, AT Galiakbarov, I Y Karasev, T F Gabdrakhmanova**

Kazan Federal University, 423800, 68/19 Mira avenue, Naberezhnye Chelny, Russia

misharin_82@mail.ru

**Abstract**. The paper presents the results of studies of polymer compositions based on aluminosilicate microspheres. Polyester resin Yarkopol-20 and epoxy glue ED-20, as well as grades of aluminosilicate microspheres K 500 and K150 were chosen as the binder of aluminosilicate microspheres. The obtained thermal conductivity coefficients are comparable with the thermal conductivity coefficients of materials such as fiber asbestos, wood, expanded clay concrete, etc.

**1. Introduction** Aluminosilicate hollow microspheres (ASPM) - glass-crystalline aluminosilicate balls which are formed during high-temperature flaring of coal. They are the most valuable components of ashes from thermal power plants. They are hollow, almost perfect shaped silicate balls with a smooth surface, with a diameter of 10 to several hundred micrometers, an average of about 100 microns. The walls are continuous non-porous with a thickness of 2 to 10 microns, melting point 1400-1500 ° C, density 580-690 kg / m³. The internal cavity of the particles is filled mainly with nitrogen and carbon dioxide.

ASPM content in ash is usually very small, tenths of a percent, however, at large thermal power plants, their "production" can reach several thousand tons per year.

In the field of composite materials [1-5], interest in ultralight energy-absorbing materials and structures that contain hollow particles is growing.

The paper presents the results of studies of polymer compositions based on aluminosilicate microspheres. The polymeric resin Yarkopol-20 and the ED-20 epoxy adhesive were chosen as a binder for ASPM.

## 2. Experimental studies

For experimental studies, aluminosilicate microspheres of the K 500 and K150 grades were used.
The physical characteristics of the microspheres are presented in table 1.
The chemical composition of the microspheres is presented in table 2.

**Table 1.** Physical characteristics of microspheres

| Physical characteristics | K500 | K150 |
|---|---|---|
| Form, appearance | Sphere Floating Bulk | Sphere Floating Bulk |
| Particle size | 0-500 micron | 0-150 micron |
| Density | 0.6-0.8 g/cc (42 lbs/ft3) | 0.69-0.8 g/cc (42 lbs/ft3) |
| Bulk density | 0.35-0.45 g/cc (24 lbs/ft3) | 0.39-0.49 g/cc (24 lbs/ft3) |
| Specific gravity | 0.45-0.75 | 0.45-0.75 |
| Compressive strength | 2500-5000 psi | 2500-5000 psi |
| Mohs hardness | 5-7 | 5-7 |

| | | |
|---|---|---|
| Melting point | 1350 - 1500°C | 1350 - 1500°C |
| Thermal conductivity | 8 x $10^{-6}$(°K) | 8 x $10^{-6}$(°K) |
| Humidity | 0,5 % max | 0,5 % max |
| Precipitation | 5% max | 5% max |
| Wall thickness, from | 5%-10% | - |
| Microsphere diameter | 6 - 7 | - |

**Table 2.** The chemical composition of microspheres

| Element | K500 | K150 |
|---|---|---|
| Silica ($SiO_2$) | 58-68% | 65-70% |
| Alumina ($AL_2O_3$) | 32-38% | 15-21% |
| Iron ($Fe_3O_2$) | 1.4 - 2 % | 1-7 % |
| CaO | 1,9% | 1-4 % |
| MgO | 1% | 0,5-2% |
| $K_2O$+$Na_2O$, no more | 1,5% | 0,1-2% |

The appearance of the microspheres is shown in Figure 1:

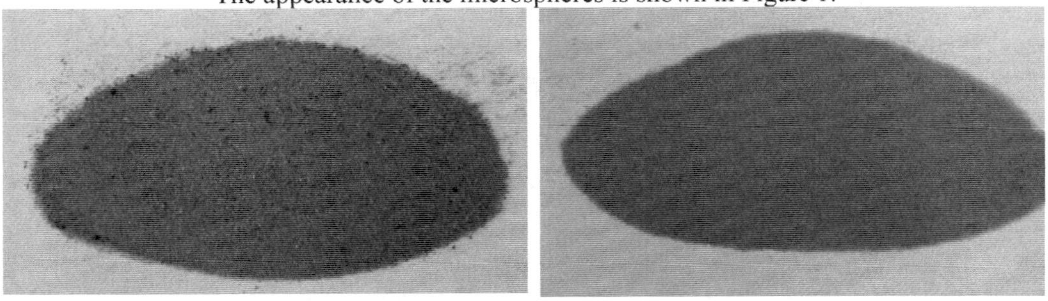

a) b)

**Figure 1.** The appearance of the microspheres: a) K500; b) K150

Mixing was carried out in the ratio of volumes 50 to 50 and 70 to 30, resins and microspheres, respectively. Full setting occurs in 24 hours. Some obtained experimental samples are presented in Figure 2.

a) b)

**Figure 2.** Samples of composite materials: a) Binding agent - Yarkopol -110, K500 microspheres, ratio 70/30, weight 163 g; b) Binding agent - Yarkopol -110, K150 microspheres, 50/50 ratio, weight 154.5 g.

Measurements to determine the thermal conductivity of the samples were carried out on a laboratory setup "Calorimetric chamber", the measurement method is a single plate [6-8].

The amount of heat that passes through a sample of material over time $\Delta\tau$ is proportional to the temperature difference $\Delta T$ between the front and back sides of the plate, the surface area of the plate S and inversely proportional to the plate thickness *d*.

$$\frac{\Delta Q}{\Delta\tau} = \lambda \cdot \frac{S}{d} \cdot \Delta T$$

where $\lambda$ is called the thermal conductivity coefficient.

In the unit plate method, thermal conductivity is determined by direct measurement of the heat flux $\frac{\Delta Q}{\Delta\tau}$.

$$\lambda = \frac{\Delta Q}{\Delta\tau} \cdot \frac{d}{S \cdot \Delta T}$$

The calculation results of various samples are presented in table 3

Thus, the obtained thermal conductivity coefficients are comparable with the thermal conductivity of such materials as fibrous asbestos, wood, expanded clay concrete, etc.

**Table 3.** The calculation results of $\lambda$

| № | Binding Agent | Microsphere brand | Volume ratio | Sample area, $м^2$ | The average thickness of the sample, mm | W, Watt | $\Delta T$ K | $\lambda, \frac{Вт}{м \cdot К}$ |
|---|---|---|---|---|---|---|---|---|
| 1 | Yarkopol-110 | K150 | 50/50 | 0,0225 | 7,187 | 10 | 19,7 | 0,162 |
| 2 | Yarkopol-110 | K150 | 70/30 | 0,0225 | 7,212 | 10 | 14,2 | 0,225 |
| 3 | ED-20 | K150 | 50/50 | 0,0225 | 8,162 | 10 | 15,6 | 0,232 |
| 4 | ED-20 | K150 | 70/30 | 0,0225 | 8,275 | 10 | 17 | 0,216 |
| 5 | Yarkopol-110 | K500 | 70/30 | 0,0225 | 8,575 | 10 | 20 | 0,190 |
| 6 | Yarkopol-110 | K500 | 50/50 | 0,0225 | 7,612 | 10 | 25,9 | 0,130 |

## 3. The conclusion

In the course of the work, the conditions for the formation of aluminosilicate microspheres were studied, the scale was estimated, and promising directions for development were considered. Ash and slag waste is really a danger to the environment. One of the ways to solve this problem is the use of coal energy wastes for useful purposes. Such is the extraction of aluminosilicate microspheres and their use in composite materials.

### References

[1] Vasiliev V V, Protasov V D, Bolotin V V and other 1990 *Composite materials: Handbook* (Moscow: mechanical engineering) p 512.

[2] Bazhenov S L 2014 *Mechanics and technology of composite materials: scientific publication* (Dolgoprudny: publishing House "Intellect") p 328.

[3] Galimov E R, Samoylov V M, Danilov E A, Fedyaev V L, Galimova N Ya and Orlov M A 2017 Production of thermally conductive carbon foams and their application in automobile transport // *IOP Conf. Series: Materials Science and Engineering* vol 240 issue 1 p 1-4.

[4] Basov N, Kazankov Yu and Ljubartovich V A 1986 *A calculation and design of equipment for production and processing of polymeric materials: textbook for universities* (Moscow: Chemistry) p 488.

[5] Matveenko V P, Smetannikov O Yu, Trufanov N Ah and Shardakov I N 2009 *Thermomechanics of polymeric materials in conditions of relaxation transition* (Moscow:

FIZMATLIT) p 176.

[6] L.N. Shafigullin, N.V. Romanova, A.T. Gabrakhmanov. Research of polyethylene in products of autocomponents // Materials Science Forum. International Russian Conference on Materials Science and Metallurgical Technology; Vladivostok; Russian Federation; October 1 - October 4, 2018. Volume 946 MSF, 2019, Pages 103-108.

[7] Gabdrakhmanov A T, Galiakbarov A T, Samigullin A D, Galiakbarov R T 2016 The calculation of a thermal field in the surface of a processed partunder the influence of a low-temperature plasma *IOP Conference Series: Materials Science and Engineering* Volume 134, Issue 1, Article number 012040. DOI: 10.1088/1757-899X/134/1/012040.

[8] Gabdrakhmanov A T, Israphilov I H, Galiakbarov A T 2014 The study the erosion of the electrodes under the influence moving electric arc *Journal of Physics: Conference Series* Volume 567, Issue 1, Article number 012013. DOI: 10.1088/1742-6596/567/1/012013.

IOP Conf. Series: Materials Science and Engineering **915** (2020) 012013 doi:10.1088/1757-899X/915/1/012013

# Application of individual educational trajectories and adaptive content in engineering education

**R D Galeev[1], I R Gilmanshin[1,2], N F Kashapov[1], S I Gilmanshina,[1]**

[1]Kazan Federal University, Kazan, Russia, 420008
[2] Kazan National Research Technical University named after A N Tupolev, Kazan, Russia

is-er@yandex.ru

**Abstract.** The article explores the educational technology of personalized training in the training of engineers. A hypothesis is presented on the relationship between the quality of student preparation and the individual educational trajectory reinforced by feedbacks. The use of new pedagogical technologies based on machine learning and big data is proposed. An algorithm for creating personalized educational systems is defined.

The main factor in the development of the competitive advantages of the Engineering Institute is the connection with industry, more precisely, the introduction of the achievements of fundamental science and advanced methods of development, design, production, testing and testing of products and processes at enterprises [1,4]. A continuous increase in the requirements for engineering education necessitates the intensification of the learning process, increasing its effectiveness, which is especially important given that for the study of engineering disciplines to ensure the development of not only subject-professional, but also key, basic competencies.

Currently, there is a shortage of engineering, design and research personnel capable of bringing to the industrial practice the modern results of the development of science, engineering and technology. A number of researchers believe that the main means of achieving these goals is to increase the role and importance of information technology in the educational process. The construction of intelligent learning systems allows achieving flexibility and adaptability at all stages of the organization of the learning process. Adaptive learning is considered one of the most promising and dynamically developing areas in the field of e-learning (E-Learning).

At the same time, as part of full-time studies, adaptability is often understood as the expansion of the rights of university students in the selection of the subjects studied in order to achieve a compromise between the hard and soft version of the educational program. The main methodological aspects of the algorithm of such training can be found in [5,8,9]. Such approaches to the creation of adaptive educational programs to some extent adapt to the apparent or imaginary needs of the student, but contribute little to the achievement of the programmed learning outcome in the form of obtaining the necessary competencies, that is, the trajectory itself becomes adaptive, but not the result. Gamukin V.V. [3,10] bluntly says that "this is just one of the forms of the learning process, weakly associated with the substantive content of the disciplines and their relevance in the real life of the graduate." With this approach, the conclusion arises that "embarking on a trajectory" will mean for the subject a loss of will to change something at the next stage of the educational process. " In this interpretation, the educational trajectory resembles a Procrustean bed in which the educational institution puts the student, depriving him of the opportunity to change anything.

A different understanding of adaptability came about as a result of the massive adoption of adaptive testing technologies. According to some authors, traditional testing, which is implemented using standardized tests, gradually loses its relevance and turns into more effective intelligent forms of adaptive testing [4,7]. Intelligent forms of knowledge diagnostics are based on different from traditional theoretical and methodological foundations and other technologies for constructing and reproducing tests. The adaptive test allows you to determine the level of knowledge much faster, because subjects with a high level of training and subjects with a low level of preparation will solve completely different subsets of tasks, where the first one will see a greater number of difficult questions, and the second - simple ones [11,17]. Modern forms of adaptive testing make it possible to more accurately build a model of students' mastered competencies necessary for the educational information system of a university.

Since the adaptive testing system adapts to the user level directly during the testing process, more complex adaptive educational systems should inevitably begin to appear. Thanks to flexible adaptive mechanisms, the system can determine exactly which question and with which coefficient of difficulty should be presented to the test subject at any given moment in time, as well as what material he needs to be given at a given moment in time.

Despite the fact that leaps and bounds in their development, adaptive testing systems are required to online education, we believe that it is advisable to use the best practices for the introduction of adaptive learning and individual educational trajectories in full-time education. Digitalization should provide support for decision-making in all forms of education, not excluding, of course, the distance form, where in fact the influence of information technology on the process of mastering knowledge has become obvious. The focus on digitalization in full-time and blended education is aimed at creating an information technology basis for creating individual educational paths using machine learning and big data. It is these two technologies that significantly increase the possibilities of true personalization of education.

In this regard, we plan to introduce into the work of the Engineering Institute a comprehensive methodology for quality control and monitoring of individual training paths. The main objective of this system is the generation of a personalized educational learning path for each student and the control of factors that influence academic success, as well as preventing students from falling into risk groups by timely notification of teachers and curators about knowledge gaps and problematic places that have formed. By selecting the optimal material format for each student, the created system tries to constantly maintain interest in teaching at a high level. The system will be built on the methodology of Education data engineering (EDE) - the adjustment of educational programs optimized for the collection and analysis of data in order to increase the effectiveness of educational activities. A workable support system for individual educational paths necessarily includes the following tasks - modular elements: 1) building together with employers fields of demanded competencies; 2) hypothesis about optimal individual educational trajectories; 3) the development of adaptive educational content using the achievements in the field of pedagogical design; 4) analysis of the digital footprint obtained from the available training formats; 5) the proof and refutation of hypotheses related to the development and training of students.

Let us consider in more detail each task - a module.

Building together with employers fields of demanded competencies. This problem can be most successfully solved in training target groups in cooperation with enterprises. Technologies make it possible to formalize the needs of large employers based on a common understanding of the situation on the labor market in terms of specific skill sets and gaining the ability to form the necessary skills. The whole process of adaptive learning must be built on the basis of the ultimate goal - the student getting the necessary set of skills at a certain point in time. There are examples of creating a methodology for such a formalization of university results.

IOP Conf. Series: Materials Science and Engineering **915** (2020) 012013 doi:10.1088/1757-899X/915/1/012013

At 2035 University, using the Natural Language Processing (NLP) method, they received the following large blocks of key competencies not tied to a specific industry for the 50 largest enterprises in Russia: thinking; work with systems and processes; self-regulation; human interaction; work with novelty and changes; communication; work with information; entrepreneurship; value regulatory competencies.

Each of these core competencies is decomposed right down to measurable skills. The specified set of competencies will be supplemented by industry-specific professional competencies that will be formalized in work in project teams using specialized equipment and software, including tool ownership, understanding of the subject area, and transition to product logic. The introduction of students' work in project teams, using practical cases of enterprises, will make it possible to obtain a significant amount of data for understanding the process of forming the necessary competencies.

Within the framework of the Faculty of Engineering of KFU, the issue of purchasing specialized software with which you can implement team projects to create digital doubles of products and processes is discussed. Team work of students in the framework of automated CAD / CAE-PLM systems with the parallel use of digital footprint systems will complement the competency profiles of students. In addition, working on real systems that work in industry will constantly generate a stream of data on the necessary skills. As part of the work, the personnel services of enterprises should formalize the specified sets of skills and abilities that they want to receive from graduates and give them to graduating departments.

Creating an expanded profile of competencies will not only allow you to adjust the individual educational path, but will also give an understanding of the patterns of behavior of a specialist in certain conditions, which will make him a much deeper tool compared to profiles in existing online recruitment systems. Over the years of training, a digital competency profile will be obtained, which, at the request of the graduate, can be used further. The array of student information available in the university's information system during training can be a powerful help in the labor market. For this, a graduate can be given access to the interface of his profile with the ability to add new data there related to his professional activities and further training. With a high probability, the launch of such a service for former university students will trigger a positive feedback - it is the best graduates who will be interested to open the maximum amount of information about themselves to potential employers, and employers, in turn, will be interested in maximum knowledge about their potential employees. Indirectly, this should also increase the demand for additional education services, advanced training, etc. I believe that universities are losing very significant revenue, not trying to take an active position in the recruitment market.

Each of these core competencies is decomposed right down to measurable skills. The specified set of competencies will be supplemented by industry-specific professional competencies that will be formalized in work in project teams using specialized equipment and software, including tool ownership, understanding of the subject area, and transition to product logic. The introduction of students' work in project teams, using practical cases of enterprises, will make it possible to obtain a significant amount of data for understanding the process of forming the necessary competencies.

Within the framework of the Faculty of Engineering of KFU, the issue of purchasing specialized software with which you can implement team projects to create digital doubles of products and processes is discussed. Team work of students in the framework of automated CAD / CAE-PLM systems with the parallel use of digital footprint systems will complement the competency profiles of students. In addition, working on real systems that work in industry will constantly generate a stream of data on the necessary skills. As part of the work, the personnel services of enterprises should formalize the specified sets of skills and abilities that they want to receive from graduates and give them to graduating departments.

Creating an expanded profile of competencies will not only allow you to adjust the individual educational path, but will also give an understanding of the patterns of behavior of a specialist in certain conditions, which will make him a much deeper tool compared to profiles in existing online recruitment systems. Over the years of training, a digital competency profile will be obtained, which, at the request of the graduate, can be used further. The array of student information available in the university's information system during training can be a powerful help in the labor market. For this, a graduate can be given access to the interface of his profile with the ability to add new data there related to his professional activities and further training. With a high probability, the launch of such a service for former university students will trigger a positive feedback - it is the best graduates who will be interested to open the maximum amount of information about themselves to potential employers, and employers, in turn, will be interested in maximum knowledge about their potential employees. Indirectly, this should also increase the demand for additional education services, advanced training, etc. I believe that universities are losing very significant revenue, not trying to take an active position in the recruitment market.

Hypothesis about optimal individual educational trajectories. To solve this problem, it is necessary to design training goals and compare them both with the content (content) of the discipline and with assessment tools for diagnosing the achievement of goals. The task of designing learning goals is to clearly, explicitly describe them through the learning outcomes expressed in students 'actions that can be further recognized by the corresponding information system algorithm.

In modern conditions, the results of education are the expected and measurable specific achievements expressed in the language of knowledge, skills, abilities, competencies that a student or graduate will be able to demonstrate upon completion of the educational program or part thereof [15,16].

The adaptive learning process is a consistent presentation of educational modules to a student - logically completed pieces of knowledge presented in textual, graphic or media form and designed to reveal a specific topic or concept [7, 8]. In addition, the module can be practical in nature and intended to develop a student with certain skills or abilities - it can be an interactive simulator or demonstration, test tasks, exercises, etc. The knowledge and skills formed after mastering the module are the output competencies of the module. The input will be those competencies that are necessary for the student to master the module. Presentation of the next module to a student depends on the current level of his knowledge on one or another competence - the algorithm will select the module whose passage at the current time will give the best result by the end of the training.

The competencies that a student should possess based on the results of studying a particular course are formulated in the most general form in state educational standards [2]. To create a workable competency graph, it is necessary to decompose or disaggregate the basic competencies of the Federal State Educational Standard into composite subcompetencies [7, 8]. These subcompetencies are disaggregated up to specifically measurable learning outcomes, which the teacher will be able to evaluate using selected methods, for example, Bloom's taxonomy [16, etc.]. Each formulated learning outcome is decomposed into elementary skills, knowledge and skills. Decomposition is carried out to the level where each elementary skill can be evaluated with one single evaluative action.

In the created training column, competencies, subcompetencies and specific measurable skills and knowledge will be the vertices, and adaptive educational content and the process of obtaining knowledge and skills will be edges. Having defined the list of competencies, subcompetencies and skills, it is necessary to describe the so-called prerequisites for each future educational module, as well as assign existing and existing competencies to the existing modules in the column, or create a module for each open competency, or create a module for any discipline and assign him competence from the column of competencies, while tracking information about which modules implement the assigned competency. Accordingly, based on this list, the information system algorithms will search for covering modules from the database and calculate the primary trajectory of their passage, which will be adjusted based on the student's learning results.

Translation of learning outcomes into the language of action and ensuring the uniqueness of this translation requires the construction of a hierarchical system of goals, within which their categories and successive levels are distinguished, as well as the actual formulation of goals in the most clear, concise, formalized and verified language. These problems are especially relevant in engineering education, because by its nature engineering education organically includes both methods of general engineering disciplines and the apparatus of mathematics, physics, chemistry and other natural sciences. The processes of analysis, evaluation, generation of technical solutions in engineering disciplines are often poorly formalized, which makes it difficult to control students' achievements in these areas [15,16].

The graduating departments should formulate what skills and knowledge students should receive in previous courses (preferably, in the context of each subject). Based on the data about the applicant, the system will have to build assumptions and choose the proposed route inside the training graph.

The development of adaptive educational content using the achievements in the field of pedagogical design.

In the context of global informatization, the most important factor in the development and competitiveness of engineering education is modern information and communication technologies aimed at creating, storing, processing and providing effective ways of presenting information to students [13: 7]. In the current conditions of media content dominance, the success of becoming a competent engineering specialist largely depends on students getting information about the latest scientific and technological achievements in the field of core and related professions.

The process of designing training materials is a sequence of clearly defined procedures that are grouped into a number of stages and have specific tasks and methods for solving them. Most often, in the development of pedagogical design, stages such as analysis, design, development are used.

Analysis of the digital footprint obtained from the available training formats. It is advisable to begin the analysis of the digital trace already at the stage of work with applicants. Software products based on solutions in the field of analytics have begun to appear, which, using the digital trail of high school students in social networks, diagnose their cognitive and psychological properties with high accuracy. As a data source for the analysis of the digital footprint, subscriptions and likes on social networks are mainly used, less often text samples (modern high school students leave little text material). Depending on what sites the user went to, what he read, watched and “liked”, modern technology allows him to build his virtual profile. This approach has been used for several years in Western educational institutions, in Russia it was tested at TSU. There was introduced a system that works in several stages.

We believe that the use of this approach at the KFU Engineering Institute, taking into account a much wider geography of interests than TSU, will make it possible to significantly reduce the cost of recruiting applicants, including foreign ones, by 10-20 points to increase the average performance on the exam, already at this stage take a step towards personalization learning. Of course, in order for this approach to work in KFU, it is necessary to significantly refine it, both in the number of social networks and other electronic resources that are monitored, and in the training sample of students for a neural network. To this service it will be possible to add analytics for participation in competitions, passing tests and other work with the university portal, and create a full-fledged digital profile of a university student. Based on the created profile, KFU will be able to make the most suitable proposal for him to submit documents to a particular faculty.

After the applicant enters the Engineering Institute, his digital profile will be formed and constantly updated. This work can be started to build from existing data for monitoring performance - these are attendance data and current grades in practical exercises. The collection of this data is quite easily automated through the introduction of an electronic journal, to which you can connect an identification system according to the data of mobile telephones (for example, via Bluetooth channel, IFIT authorization).

Next, the student’s digital profile management system needs to be developed on the part of predictive analytics, to predict events based on a comparison of data for the past and current period. This will allow you to pre-identify students who are inclined to abandon their studies and therefore need special attention from the teacher and to develop an alert system that tracks the level of student involvement in the development of courses. It is important that the conclusion is made not on the basis of assessments, but on the basis of how much time the student spends on tasks, what types of exercises he does better and faster, and which ones he throws. Using this data, potentially "falling out" students are determined several months before the end of the course, when the situation can still be corrected.

Once organized into the system of obvious evidence, you can begin to consider more exotic methods of analyzing the digital footprint. For example, the speech analysis algorithm contains the following steps: recording audio with reference to the speaker; audio transcription and semantic analysis; allocation of frequency indicators by terms and semantic kernels; comparison of the obtained indicators with the reference on this topic in the framework of the general taxonomy.

The proof and refutation of hypotheses related to the development and training of students. This modular subsystem can be started with simple but efficient tree-like methods for constructing trajectories, and when the system works and accuracy will be important - begin to introduce analysis technologies based on machine learning algorithms that will be able to adapt electronic courses at any time and automatically adjust the learning path , and also provide teachers with collected analytical data for each student. In addition to statistics, the adaptive learning platform generates recommendations for teachers based on the information gathered about the student and suggests what materials the student should show next.

Genuine adaptive learning should respond in real time to the results of an individual student and his actions in the system. This approach increases the likelihood that a student will receive holistic, problem-oriented content at the right time and achieve their goals. The system will allow you to dynamically adjust the level and type of courses depending on the individual knowledge and abilities of the student. Optimization of the level of complexity of training, in turn, positively affects the internal motivation of students and their interest in acquiring new knowledge. Collecting this kind of data about students will allow not only to adapt electronic courses on the fly, but also to use the data collected by tutors and teachers.

**References**

[1] Akchelov E.O., Pokholkov Yu.P. The role of the education system in the economic development of the region (on the example of Singapore) // Vocational education in Russia and abroad. - 2016 .-- No. 4 (24). - S. 221 - 226.

[2] Baidenko V.Yu. Identification of the competencies of university graduates as a necessary stage in the design of new generation of state educational standards of higher professional education: method. allowance. M .: Research Center for the Problems of Quality of Training of Specialists, 2006. 72 p.

[3] Gamukin VV Individual educational routes at the university // Engineering Education. - 2019.- No. 25. - S. 28 - 36.

[4] Glova V.I., Duplik S.V. Models of pedagogical testing of students // Bulletin of Kazan. state tech. University of them. A.N. Tupolev. - 2003. - No. 2. - S. 74 - 79.

[5] Zhdanko T.A., Zhivokorentseva T.V., Chuprova O.F. Designing individual educational routes (IOM) for students at a university // Scientific and Pedagogical Journal of Eastern Siberia Magister Dixit. - 2014. - No. 1 (13). - S. 140 - 146.

[6] Zakharova A.A., Lazareva A.N. Support for the selection of an educational trajectory of an individual based on ontology // Scientific and Practical Journal "New Research in the Development of Engineering and Technology". - 2017. - No. 2. - S. 61 - 68.

[7] Krechetov I.A., Dorofeeva M.Yu., Degtyarev A.V. We reveal the potential of adaptive learning: from development to implementation. // Mat. Int. conf. ELearning Stakeholders and Researchers Summit 2018, December 05 - 06, 2018 - M. - P.76 - 88.

[8] Lushnikov A.V. An algorithm for choosing optimal learning paths in a two-level education system // News of higher educational institutions. Volga region. Technical science. - 2014. - No. 4 (32). - S. 32 - 40.

[9] Mitsel A.A. Chernyaeva N.V. Dynamic model of individual student learning path management // Bulletin of the Voronezh State University of Engineering Technologies. - 2015. - No. 2. - S. 77 - 81.

[10] Myasoedova V.A., Lebedeva N.A. Problems of modifying the activities of Russian universities on the basis of world practice in the development of educational institutions // Modern Education. - 2016. - No 4. - S. 29 - 37.
[11] Nikiforov, O.Yu. The use of a computer testing environment based on the test database in a test form in a quality control system of a higher educational institution // Quality management of continuing education: Collection of articles of the 2nd interregional scientific and practical conference // Otv. Editor E.Yu. Bakhtenko; Ministry of Education and Science of the Russian Federation; Depart. Education Volog. reg .; Vologod. state ped un-t - Vologda: Voronezh State Pedagogical University, 2011 .-- 192 p. - S. 72 - 79.
[12] Sandler D.G., Suschenko A.D. Economic motivation of choice by graduate students of educational, scientific and professional paths // Economy of the region. - 2016. - No. 2. - S. 547 - 559.
[13] Uvarov A.Yu. Pedagogical design // Informatics. - 2003. - No. 30. - C.2 - 31.
[14] Shatalov M.A., Mychka S.Yu. Features of the implementation of the individual educational route (IOM) in preparing students for higher education programs // St. Petersburg Educational Bulletin. - 2017. - No. 1 (5). - S. 18 - 21.
[15] Tsvetkov Yu.B. Features of the design of educational goals of disciplines of engineering educational programs // Science and Education. MSTU named after N.E. Bauman. Electron Journal. - No. 201. - C.1 - 14.
[16] Anderson L.W., Krathwohl D.R., eds. A taxonomy for learning, teaching, and assessing: A revision of Bloom's taxonomy of educational objectives. New York: Longman, 2001.

IOP Conf. Series: Materials Science and Engineering **915** (2020) 012014 doi:10.1088/1757-899X/915/1/012014

# Experience in the application of nitriding in a glow discharge to increase the efficiency of gear processing by the method of grinding

**V N Gilman[1], A I Fashutdinov[1, 2] and I P Balabanov[1, 2]**

[1] Kazan Federal University, 18 Kremlevskaya str., Kazan, 420008, Russian Federation
[2] Kazan National Research Technical University named after A. N. Tupolev - KAI, 10 K. Marx street, Kazan, 420111, Russian Federation

tim_vg@mail.ru, balabanovip@mail.ru, airat.fashutdinov@kamaz.ru

**Abstract.** The article presents the experience of increasing the resistance by nitriding the gear tool using the method of grinding, using the Skywing technology. Tool material P6M5K5 with hardness 62-64HRC (analog: 1.3243, HS6-5-2-5 – Germany, W6Mo5Cr4V2Co5-China). Gear material is 20crnimoh structural alloy steel (analog 1.6523, 21NiCrMo2 – Germany). Nitriding was carried out by ionoplasmic nitriding, in an ammonia gas environment, at a temperature of 450 degrees Celsius, for 40 hours, to a depth of 0.6 mm. In addition, a wear-resistant coating based on alumotite carbonitrides was applied, with the activation of the surfaces by a hydroabrasive installation based on aluminum oxide. Washing was carried out in an environment of alkaline surfactants. This experience led to rapid destruction of the cutting edges. Experience has shown that the saturation of tool steel with nitrogen did not yield the expected dispersed nitrides. The material has become fragle. This is due to the oversaturation of tool steel with alloying elements. In continuation, it is recommended to conduct a new experiment with heating to 500-520 degrees Celsius and nitrogen in a glow discharge lasting 2-3 hours for an expected layer of 20-40 microns. In order to avoid loss of adhesion of the wear-resistant coating for some nitrogen etching, it is assumed that an argon volume is introduced into the gas mixture.

Improving the efficiency of the cutting tool can be achieved in two main ways: reducing the cost of the tool, or improving its characteristics [1,2]. Under the characteristics, it is necessary to take the ability of the tool to work at more extreme modes (with higher performance) with equal resistance or a direct increase in resistance without changing the processing modes [3,4,5]. Both the first and the second can be achieved by selecting the optimal geometry of the cutting wedge, the state of the structure and chemical composition of the tool material, its physical characteristics, the presence and type of wear-resistant coating.

For processing the internal gear, a special tool for tooth-grinding is used applying the Skywing technology. After grinding, the tool is prepared according to a standard technological process that includes deburring, rounding the cutting edge, cleaning, ion-plasma application of a multi-layer wear-resistant coating in a vacuum based on titanium and aluminum nitrides, and polishing [6,7,8,9]. When comparing tools made of ASP2030 powder steel with a hardness of 65-67HRC and p6m5k5 high-speed steel with a hardness of 62-64HRC, a difference in durability was revealed by 15-20% in favor of the former. In addition, the analysis of the wear of the cutting edges showed insufficient strength of the cutting edge associated, in our opinion, with local staining of the material.

Figure 1 shows the stages of wear recorded during the life cycle of the tool. It is obvious that microschips gradually grow (Fig. 1a, b, c) and, at a certain moment, an avalanche-like process of critical wear occurs (Fig. 1d). The main problem of the observed process is instability of critical wear. That is, the transition from the stage of an allowable defect on the cutting edge of the tool to 0.3 mm to a critical value of more than 0.3 mm is not predictable.

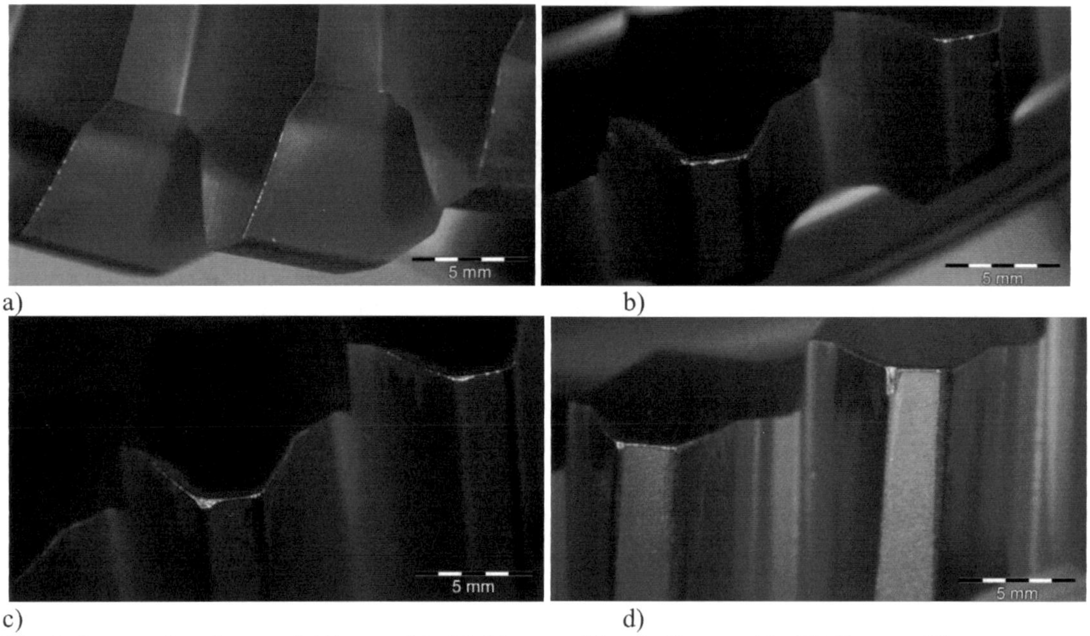


Fig. 1 Stages of tool wear: 5-10 parts (a), 20-25 parts (b), 40-50 parts (c), 55-70 parts (d)

Taking into account somewhat better indicators of tool life with higher hardness, the likely cause of premature failure [11,12,13,14] were considered physical characteristics of the tool. To improve them, it was decided to conduct an experiment on hardening the tool material with the process of ion-plasma nitrogenation.

It is known that correctly selected nitriding regimes make it possible to saturate the surface layer of steel to a depth of 0.6 mm [15], thereby changing the state of the structure, static and dynamic strength, wear resistance, deformability, and corrosion properties [16,17,18,19,20]. Nitriding was carried out at the "Kleckner" installation in a gaseous environment of ammonia at a temperature of 450 degrees Celsius for 40 hours.

In addition to ion-plasma nitration, a wear-resistant coating based on aluminotitanium carbonitrides was applied. The issue of insufficient coating adhesion to the tool material was solved by deep cleaning and surface activation by a waterjet installation, where aluminum oxide was used as a working material. At the final stage, ultrasonic washing was performed in an alkaline surfactant medium.

The results of the subsequent destructive testing of the tool for tilling showed a supersaturation of the volume with nitrogen to a depth of 0.4 mm, an extremely high hardness of the surface layer. It ultimately led to a negative result when testing the tool. The destruction of the cutting edges occurred almost immediately after contact with the workpiece material (Fig. 2).

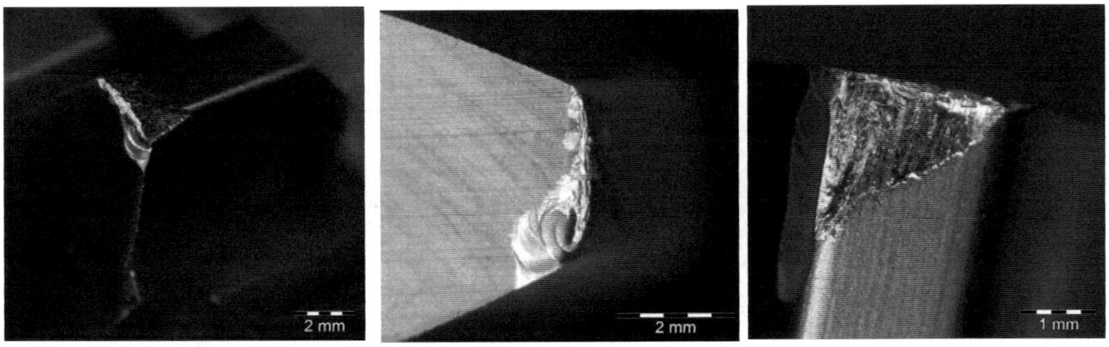


Fig. 2 Destruction of the cutting edges of the tool

It is obvious that the selected modes of chemical-thermal treatment completely did not correspond to this material and adversely affected the physical characteristics. In fact, we got a hard and very fragile metal.

Further work in this direction could be the introduction of the ionoplasmic nitration process in the process of vacuum ionoplasmic coating. In the process of which, after heating to 500-520 degrees Celsius and cleaning the tool surface with a glow discharge, a process of nitriding in a glow discharge lasting 2-3 hours could follow, allowing to saturate a very thin (20-40mkm) surface layer of the tool material. In order to avoid loss of adhesion of the wear-resistant coating for some nitrogen bleed, it is assumed that a volume of argon to be introduced into the gas mixture.

**References**

[1] Kondrashov, A. G., Safarov, D. T., Faskhutdinov A. I. and Davletshina G. K. Single-Turn Worm Mills for Conical Round-Tooth Gears / Russian Engineering Research, 2017, Vol. 37, No. 9, pp. 812–813. reference

[2] Weiss Denis, The Effect of Cutting Edge Rounding Geometry on Tool Wear, Machine tool Park № 12,2016.

[3] Kasyanov, S.V., Kondrashov A. G. and Safarov D. T. Research of characteristics of wearproof coating for cutting tools / INTERFINISH-SERIA 2014: International Conference on Surface Engineering for Research and Industrial Applications. 2014. P. 62.s

[4] Faskhutdinov A. I., Emelyanov D. V., Blinova A. S. Influence of the size of the cut layer on the productivity of mechanical processing / Proceedings of scientific articles of the 2nd international youth scientific and practical conference "Progressive technologies and processes" September 24-25, 2015 in 3 volumes. Kursk.: ZAO Universitetskaya kniga. - 2015. Pp. 111-115.

[5] Chatterjee-Fisher, R., Eisell, F. et al., Nitriding and carbonitriding, TRANS./ edited by Supov A.V. M.: metallurgy, 1990. 280s.

[6] Denkenaa, B., Koehlera J. and Rehe M. Influence of the Honed Cutting Edge on Tool Wear and Surface Integrity in Slot Milling of 42CrMo4 Steel. Procedia CIPR (2012) 190-195

[7] Khusainov R. M., Golovko A. N., Petrov S. M., Yurasov S. Yu., Balabanov I. P., grechishnikov V. A., Romanov V. B., Pivkin P. M. / / Determination of tool parameters in technological cutting systems / STIN. 2016. No. 10. Pp. 17-20.

[8] Kasyanov, S.V., Kondrashov A. G. and Safarov D. T. Rapid Assessment of Wear-Resistant Tool Coatings / Russian Engineering Research, 2017, Vol. 37, No. 11, pp. 969–973.

[9] Khusainov R. M and Khaziev R. R. Modeling of forming technological errors in processing by gear shaping machine // IOP Conference Series: Materials Science and Engineering. - 2017. - Vol.240, Is.1. - Art. № 012045.

[10] Alanou M.P., Evans H..P, Snidle R.W. Effect of different surface treatments and coatings on the scuffing performance of hardened steel discs at very high sliding speeds. Tribology International. 2004;37:93-102.

[11] Bell T, Morton PH, Bloyce A. TOWARDS THE DESIGN OF DYNAMICALLY LOADED TITANIUM ENGINEERING COMPONENTS. Materials Science and Engineering a-Structural Materials Properties Microstructure and Processing. 1994;184:73-86.

[12] Boniardi M, D'Errico F, Tagliabue C. Influence of carburizing and nitriding on failure of gears - A case study. Engineering Failure Analysis. 2006;13:312-39.
[13] Bull SJ, Evans JT, Shaw BA, Hofmann DA. The effect of the white layer on micro-pitting and surface contact fatigue failure of nitrided gears. Proceedings of the Institution of Mechanical Engineers Part J-Journal of Engineering Tribology. 1999;213:305-13.
[14] Cardoso NFR, Martins RC, Seabra JHO, Igartua A, Rodriguez JC, Luther R. Micropitting performance of nitrided steel gears lubricated with mineral and ester oils. Tribology International. 2009;42:77-87.
[15] Leushin, I.O. Technologies for Use in the Formation of a Differentiated Structure in Iron Billets Used in Glass Molds / I.O. Leushin, D.G. Chystyakov // METAL SCIENCE AND HEAT TREATMENT. - 2016. - №5-6. - T. 58. - p. 299-302.
[16] Leushin, I.O. Creep of the investment mold material / I.O. Leushin, L.I. Leushina, O.S. Koshelev // Russian Metallurgy. - 2017. - № 13 (DEC. 2017). - C. 1092-1095.
[17] Karamis MB. EXPERIMENTAL-STUDY OF THE ABRASIVE WEAR BEHAVIOR OF PLASMA-NITRIDED GEARING STEEL. Wear. 1993;161:199-206.
[18] Ohue Y, Matsumoto K. Sliding-rolling contact fatigue and wear of maraging steel roller with ion-nitriding and fine particle shot-peening. Wear. 2007;263:782-9.
[19] Rolinski E, Amer J, Sharp G. Negative effects of reactive sputtering in an industrial plasma nitriding. Journal of Materials Engineering and Performance. 2005;14:343-50.
[20] A.V. Shaparev and I.A. Savin, Influence of the State of the Contact Surfaces on the Formation of the Joint of Steel and Brass during Cold Cladding. Solid State Phenomena, Vol. 284, pp. 319-325, 2018 DOI: https://doi.org/10.4028/www.scientific.net/SSP.284.319
[21] Fabre A, Evans HP, Barrallier L, Sharif KJ, Desvignes M. Prediction of microgeometrical influences on micropitting fatigue damage on 32CrMoV13 steel. Tribology International. 2013;59:129-40.
[22] Fujii M, Kumar MA, Yoshida A. Influence of DLC coating thickness on tribological characteristics under sliding rolling contact condition. Tribology International. 2011;44:1289-95.
[23] Goret V, Fabre A, Barrallier L, Vardon P. Evaluation by synchrotron radiation of shape factor effects on residual stress in nitrided layers. In: Reimers W, Quander S, editors. Residual Stresses Vii. Materials Science Forum. 524-5252006. p. 285-90.
[24] Gavariev R.V. and Savin, I.A. Research of the Mechanism of Destruction of Compression Molds for Casting under Pressure of Color Alloys. Solid State Phenomena, Vol. 284, pp. 326-331, 2018 DOI: https://doi.org/10.4028/www.scientific.net/SSP.284.326

# Methodological features of using elements of distance learning in the training of competent specialists

**I R Gilmanshin[1,2], S S Kosmodemyanskaya[1], A I Galeeva[3], K Saduakaskyzy[4]**

[1]Kazan Federal University, Kazan, Russia, 420008
[2] Kazan National Research Technical University named after A N Tupolev, Kazan, Russia
[3]Kazan State Power Engineering University, Kazan, 420066, Russian Federation
[4]Korkyt Ata Kyzylorda University, Aiteke bi, 29A; 120014, Kyzylorda, Republic of Kazakhstan

is-er@yandex.ru

**Abstract.** The paper considers the problem of using distance learning of students of engineering and chemical education in the framework of the digitalization of the learning system. Authors consider distance learning of profile disciplines as an element of a personality-oriented approach in the context of professional self-education and as a form of preparation for future professional activity. The rationale for the feasibility of variable forms of a holistic distance learning system based on methodological support is presented, the basic principles for constructing educational trajectories are determined. The authors present some results of a pedagogical experiment over a seven-year study period, confirming the effectiveness of the developed distance learning technology.

The changes that occurred at the turn of 2020 in countries of the world community made enormous corrections to all areas of human life, which, of course, entailed the educational sphere. The changes that have occurred have identified the problems of resolving new problems in the formation of a new generation of specialist engineers (for more details see [1], [2], [3]) focused on breakthrough technical research ([4], [5], [6], [7 ]). In almost every subject of the country, education has transferred to the status of distance learning using information and communication technologies (for more details see [8], [9], [10]), including the formation of 21st century competencies and environmental literacy (studies [11], [12] ) This contributes to the global informatization of the community, the distribution in educational institutions of the latest computer technology and modern software. To ensure the practical implementation of the developed major state and international programs, special training of future specialists in the field of organization and conduct of distance learning is required.

We have been studying this topic of research since 2013 when developing and adapting electronic digital resources (ESM) on the MOODLE platform for students of Kazan Federal University. A modular object-oriented dynamic learning environment is one of the most used forms of distance learning (Moodle is an abbreviation for Modular Object-Oriented Dynamic Learning Environment, a modular object-oriented dynamic learning environment).

The analysis of the existing literature and advanced pedagogical experience on the problem of training qualified specialists - university graduates in distance learning allows us to identify the leading contradiction between the needs of the modern customer (from the position of society and the employer) and the level of application of the methodology for the formation of over-subject competencies of university graduates.

Earlier in [13], we substantiated the need to form competencies of such a plan, both for future engineers and for future chemistry teachers. In the leading universities of Russia for a long time there has been a certain discrepancy in the level of competence of graduates with the expectations of the customer. We consider the interconnected components in the preparation of students - future teachers and future engineers, determined by the conditions and methodological possibilities of using distance learning with the digitalization components of the educational process.

Distance learning in [14] is considered as a technology for the purposeful and methodically organized management of the educational and cognitive activities of students who are at the time of training at a certain distance from the educational center itself. Moreover, the level of education received does not affect the learning process, which is carried out using modern means and methods of pedagogical communication between the teacher and students. According to the methods and forms of organization of the educational process, distance and distance learning have certain similar features.

To summarize the results, we used experimental data from the last 7 years of our research, where we are considering the possibility of using distance learning in various versions of school and university education. This is an independent work of students in the study of individual subjects / disciplines; expanding opportunities for working with gifted children while deepening the information component of courses; the study of teaching material in home schooling due to the occurrence of objective reasons (in the conditions of a compelled absence of a student or teacher in the classroom); the use of digital opportunities of distance learning in creating various learning situations; counseling students to fill the gaps in knowledge; the use of distance learning opportunities for students with specific educational needs, including foreigners, children with disabilities, etc.

The results of a study of students of 1-3 courses of the Chemical Institute named after A.M. Butlerov and the Institute of Fundamental Medicine and Biology of Kazan Federal University, presented in [15]. The question of the readiness of students (58 respondents) to conduct pedagogical activities with children with disabilities using remote technologies was studied. For inclusion as the best option for co-education of children, 60% of freshmen and 67% of third-year students identified. Mainstreaming was supported by 40% of students in both first and third year students. And the share of distance learning for children with disabilities is 28% and 53% of respondents, respectively. A comparative analysis of the answers of 3-4 year students about the correct use of distance learning chemistry for children with disabilities showed a difference in relation to this of chemical students (53%) and future chemistry-biology teachers (88%).

As you increase your subject and subject competencies, a chemistry teacher (for more details, see [15]) must learn how to conduct lessons using distance technologies, which include their application at a certain stage and / or throughout a chemistry lesson. To implement such a learning process, subjects of training (chemistry teachers and students) should have a standard set of equipment, including: a PC (or any device) with a web camera, it is also desirable to have a printer, scanner, and other devices to facilitate the process. We analyzed the sites of chemistry teachers of the Russian Federation and the Republic of Tatarstan and developed recommendations for students of pedagogical education on how to optimally create an individual page for a chemistry teacher (WEB page). After analyzing the best pedagogical experience on this topic, we came to the conclusion that the following information and communication competencies should be formed for graduates of pedagogical universities: the ability to use specialized programs starting with MS Word (this is the main means of conducting the lesson); ability to work with a virtual chemical laboratory; ability to work with electronic technologies, etc .; the ability to work on the Internet to find information, develop materials for the lesson, conduct distance learning, etc.

Many educational institutions are actively implementing distance learning, according to the program "E-learning with the use of information and telecommunication networks." In educational institutions, the classes are equipped with the equipment necessary for distance learning technologies, for example, via the Skype network. Various specialized virtual educational environments also exist and are gradually being used, such as Moodel, E-STADI, ATutor, Eliademy, Forma LMS, Dokeos, ILIAS, Opigno, OLAT. These distance learning systems are the most popular, while being distributed free of charge and may interest chemistry teachers, as well as increase their level of methodological competence. The virtual educational environment itself is a system of information resources that help teachers manage student learning.

The possibilities of creating and adapting video lectures on methodological disciplines were analyzed earlier in [16], where the following conclusions were made. Firstly, it is more convenient for some students to perceive the theoretical material in this format of “conditionally real communication”. Secondly, today it is one of the relevant forms of training. Thirdly, this type of presentation of the material forms one of the important competencies of modern education - information.

Below are the results of studies analyzing the activities of students in the context of a complete transition to a remote form of learning. The study (May, 2020) was attended by 40 students - 1st year students of the direction 44.03.01. Pedagogical education (chemistry) and undergraduates of the 1st year of study direction 44.04.01. Pedagogical education (chemical education).

Respondents indicate almost the same features of the use of new technologies and training opportunities - 36% and 38%, respectively. The ability to work at any time and in a relaxed atmosphere suits 45% and 38% of respondents. Saving time for moving (31% of students and 64% of undergraduates) does not reduce collaboration with documents and demonstration of various materials for the lesson (31%.). The lack of a real experiment worries 64% of undergraduates and 41% of students. 55% and 69%, respectively, indicate problems with technical equipment and a weak Internet signal. The need for additional training of students and teachers in the field of information and communication competencies for distance learning is indicated by 45% of undergraduates and 17% of students.

Findings. The distance form of training is quite relevant and most effective for the planned immersion of students in preparation for professional activities as highly qualified future specialists.

Open access to the site of the Kazan Federal University provides links to the largest foreign electronic educational resources. For example, Coursera, The Open university, EDX, etc. The university also presents electronic versions of libraries: “eLIBRARY.RU”, “Doe” Electronic Library System, National Electronic Library and others. When working on these distance systems or on the developed personal website, the joint work of the teacher and tutor of distance education is possible.

Thus, work on distance learning allows future engineers and chemistry teachers to form the subject and over-subject competencies of highly qualified specialists.

**References**

[1] Kosmodemyanskaya S.S, Gilmanshin I.R, Azimov Y.I. (2019). Features of the formation of interdisciplinary competencies in engineering education. IOP Conference Series: Materials Science and Engineering. 2019. Vol. 570, Is.1. Art. № 012063.

[2] Gilmanshin, I.R., Gilmanshina, S.I., Galeeva, A.I., Gadirova, E.M., Galeev, I.A. (2019). Methods of training specialists in the field of sustainable resource management and waste management. IOP Conference Series: Materials Science and Engineering, Volume 570, Number 1. https://iopscience.iop.org/article/10.1088/1757-899X/570/1/012025

[3] Gilmanshin I.R., Gilmanshina S.I. (2017). Competence formation of engineering directions students in the field of energy saving as a way to create new generation technologies. IOP Conference Series: Materials Science and Engineering, Volume 240, Number 1, 2015, pp. 012022. doi:10.1088/1757-899X/240/1/012022.

[4] Kashapov N.F., Gil'manshin I.R., Konahina I.A. et al. (2014). System analysis of the energy complex of engineering enterprise as a basic tool of effective energy management. IOP Conference Series: Materials Science and Engineering, Vol. 69, 2013. DOI: 10.1088/1757-899X/69/1/012024.

[5] Sabirzyanov, R.G., Gilmanshin, I.R., Kashapov, N.F., (...), Galeev, I.A., Krainova, D.R. (2018). Material and technical base of the laboratory for the thermal destruction of production and

consumption waste for the training of specialists in the field of waste management. IOP Conference Series: Materials Science and Engineering, 412, (2018) 012061. doi:10.1088/1757-899X/412/1/012061. http://iopscience.iop.org/article/10.1088/1757-899X/412/1/012061.

[6] Gilmanshin I.R., Kashapov N.F., Gilmanshina S.I., Galeeva A.I. (2016). Landfill energy complex based on the renewable energy installations. IOP Conference Series: Materials Science and Engineering, Volume 134, Number 1, 2015, pp.12007. DOI: 10.1088/1757-899X/134/1/012007.

[7] Duarte, A.J., Malheiro, B., Arno E., Perat, I., Silva, M.F., Fuentes-Dura, P., et al. (2020). Engineering Education for Sustainable Development: The European Project Semester Approach. IEEE Transactions on Education. Vol. 63(2), P. 108-117.

[8] Gilmanshin I.R., Gilmanshina S.I. (2018). Technology of e-learning in the university education. IOP Conference Series: Materials Science and Engineering, 412, (2018) 012022. doi:10.1088/1757-899X/412/1/012022. https://iopscience.iop.org/article/10.1088/1757-899X/412/1/012022.

[9] Gilmanshina, S.I., Gilmanshin, I.R., Bendjukevich, K.G. (2019). Digital learning technologies in the University education system. IOP Conference Series: Materials Science and Engineering, Volume 570, Number 1. https://iopscience.iop.org/article/10.1088/1757-899X/570/1/012027.

[10] Kosmodemyanskaya S.S., Nizamov I.D., Yarullin I.F. (2017). Information and educational environment in the system of practices of the future teacher of chemistry. QUID-INVESTIGACION/QUID 2017, pp.373-378, Special Issue N°1-ISSN:1692-343X. http://revistas.proeditio.com/iush/quid/article/view/1794/1792

[11] Gilmanshina, S.I., Gilmanshin, I.R., Dyatlova, E.V. (2019). Formation of competencies of the 21st century by means of project activities. IOP Conference Series: Materials Science and Engineering, Volume 570, Number 1. https://iopscience.iop.org/article/10.1088/1757-899X/570/1/012026.

[12] Gilmanshina S.I., Gayfullina A.Z., Fedotova N.R., Gilmanshin I.R. (2018). Building the Environmental Literacy Through School Natural Science Education. Astra Salvensis, VI (2018), Special Issue, p. 245-253. https://astrasalva.files.wordpress.com/2018/07/astra-salvensis-vi-2018-special-issue.pdf.

[13] Cosmodemyanskaya S.S. (2017). Electronic educational resource and case technologies as a means of becoming a future chemistry teacher. Bulletin of Kazan Technological University. Kazan, T. 17, No. 17. - 2014.S. 281-284.

[14] Ponomareva M.N. (2019). Digital educational environment of a professional educational organization: directions of development. Innovative development of vocational education. 2019.No1 (21). S. 59-65.

[15] Voronina A.V. (2018). The use of distance technologies in chemistry lessons with a personality-oriented approach. Modernization of professional training of specialists in the field of science education. Minsk: BSPU, 2018.P.67-69.

[16] Darzemanova D.L., Kosmodemyanskaya S.S. (2018). Methodological features of teaching chemistry to foreign students of non-chemical specialties at the university. Fundamental and applied sciences today XIV: Proceedings of the Conference. North Charleston, February 20-21, 2018, Vol. 3. North Charleston, SC, USA: CreateSpace, 2018, p. 123, 44-46 p.

IOP Conf. Series: Materials Science and Engineering **915** (2020) 012016 doi:10.1088/1757-899X/915/1/012016

# Pedagogical conditions for training science teachers for sustainable development of society

**S I Gilmanshina[1], R N Sagitova[1], I R Gilmanshin[1,2], A I Galeeva[3] and I A Galeev[1]**

[1]Kazan Federal University, Kazan, Russia, 420008
[2] Kazan National Research Technical University named after A N Tupolev, Kazan, Russia
[3]Kazan State Power Engineering University, Kazan, 420066, Russian Federation

is-er@yandex.ru

**Abstract.** The concept of "science education for sustainable development" is considered, including environmental education, social and economic components. Emphasis is placed on the environmental component. The pedagogical conditions for the preparation of teachers of natural sciences for the sustainable development of society are identified and justified. The stages of their implementation are considered on the example of chemical education.

In modern socio-economic conditions of energy conservation and sustainable management of resources and waste (for more details see [1], [2], [3]), the need to anticipate threats and anticipate responses to them, there is a need to train teachers of natural sciences who are familiar with the concept of sustainable development and capable of conscious implementation of education in the interests of sustainable development of society. The teacher's readiness for the formation of environmental literacy (research [4], [5]), assessing the creativity of students in achieving the goals of the concept of sustainable development in the context of digitalization of education is relevant. These circumstances require a new look at the preparation of the teacher of natural sciences and the formation of professional thinking (research [6], [7], [8], [9]) in modern conditions of project training (for more details see [10]) and digitalization of education ( more details in studies [11], [12]).

The concept of "education for sustainable development" was formed relatively recently and includes environmental education, social and economic components. Today, when a common systemic concept of education is being formed in the interests of sustainable development, designing the appropriate educational process for teacher training is becoming one of the urgent tasks.

The training of teachers of natural sciences in the interests of sustainable development of society is based on the upbringing of a set of personal qualities, the formation of professionally significant competencies and is of a systemic nature with an emphasis on futurization of teacher education (studying and modeling the future).

Research problem: what are the pedagogical conditions for the training of teachers of science in the interests of sustainable development of society in the system of university education.

Purpose: to identify the pedagogical conditions for the training of teachers of science in the interests of sustainable development of society on the example of university chemical and pedagogical education.

Methodological basis: practice-oriented and integrative approaches.

The studies conducted allowed us to identify four pedagogical conditions for the preparation of chemistry teachers in the interests of sustainable development of society.

The first condition is related to the adjustment of the curriculum in accordance with the environmental component of the concept of sustainable development. For example, in the curriculum of teacher training in the direction of "Pedagogical education, profile chemistry" introduced chemical and environmental disciplines.

These are “Environmental Chemistry”, “School Ecological and Analytical Monitoring”, “Chemical Systems and Ecological Risk”, “Problems of Ecological and Analytical Monitoring of Environmental Pollution” and others (taking into account the shift in emphasis on the study and modeling of the future, which corresponds to the futurization of education )

For example, in the process of studying the course “School environmental and analytical monitoring” the topics are considered: “Environmental and analytical monitoring as an integral part of environmental monitoring”, “School environmental and environmental and analytical monitoring”, “Methodology of school environmental and analytical monitoring”, where 36 hours for a laboratory workshop are allotted 24 hours. The course “Chemical systems and environmental risk” includes topics such as “Basic concepts of environmental risk”, “Man-made accidents as sources of negative factors in the technosphere”, “Environmental safety management in the chemical industry” and others (from the 36 hours contact the laboratory workshop is 18 hours )

The second condition is the transformation of laboratory work in chemistry as an important factor in the formation of students' skills to achieve the goals of the concept of sustainable development. For example, the course of colloid chemistry includes work to eliminate the turbidity of the solution using various coagulants, the laboratory workshop on physical and organic chemistry is also significantly transformed taking into account the ideas of sustainable development of society. Small amounts of reagents are used, organic substances for a laboratory workshop are selected taking into account their minimum environmental load, thermal analysis in physical chemistry is carried out in sealed capsules, etc.

The studies conducted allowed us to identify four pedagogical conditions for the preparation of chemistry teachers in the interests of sustainable development of society.

The first condition is related to the adjustment of the curriculum in accordance with the environmental component of the concept of sustainable development. For example, in the curriculum of teacher training in the direction of "Pedagogical education, profile chemistry" introduced chemical and environmental disciplines.

These are “Environmental Chemistry”, “School Ecological and Analytical Monitoring”, “Chemical Systems and Ecological Risk”, “Problems of Ecological and Analytical Monitoring of Environmental Pollution” and others (taking into account the shift in emphasis on the study and modeling of the future, which corresponds to the futurization of education )

For example, in the process of studying the course “School environmental and analytical monitoring” the topics are considered: “Environmental and analytical monitoring as an integral part of environmental monitoring”, “School environmental and environmental and analytical monitoring”, “Methodology of school environmental and analytical monitoring”, where 36 hours for a laboratory workshop are allotted 24 hours. The course “Chemical systems and environmental risk” includes topics such as “Basic concepts of environmental risk”, “Man-made accidents as sources of negative factors in the technosphere”, “Environmental safety management in the chemical industry” and others (from the 36 hours contact the laboratory workshop is 18 hours )

The third condition is the development of students' creativity and systemic thinking through the implementation of complex chemical and environmental projects and the use of smart card technology (the problem to be solved is placed in the center, from which there are branches that demonstrate possible solutions (for details, see [13]). Students to the future Group and individual projects are offered to bachelors of teacher education in chemistry, projects are organized in three blocks related to pollution monitoring and cleaning of environmental objects, physicochemical and chemical analysis of food products and medicines, and study of household chemicals.

The fourth condition is the development of the content of a special topic that reveals the essence of the concept of "education in the interests of sustainable development" in one of the courses of the block of methodological disciplines. For example, in the scientific and methodological course "Didactics of Chemistry" a new topic "Chemical and pedagogical education for the sustainable development of society" is included. This topic involves the consideration of the concept of sustainable development, the concept of education for sustainable development, the specificity and essence of the concept of chemical and pedagogical education in the interests of sustainable development of society.

These pedagogical conditions are implemented in different ways in the preparation of chemistry teachers at the undergraduate level in the system of pedagogical education of Kazan Federal University.

The first pedagogical condition - curriculum adjustment in accordance with the environmental component of the concept of sustainable development - has been implemented since 2011.

Initially, only two disciplines "Chemical fundamentals of ecology" and "Chemical processes in the environment" were included in the curriculum. Later, at the second stage of adjustment, the curriculum was replenished with the courses "Soil Chemistry", "Chemical Systems and Environmental Risk", "Problems of Ecological and Analytical Monitoring of Environmental Pollution".

The third stage of curriculum adjustment in the direction of "Pedagogical education, profile chemistry" is associated with the introduction of the requirements of the new generation of federal state educational standards in the educational process. This is the need to improve the quality of environmental education and training of students of general educational institutions at the lessons of the natural science block of disciplines. Since the chemistry teacher will also carry out this activity, it became necessary to include in the curriculum such courses as "Physico-chemical methods of analysis in school project activities" and "School environmental and analytical monitoring". The work programs of both disciplines are designed with a clear focus on the capabilities of the school chemistry laboratory and the implementation of the requirements of the new generation of federal state educational standards.

Initially, only two disciplines "Chemical fundamentals of ecology" and "Chemical processes in the environment" were included in the curriculum. Later, at the second stage of adjustment, the curriculum was replenished with the courses "Soil Chemistry", "Chemical Systems and Environmental Risk", "Problems of Ecological and Analytical Monitoring of Environmental Pollution".

Moreover, the need to develop the content of these disciplines arose according to the wishes of chemistry teachers who annually undergo advanced training at the Department of Chemical Education of Kazan Federal University. As a result of their repeated questioning, the following results were obtained. To the question "Do you consider it necessary to introduce an environmental component in the school chemistry course", the majority of respondents, regardless of their teaching experience, answered "Yes" (92% of 164 respondents).

At the same time, to the question "Where is it more expedient to implement the environmental component in chemistry education - in the classroom or in extracurricular activities?" teachers' answers were divided depending on the teaching experience. 50% of teachers with experience of up to 5 years, consider that the environmental component is necessary in chemistry classes. This opinion is shared by 30% of teachers with an experience of 5 to 15 years and only 20% of teachers with an experience of more than 15 years. More experienced teachers prefer to introduce the environmental component into extracurricular activities, explaining that, at the time, in the process of studying at a higher school, they taught chemical disciplines without an environmental component. In general, teachers were asked to bring the laboratory workshop of environmental-oriented university disciplines closer to school conditions.

Thus, a survey of existing teachers confirms the need to adjust the curriculum for the preparation of chemistry teachers in accordance with the environmental component of the concept of sustainable development.

IOP Conf. Series: Materials Science and Engineering **915** (2020) 012016 doi:10.1088/1757-899X/915/1/012016

The second pedagogical condition - the transformation of laboratory work in chemistry as an important factor in the formation of students' skills to achieve the goals of the concept of sustainable development - has been systematically implemented in courses of physical, analytical and organic chemistry for a rather long time (since 2011).

The third pedagogical condition - the development of students' creativity and systemic thinking through the implementation of complex chemical and environmental projects and the use of smart card technology has been performed relatively recently.

Students - future chemistry teachers carry out chemical and environmental projects in the course "Didactics of chemistry" when studying the topic "Development and education of students using the subject" chemistry "in the section" Formation of chemical and environmental culture of students ". At the same time, ecological knowledge is considered as a component of worldview knowledge and the conditions for their formation in chemistry lessons.

At the same time, the results of the survey of graduate students and current teachers of chemistry indicate the need to strengthen the work on promoting the ideas of sustainable development and more fully implementing the concept of "education for sustainable development" in the system of university teacher education. This served as an important factor in the development and implementation of the fourth pedagogical condition in the 2019-2020 academic year.

The third pedagogical condition - the development of students' creativity and systemic thinking through the implementation of complex chemical and environmental projects and the use of smart card technology has been performed relatively recently.

Students - future chemistry teachers carry out chemical and environmental projects in the course "Didactics of chemistry" when studying the topic "Development and education of students using the subject" chemistry "in the section" Formation of chemical and environmental culture of students ". At the same time, ecological knowledge is considered as a component of worldview knowledge and the conditions for their formation in chemistry lessons.

The fourth pedagogical condition is the development of the content of a special topic that reveals the essence of the concept of "education in the interests of sustainable development" in one of the courses of the block of methodological disciplines. A chemistry didactics course was selected. Currently, as part of this course, a new theme is being developed: "Chemical-pedagogical education for the sustainable development of society". Lecture presentations "Concepts of Sustainable Development" and "Education for Sustainable Development" were developed. Work is underway on the third presentation, "Chemical and Pedagogical Education for the Sustainable Development of Society," a glossary and a bank of control questions, and topics are selected for the corresponding design work for students. It is planned to introduce it into the educational process in the 2020–2021 academic year.

Thus, the pedagogical conditions for the preparation of chemistry teachers in the interests of the sustainable development of society in the system of university education have been identified and are being tested. This is the inclusion in the curriculum of environmental disciplines; transformation of laboratory work with the aim of forming students' skills to achieve the goals of the concept of sustainable development; inclusion in the content of methodological disciplines topics that reveal the essence of the concept of "education in the interests of sustainable development"; development of creativity and systems thinking through the implementation of complex projects and the use of smart card technology.

**References**

[1] Kashapov N.F., Gil'manshin I.R., Konahina I.A. et al. (2014). System analysis of the energy complex of engineering enterprise as a basic tool of effective energy management. IOP Conference Series: Materials Science and Engineering, Vol. 69, 2013. DOI: 10.1088/1757-899X/69/1/012024.

[2] Gilmanshin I.R., Kashapov N.F., Gilmanshina S.I., Galeeva A.I. (2016). Landfill energy complex based on the renewable energy installations. IOP Conference Series: Materials

Science and Engineering, Volume 134, Number 1, 2015, pp.12007. DOI: 10.1088/1757-899X/134/1/012007.

[3] Gilmanshin, I.R., Gilmanshina, S.I., Galeeva, A.I., Gadirova, E.M., Galeev, I.A. (2019). Methods of training specialists in the field of sustainable resource management and waste management. IOP Conference Series: Materials Science and Engineering, Volume 570, Number 1. https://iopscience.iop.org/article/10.1088/1757-899X/570/1/012025

[4] Gilmanshina S.I., Gayfullina A.Z., Fedotova N.R., Gilmanshin I.R. (2018). Building the Environmental Literacy Through School Natural Science Education. Astra Salvensis, VI (2018), Special Issue, p. 245-253. https://astrasalva.files.wordpress.com/2018/07/astra-salvensis-vi-2018-special-issue.pdf.

[5] Sibgatova K.I., Gilmanshina S.I., Khalikova F.D., Gilmanshin I.R., Akchurina I.R., Shchaveleva N.G., Fassakhova G.R. (2015). Peculiarities of pupils and vocational college students' career guidance modeling in the integrated system "school – college –enterprise". Asian Social Science. Vol. 11. No. 1. P. 386-391. DOI: 10.5539/ass.v11n1p386.

[6] Sakhieva R.G., Gilmanshina S.I., Gilmanshin I.R., Kosmodemyanskaya S.S., Akchurina I.R., Sagitova R.N. (2015). A Portfolio as an Alternative Means of Presenting the University Students Achievements // Asian Social Science; Vol. 11, No. 3; 2015. P. 162-167. ISSN 1911-2017 (Print) E-ISSN 1911-2025 (Online). DOI: 10.5539/ass.v11n3p162-167.

[7] Gilmanshin I.R., Gilmanshina S.I. (2016). The formation of students' engineering thinking as a way to create new techniques, technologies, materials. 2016 IOP Conf. Ser.: Mater. Sci. Eng. 134 012006. DOI: 10.1088/1757-899X/134/1/012006.

[8] Gilmanshina S.I., Sagitova R.N. and et. (2015). Professional Thinking Formation Features of Prospective Natural Science Teachers Relying on the Competence-Based Approach. Review of European Studies, Vol 7, No 3 (2015) P. 341-349 DOI: 10.5539/res.v7n3p341.

[9] Samigullina G.S., Gilmanchina S.I., Gaisin I.T., Gilmanshin I.R., Akchurina I.R. (2015). Professional and Creative Development of Natural Geographic Course Teachers within the Process of Professional Retraining. International Education Studies, 8(4), 2015. http://dx.doi.org/10.5539/ies.v8n4p159.

[10] Gilmanshina, S.I., Gilmanshin, I.R., Dyatlova, E.V. (2019). Formation of competencies of the 21st century by means of project activities. IOP Conference Series: Materials Science and Engineering, Volume 570, Number 1. https://iopscience.iop.org/article/10.1088/1757-899X/570/1/012026.

[11] Bergdahl, N., Nouri, J., Fors, U. (2020). Disengagement, engagement and digital skills in technology-enhanced learning. Education and Information Technologie, Volume 25(2). P. 957-983.

[12] Gilmanshina, S.I., Gilmanshin, I.R., Bendjukevich, K.G. (2019). Digital learning technologies in the University education system. IOP Conference Series: Materials Science and Engineering, Volume 570, Number 1. https://iopscience.iop.org/article/10.1088/1757-899X/570/1/012027.

[13] Tony Buzan. Mind Map Mastery: The Complete Guide to Learning and Using the Most Powerful Thinking Tool in the Universe, Watkins Publishing, 2018. 224 p.

IOP Conf. Series: Materials Science and Engineering **915** (2020) 012017 doi:10.1088/1757-899X/915/1/012017

# Digital educational environment: training future teachers to design additional education

**S I Gilmanshina[1], V A Minnakhmetova[1], I R Gilmanshin[1,2], A I Galeeva[3], G M Abyzbekova[4]**

[1]Kazan Federal University, Kazan, Russia, 420008
[2] Kazan National Research Technical University named after A N Tupolev, Kazan, Russia
[3]Kazan State Power Engineering University, Kazan, 420066, Russian Federation
[4]Korkyt Ata Kyzylorda University, Aiteke bi, 29A; 120014, Kyzylorda, Republic of Kazakhstan

is-er@yandex.ru

**Abstract.** The concepts of further education - care and development are considered. Additional general educational programs are justified taking into account age and individual characteristics of children, adolescents and youth. The possibilities of a digital educational environment for teaching future teachers to design additional chemical education in primary and secondary schools are considered.

Today, along with the relevance of research in the field of efficient energy management [1], renewable energy ([2], [3]) productive resource and waste management ([4], [5], [6]), as well as digitalization of education (more in [7], [8]), project training (research [9], [10]) and environmental literacy (for more details in [11]), there is an acute question of additional education of students as an important factor in revealing and developing abilities and professional self-determination .

In European countries, the requirements for further education are determined by two of its concepts. This is the concept of care (helping children and adolescents with serious problems) and the concept of development. The concept of continuing care is widely represented in the Netherlands and Poland. The development concept is the basis of further education in Eastern and Central Europe, where the emphasis is on the interests of students who do not have physical, psychological and social problems.

Additional education in Russia, as noted in [12], is aimed at the formation and development of creative abilities, adaptation to life in society and the vocational guidance of students, as well as at identifying and supporting children with outstanding abilities.

Additional general education programs are divided into general development and pre-vocational programs. In any case, additional education must be carried out taking into account the age and individual characteristics of children, adolescents and youth.

Age-related periodization has been widely studied in the works of Russian psychologists ([13], [14]). According to L.S. Vygotsky, the transition from one age to another is revolutionary, which determines the presence of "crises of age development", when there is a transition from one age to another [13]. The adolescent and youthful ages, which are considered the main in the formation of the psychological and social well-being of a person, are especially distinguished.

Briefly consider the age characteristics of children 13-17 years old. Eighth-graders are characterized by a deterioration in educational activity (non-fulfillment of tasks, negligence and sloppiness in school records, a change in emphasis) compared to the seventh grade. At the same time, many students are interested in non-standard tasks of intelligence, requiring persistence and independence.

The peculiarity of ninth-graders is associated with the integration of their teenage and youthful age traits, which often serves as a reason for the protest nature of interactions with adults and teachers as well. Some of them lose interest in academic subjects, begin to poorly navigate in educational activities. During this period, design and research activities are significant.

In the tenth grade, students often show signs of fatigue from learning activities, increasing interest in a variety of extracurricular activities. At the same time, by the tenth grade, most students are determined with the direction of their future professional activity, which requires increased attention to specialized subjects.

In grade 11, the priority is the correction of the choice of a future profession and the construction of an individual learning path. Eleventh graders begin to differentiate subjects on the basis of the principle that "the subject is needed to enter the university - not needed."

Of course, the noted age-related features of students must be taken into account both in teaching chemistry in the classroom and in the implementation of additional chemical education.

When preparing undergraduate students for their future professional and pedagogical activities, a chemistry teacher also raises the question of additional chemical education, which is not always given due attention.

In the curriculum of the Kazan Federal University for the preparation of bachelors of teacher education in chemistry, there is no special discipline in additional education. Issues of additional chemical education are considered in the courses "Methodology for solving problems of increased complexity", "Methodology for solving problems in physical colloid chemistry in a school chemistry course", "Methodology for solving olympiad problems", "Didactics of chemistry", "Modern educational technologies", "Modern pedagogical technologies training."

In general, at the Department of Chemical Education of Kazan Federal University, the preparation of undergraduate students for the implementation of additional chemical education in primary and secondary schools is associated with solving problems related to familiarization with:

1) the solution of the calculated olympiad problems;
2) solving experimental olympiad problems;
3) theoretical and experimental training of highly motivated adolescents of Kazan at the Small Chemical Institute;
4) practice-oriented training of lyceum students in the laboratories of a specialized institute;
5) the use of show technology in additional chemical education.

The first task - to acquaint with the solution of the calculated olympiad problems - is carried out in courses of special methodological disciplines, such as "Methods for solving olympiad problems" and "Methods for solving problems in physical colloid chemistry in a school chemistry course." An electronic educational resource has been developed to help study the last discipline, where the topics are presented: “Methods of thermochemical calculations and thermodynamic substantiation of the possibility and limit of chemical reactions,” “Methods for calculating chemical equilibrium,” “Methods for calculating equilibria in electrolyte solutions,” “Methods for kinetic calculations” , "Methodology for the calculation of electrochemical systems." For each topic, the theoretical foundations, the methodology for calculating typical problems, the methodology for solving olympiad problems, questions for self-monitoring on theoretical material, computational problems (for all topics about 380) and current control for solving olympiad problems are presented.

The second task (to familiarize with the solution of experimental olympiad problems) and the fourth task (to familiarize with practice-oriented training of lyceum students based on laboratories of the profile institute) are carried out in the course "Didactics of chemistry (the electronic educational resource of the same name has been developed). At this course, students complete the tasks of the experimental tour of chemical olympiads of various levels and get acquainted with the programs of practice-oriented training (with a focus on the integration of knowledge). The fifth task - to acquaint with show technologies in additional chemical education - is solved in the course "Modern educational technologies."

Let us dwell in more detail on the third task (to acquaint with the theoretical and experimental training of highly motivated adolescents in Kazan at the Small Chemical Institute), which is being solved in the course "Didactics of Chemistry". To help study this discipline, the electronic educational resource of the same name (ESM) has also been developed and introduced, which allows students to optimize their independent scientific and methodological work. This resource, in addition to lecture and other methodological material, contains the electronic resource "Book" and the element of the electronic resource "Seminar". The electronic resource "Book" allows us to present the systematized material available at the department according to programs of additional education, depending on the age characteristics of students. And the "Workshop" element is oriented towards a practical lesson on the development by students of an author's program of additional chemical education for a certain age group of schoolchildren. The invariant is the program structure, which contains sections: general characteristics of the program, goal and objectives, planned learning outcomes, curriculum, calendar training schedule, organizational and pedagogical conditions, certification form, assessment tools, subsequent support.

Thus, the system of preparing students for the implementation of additional chemical education includes subsystems for familiarizing themselves with the solution of design and experimental olympiad problems; theoretical and experimental training of highly motivated adolescents after school hours at the university; practice-oriented teaching of schoolchildren on the basis of laboratories of the relevant institute; application of show technologies in additional chemical education in order to increase motivation. At the same time, the digital educational environment of the federal university is widely used.

**References**

[1] Kashapov N.F., Gil'manshin I.R., Konahina I.A. et al. (2014). System analysis of the energy complex of engineering enterprise as a basic tool of effective energy management. IOP Conference Series: Materials Science and Engineering, Vol. 69, 2013. DOI: 10.1088/1757-899X/69/1/012024.

[2] Gilmanshin I.R., Kashapov N.F., Gilmanshina S.I., Galeeva A.I. (2016). Landfill energy complex based on the renewable energy installations. IOP Conference Series: Materials Science and Engineering, Volume 134, Number 1, 2015, pp.12007. DOI: 10.1088/1757-899X/134/1/012007.

[3] Gilmanshin I.R., Gilmanshina S.I. (2017). Competence formation of engineering directions students in the field of energy saving as a way to create new generation technologies. IOP Conference Series: Materials Science and Engineering, Volume 240, Number 1, 2015, pp. 012022. doi:10.1088/1757-899X/240/1/012022.

[4] Sabirzyanov, R.G., Gilmanshin, I.R., Kashapov, N.F., (...), Galeev, I.A., Krainova, D.R. (2018). Material and technical base of the laboratory for the thermal destruction of production and consumption waste for the training of specialists in the field of waste management. IOP Conference Series: Materials Science and Engineering, 412, (2018) 012061. doi:10.1088/1757-899X/412/1/012061. http://iopscience.iop.org/article/10.1088/1757-899X/412/1/012061.

[5] Azimov Yu.I., Gilmanshin I.R., Gilmanshina S.I. (2016). Modern technologies of waste utilization from industrial tire production. IOP Conference Series: Materials Science and Engineering, Volume 134, Number 1, 2015, pp.12003. DOI: 10.1088/1757-899X/134/1/012003.

[6] Gilmanshin, I.R., Gilmanshina, S.I., Galeeva, A.I., Gadirova, E.M., Galeev, I.A. (2019). Methods of training specialists in the field of sustainable resource management and waste

management. IOP Conference Series: Materials Science and Engineering, Volume 570, Number 1. https://iopscience.iop.org/article/10.1088/1757-899X/570/1/012025.

[7] Gilmanshin I.R., Gilmanshina S.I. (2018). Technology of e-learning in the university education. IOP Conference Series: Materials Science and Engineering, 412, (2018) 012022. doi:10.1088/1757-899X/412/1/012022. https://iopscience.iop.org/article/10.1088/1757-899X/412/1/012022.

[8] Bergdahl, N., Nouri, J., Fors, U. (2020). Disengagement, engagement and digital skills in technology-enhanced learning. Education and Information Technologie, Volume 25(2). P. 957-983.

[9] Gilmanshina, S.I., Gilmanshin, I.R., Dyatlova, E.V. (2019). Formation of competencies of the 21st century by means of project activities. IOP Conference Series: Materials Science and Engineering, Volume 570, Number 1. https://iopscience.iop.org/article/10.1088/1757-899X/570/1/012026.

[10] Khalikova, F.D., Gilmanshina, S.I. (2016). Practical-oriented teaching of gifted youth in the field of natural sciences. IOP Conf. Series: Materials Science and Engineering. Vol. 240 (2017) 012042 doi:10.1088/1757-899X/240/1/012042. http://iopscience.iop.org/article/10.1088/1757-899X/240/1/012042. http://apps.webofknowledge.com/Search.do?product=WOS&SID=C4x3B9OxRwnlKplUbLj &search_mode=GeneralSearch&prID=4d9b09c1-6a8f-4f3b-a1cd-0e5aa9565462.

[11] Gilmanshina S.I., Gayfullina A.Z., Fedotova N.R., Gilmanshin I.R. (2018). Building the Environmental Literacy Through School Natural Science Education. Astra Salvensis, VI (2018), Special Issue, p. 245-253. https://astrasalva.files.wordpress.com/2018/07/astra-salvensis-vi-2018-special-issue.pdf.

[12] Federal Law of December 29, 2012 N 273-ФЗ (as amended on February 6, 2020) “On Education in the Russian Federation”. Article 75. Further education of children and adults.

[13] Vygotsky, L.S. Psychology of child development. M .: Publishing house "Sense". Eksmo Publishing House, 2004.512 p.

[14] Leontyev, A.N. Lectures on general psychology. M .: Publishing house Sense, 2019.511 s.

# The digitalization of training of bachelors - the future science teachers

**S I Gilmanshina[1], I R Gilmanshin[1,2], A I Galeeva[3], G M Abyzbekova[4]**

[1]Kazan Federal University, Kazan, Russia, 420008
[2] Kazan National Research Technical University named after A N Tupolev, Kazan, Russia
[3]Kazan State Power Engineering University, Kazan, 420066, Russian Federation
[4]Korkyt Ata Kyzylorda University, Aiteke bi, 29A; 120014, Kyzylorda, Republic of Kazakhstan

is-er@yandex.ru

**Abstract.** The types of digitalization in the training of bachelors - the future science teachers are identified. The peculiarity of digitization of methodological training of future teachers is considered on the example of the author's digital educational resource of chemical education. The technology of working with the developed digital resource is presented.

The development of digitalization is important for all spheres of the national economy and for education as well. Due to the rapid aging of the knowledge gained at the university for solving the practical problems of the post-industrial society (for more details see [1], [2], [3], [4]), there is a need for digital educational resources in education. This is due to the need of society for modern creative specialists with developed professional thinking (research [5], [6], [7], [8]), educators who are able to creatively adapt digital teaching technologies to the needs of students in the context of educational transformation (research [ 9], [10], [11], [12]). Digital digital educational resources are necessary, both elements of digitalization of university education (research [13], [14]), and to improve the qualifications of teachers without interruption from the educational process (in recent years, the term "digital school" has even appeared).

Today, during the pandemic, the negative aspects of large-scale digital learning have emerged, such as the replacement of real communication with peers and late socialization, the lack of practical competencies formed by a virtual laboratory workshop. In connection with the aforementioned, a mixed form of training in the form of the use of digitalization elements in the preparation of bachelors and teachers of natural science education has become relevant.

Research problem: what is the peculiarity of the use of digitalization in the preparation of bachelors and teachers of science education.

The purpose of the study: to identify the peculiarity of digitalization of the methodological training of bachelors-teachers of science education in mixed education on the example of chemical education.

The methodological basis of the study is the practice-oriented and competent approaches.

The conducted studies have allowed to establish three types of digitalization of the methodological training of bachelors and teachers of natural science education.

The first type manifests itself in full digitalization and refers to the conditions of self-learning, when the future teacher independently masters the methodology of teaching the subject through digital educational resources, including the methodology of the demonstration and student experiment. After completing the course, he passes a digital test control. Teaching practice at the school also takes place online. This is the methodological feature of this type of digitalization.

The second type is manifested in the digitalization of the student learning process, which is led by the teacher. In this case, the teacher directs the student's educational research activities, advises him through chat, email, phone. The student masters the methodological features of the demonstration and student experiment through digital educational resources, video clips presented in a digital educational resource or posted on YouTube. Teaching practice at the school also takes place online. The peculiarity of digitalization of the methodological training of bachelors-teachers of science education in this case is related to the specifics of the demonstration and student experiment, as well as on-line pedagogical practice.

The third type is manifested in a mixed form of training, when there are elements of digitalization of the preparation of the future teacher of science education. The student offline learns the methodology of teaching the subject and teaching practice in an educational institution. The digital educational resource supports methodological discipline, promotes the individualization of the learning process and the monitoring of the formation of relevant competencies.

When preparing future teachers - bachelors of teacher education (chemistry profile), Kazan Federal University implements a mixed form of training and a third type of digitalization of students' methodological training.

Consider it in more detail on the example of teaching the methodological discipline "Didactics of chemistry". To maintain it, the author's author's digital (electronic) educational resource of the same name was developed, which fully corresponds to the approved work program of the discipline and has passed the examination.

When developing a digital (electronic) resource, the emphasis was placed on the design technology for teaching chemistry, taking into account the individual educational routes of students. When compiling questions, test and case studies, both the author's material (the vast majority) and the manuals indicated in the corresponding lists of recommended literature were used.

The digital course is designed for classroom and independent work of students and includes five topics. This is "Documents reflecting the content of chemical education", "Chemical education as a didactic system. The process of teaching chemistry "," Development and education of students by means of the subject "chemistry", "Organization, management, modern technologies in chemical education", "Methodology in chemical education". Start date: October 16, 2018.

Each topic includes questions for consideration, a glossary, electronic teaching aids, recommended literature, methodological recommendations for students and teachers, lectures in the form of presentations in pdf format, topics and questions of study in practical classes, individual and group design tasks of a creative nature.

The external structure of the digital course consists of a heading, a side navigation menu and the main part of the site.

In general, a digital resource contributes to the formation of professional competencies. The student has the ability to carry out professional activities in accordance with regulatory legal acts in the field of education and standards of professional ethics; organize joint and individual educational activities of students, including those with special educational needs, in accordance with the requirements of educational standards; carry out pedagogical activities on the basis of special scientific knowledge.

As a result of mastering the digital course, the student will be able to demonstrate the ability and willingness to independently choose methods of work in the classroom, assess their capabilities when organizing various methodological situations in the lessons and choose teaching methods, develop and educate students in chemistry lessons.

This digital resource involves the following types of classes with students:

- updating and consolidating the theoretical material necessary for solving practical work ("Lecture" files), to independently deepen theoretical knowledge, the "Information Resources" files are attached (with the ability to access external sources via INTERNET according to the course as a whole and on specific topics), a glossary for the course as a whole and for each topic, as well as open electronic resources for self-testing;

- practical, personality-oriented classes for students to independently carry out practice-oriented design and experimental projects (elements of the "Practice 1, 2" course);
- practical classes for students to independently carry out projects related to the use of innovative educational technologies, real and virtual experiments, methodological approaches in teaching chemistry (elements of the "Practice 3, 4, 5" course);
- practical lesson on the implementation of the final research project (element of the course "Practice 6. Research project on chemical education");
- extracurricular counseling of students on their individual project (element of the course "Forum to discuss the topic ...")
- the results of the practical exercises on topics 3 and 4 are intellect cards developed by students, video fragments of real chemical experiments and the corresponding chemistry lessons, the best of which can be included in the Book resource, which allows you to create a digital educational like a book with the corresponding chapters and paragraphs subresource with students' creative work;
- performance of the final control - in the form of a full-time test (questions for preparation are presented in the form of a file "Sample questions for the test"); at the same time, the mark is set according to criteria established by the teacher in accordance with the point-rating system of Kazan Federal University.

The technology for working with the developed digital educational resource is given below.

1. The student must study the theoretical and practice-oriented material according to the sources indicated in the "Course Information Support", which is presented as a MOODLE object, within the time specified by the teacher. This list of sources is broken down by topic and presented as separate MOODLE objects in each topic. Some sources of information (most) are displayed on the computer screen at the specified link.
2. The student must study the first point of the source - the lecture of the teacher. Lectures are displayed in PDF format from an external source.
3. In the process of studying each topic, the student must become familiar with the glossary and open electronic resources to determine his level of mastering the material (the corresponding hyperlinks are presented).
4. The student, according to the schedule of the educational process, must complete, design and submit within the specified time frame individual and group projects and mini-projects, topics and guidelines for which are presented in the form of MOODLE objects.
5. After studying all five topics, the student (student) should submit a completed research project on chemical education, arrange it electronically, make a presentation and send it in Word or PDF format to the address of the teacher for verification and evaluation, and protect him in front of his classmates.
6. After receiving a positive assessment and familiarization with the comments, the student must independently carry out work on the mistakes made and proceed to the test certification.
7. Test certification (passing the test, conducting the final test control) can be done only by agreement with the teacher, at the time appointed by him (in person) or at the appointed time frames in absentia, remotely through a digital (electronic) educational resource.

A positive assessment is put in the gradebook and statement.

In the case of a non-positive assessment, the student (student) receives additional access to the system for studying the course through a digital resource in consultation with the teacher and in the time frame indicated by him with the subsequent delivery of the test in person.

Thus, on the example of the preparation of future bachelors of chemical education, three types of digitalization of the methodological training of bachelors-teachers of science education are identified. This is a complete digitalization in a self-learning environment; full digitalization under the guidance of a teacher; elements of digitalization in a mixed form of learning. In all cases, the peculiarity of the digitalization of the methodological training of bachelors-teachers of science education is related to

IOP Conf. Series: Materials Science and Engineering **915** (2020) 012018 doi:10.1088/1757-899X/915/1/012018

the specifics of the demonstration and student experiment. In addition, with full digitalization, there is a feature of the methodological preparation of students for teaching practice on-line.

**References**


[1] Kashapov N.F., Gil'manshin I.R., Konahina I.A. et al. (2014). System analysis of the energy complex of engineering enterprise as a basic tool of effective energy management. IOP Conference Series: Materials Science and Engineering, Vol. 69, 2013. DOI: 10.1088/1757-899X/69/1/012024.

[2] Sabirzyanov, R.G., Gilmanshin, I.R., Kashapov, N.F., (...), Galeev, I.A., Krainova, D.R. (2018). Material and technical base of the laboratory for the thermal destruction of production and consumption waste for the training of specialists in the field of waste management. IOP Conference Series: Materials Science and Engineering, 412, (2018) 012061. doi:10.1088/1757-899X/412/1/012061. http://iopscience.iop.org/article/10.1088/1757-899X/412/1/012061.

[3] Gilmanshin, I.R., Ferenets, A.V., Azimov, Yu.I., Galeeva, A.I., Gilmanshina, S.I. (2015). Innovative technologies of waste recycling with production of high performance products. IOP Conf. Series: Materials Science and Engineering Vol. 86, pp.12014-12016. doi:10.1088/1757-899X/86/1/012014. http://www.ingentaconnect.com/content/iop/mse/2015/00000086/00000001/art012014.

[4] Gilmanshin, I.R., Gilmanshina, S.I., Galeeva, A.I., Gadirova, E.M., Galeev, I.A. (2019). Methods of training specialists in the field of sustainable resource management and waste management. IOP Conference Series: Materials Science and Engineering, Volume 570, Number 1. https://iopscience.iop.org/article/10.1088/1757-899X/570/1/012025.

[5] Sakhieva R.G., Gilmanshina S.I., Gilmanshin I.R., Kosmodemyanskaya S.S., Akchurina I.R., Sagitova R.N. (2015). A Portfolio as an Alternative Means of Presenting the University Students Achievements // Asian Social Science; Vol. 11, No. 3; 2015. P. 162-167. ISSN 1911-2017 (Print) E-ISSN 1911-2025 (Online). DOI: 10.5539/ass.v11n3p162-167

[6] Gilmanshin I.R., Gilmanshina S.I. (2016). The formation of students' engineering thinking as a way to create new techniques, technologies, materials. 2016 IOP Conf. Ser.: Mater. Sci. Eng. 134 012006. DOI: 10.1088/1757-899X/134/1/012006.

[7] Gilmanshina S.I., Sagitova R.N. and et. (2015). Professional Thinking Formation Features of Prospective Natural Science Teachers Relying on the Competence-Based Approach. Review of European Studies, Vol 7, No 3 (2015) P. 341-349 DOI: 10.5539/res.v7n3p341.

[8] Samigullina G.S., Gilmanchina S.I., Gaisin I.T., Gilmanshin I.R., Akchurina I.R. (2015). Professional and Creative Development of Natural Geographic Course Teachers within the Process of Professional Retraining. International Education Studies, 8(4), 2015. http://dx.doi.org/10.5539/ies.v8n4p159.

[9] Sibgatova K.I., Gilmanshina S.I., Khalikova F.D., Gilmanshin I.R., Akchurina I.R., Shchaveleva N.G., Fassakhova G.R. (2015). Peculiarities of pupils and vocational college students' career guidance modeling in the integrated system "school – college –enterprise". Asian Social Science. Vol. 11. No. 1. P. 386-391. DOI: 10.5539/ass.v11n1p386.

[10] Mertala, P. (2020). Paradoxes of participation in the digitalization of education: a narrative account. Learning, Media and Technology, Vol. 45(2), 2020, P. 179-192.

[11] Gilmanshina S.I., Gayfullina A.Z., Fedotova N.R., Gilmanshin I.R. (2018). Building the Environmental Literacy Through School Natural Science Education. Astra Salvensis, VI (2018), Special Issue, p. 245-253. https://astrasalva.files.wordpress.com/2018/07/astra-salvensis-vi-2018-special-issue.pdf

[12] Gilmanshina, S.I., Gilmanshin, I.R., Dyatlova, E.V. (2019). Formation of competencies of the 21st century by means of project activities. IOP Conference Series: Materials Science and Engineering, Volume 570, Number 1. https://iopscience.iop.org/article/10.1088/1757-899X/570/1/012026.

[13] Gilmanshin I.R., Gilmanshina S.I. (2018). Technology of e-learning in the university education. IOP Conference Series: Materials Science and Engineering, 412, (2018) 012022. doi:10.1088/1757-899X/412/1/012022. https://iopscience.iop.org/article/10.1088/1757-899X/412/1/012022.

[14] Gilmanshina, S.I., Gilmanshin, I.R., Bendjukevich, K.G. (2019). Digital learning technologies in the University education system. IOP Conference Series: Materials Science and Engineering, Volume 570, Number 1. https://iopscience.iop.org/article/10.1088/1757-899X/570/1/012027.

IOP Conf. Series: Materials Science and Engineering **915** (2020) 012019 doi:10.1088/1757-899X/915/1/012019

# How the radius of the screw insert core influences the flow behavior of the polymer solution

**A I Kadyirov[1], J V Karaeva[1], V O Zdor[1] and E M Khusnutdinova[2]**

[1]Institute of Power Engineering and Advanced Technologies, FRC Kazan Scientific Center, Russian Academy of Sciences
2/31 Lobachevskogo Str., P.O. box 261, Kazan, 420111 Kazan
[2]Kazan (Volga Region) Federal University
18 Kremlyovskaya Str., Kazan 420008

rr-088@mail.ru

**Abstract**. In this paper the authors studied the 3-dimensional flow behavior of the polymer solution in the screw barrel. Numerical simulation was performed in the COMSOL Multiphysics® suite. An aqueous solution of polyacrylamide was regarded as a specific liquid. Demonstrated that, with a decrease in the gap between the outer tube and the screw insert core, the profile of the axial component of the velocity vector is aligned over the entire barrel cross-section, except for the region adjacent to the screw ribs.

Currently, it is difficult to achieve an exact analytical solution to the problem of the non-Newtonian flow in a screw barrel [1]. Known are only analytical solutions of the problem of the isothermal flow of ordinary Newtonian fluid in the barrel of a single-screw extruder [2]. In addition, most of the solutions to such problems were obtained with assumptions concerning the small curvature of the screw ribs and the consideration of the flow region in the rectangular form [3]. Moreover, the rotation of the screw insert means the rotation of the outer wall [4, 5, 6]. Generally, there are publications in which the power law is used to describe the rheological behavior of a polymer solution [7]. The disadvantage of this model is the infinite viscosity at zero shear rate. This paper uses the Cross-model [8, 9, 10], which takes into account the asymptotic values of the effective viscosity in threshold cases.

A consideration was given to the 3-D stationary laminar flow of a 0.5% aqueous solution of polyacrylamide in a round pipe with a CCW screw insert (Fig. 1). Mathematical modeling was performed considering that gravity is negligible. Due to this assumption, the distributions of hydrodynamic characteristics in two regions bounded by the screw insert ribs (Fig. 1) are identical. Therefore, one of these regions is regarded as the study area (area - i) further in the text.

The numerical solution was performed using the COMSOL Multiphysics® suite. The Navier-Stokes equations are taken as the basic equations:

$$\rho\left(\nabla\overline{V}\cdot\overline{V}\right)=\nabla\cdot\left(-p\mathbf{I}+\eta\left(\nabla\overline{V}+\left(\nabla\overline{V}\right)^{T}\right)\right) \quad (1)$$

$$\nabla\cdot\overline{V}=0 \quad (2)$$

where $p$ is the pressure, Pa; $\mathbf{I}$ is the unit tensor; $\eta$ is the effective viscosity, Pa·s; $\rho$ is the liquid density, kg/m$^3$; $\overline{V}=\{v_r,v_\varphi,v_z\}$ is the velocity vector.

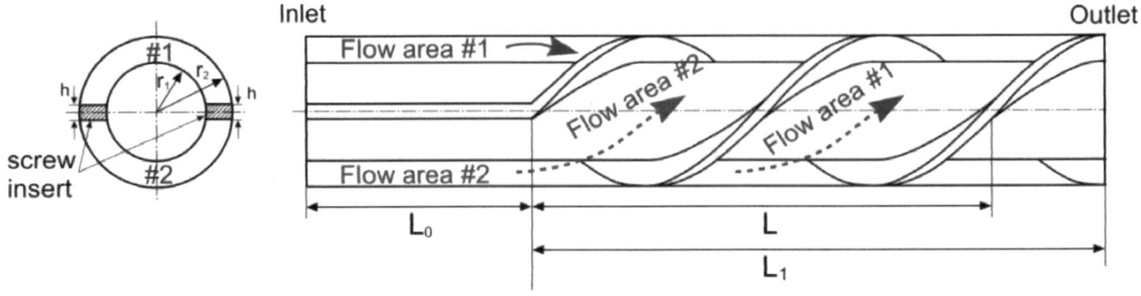


(a) (b)

Fig. 1 Transversal (b) and longitudinal (a) section of the barrel: $L_0$ is the length of the rectilinear portion; $L$ is the pitch of the screw barrel (the length of the barrel part corresponding to the 360-degree rotation of the screw ribs); $r_1$ is the radius of the inner pipe; $r_2$ is the radius of the outer pipe; # 1, # 2 are the identical fluid flow regions, limited by the outer and inner pipes and screw ribs; h is the thickness of the ribs.

The following conditions are taken as threshold conditions:

1) a developed velocity profile in the coaxial barrel is specified on the pipe inlet

$$v_z = 2U_a \frac{\left(1-(r/r_2)^2\right)\log(r_1/r_2)-\left(1-(r_1/r_2)^2\right)*\log(r/r_2)}{\left(1+(r_1/r_2)^2\right)\log(r_1/r_2)+\left(1-(r_1/r_2)^2\right)},\ v_r = 0,\ v_\varphi = 0. \quad (3)$$

where $U_a$ is the mean flow velocity.

2) the conditions for adhesion of the liquid are satisfied on the barrel walls and on the surface of the screw insert ribs;

3) the condition of equality of zero pressure is given at the pipe outlet.

For describing the rheological behavior of the aqueous solution of 0.5% polyacrylamide, the authors used the Cross-model [11] with the following parameters: $\eta_0$ = 1.08 (Pa·s), $\eta_\infty$ = 0.0023 (Pa·s) are the asymptotic values of the effective viscosity at $\dot{\gamma} \to 0$ and $\dot{\gamma} \to \infty$, respectively; $\lambda_c$ = 5 (s) is the characteristic time; $n$ = 0.8 is a coefficient that determines the viscosity reduction rate; $\dot{\gamma}$ is the shear rate, 1/s. The density of the aqueous solution of 0.5% polyacrylamide is believed to be $\rho$ = 999 kg/m$^3$.

Numerical results are presented for a stationary flow of 0.5% polyacrylamide in a barrel with the three types of screw inserts. The geometry of the screw inserts: $L/D_2 = 3$, $r_1 = 0.013$ (m), $r_2 = 0.0195$ (m), $h = 0.004$ (m). The diameter of the outer pipe $D_2 = 0.039$ (m), the length of the upstream section $L_0 = 1.5 \cdot D_2$, the length of the screw insert along the $z$-axis is equal to $L_1 = 7.5 \cdot D_2$.

As can be seen from Fig. 2, a steady-state velocity behavior is formed on the initial hydrodynamic portion of the barrel containing the screw insert. Here $A_i$ ($i = 45...360$) are the planes spaced apart from $A_0$ at the distance corresponding to the rotation of the screw insert by $i$ degrees. $A_0$ is the cross-section of the barrel from which the screw insert starts [12].

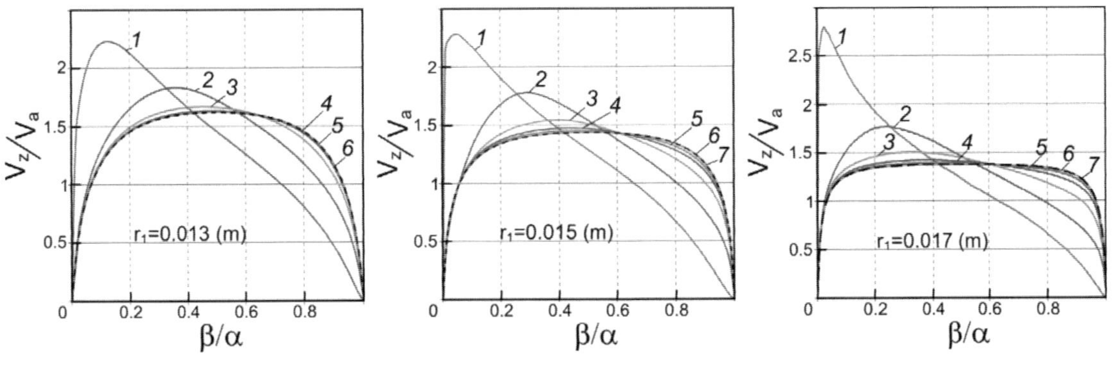


Fig. 2 Curves of the dimensionless axial component of the velocity vector in various runs of the barrel with screw insert around the circumference with radius $r = r_3$ and $U_a = 0.05$ (m/s): *1* – , *2* – $A_{45}$, *3* – $A_{90}$, *4* – $A_{135}$, *5* – $A_{180}$, *6* – $A_{360}$ (dashed black line), *7* – $A_{315}$

The velocity trends shown in Figure 3 are built around the circumference with radius $r_3 = (r_1 + r_2)/2$ (Fig. 3). The results obtained illustrate that the length of the initial hydrodynamic run does not exceed the distance limited by the plane $A_{360}$, i.e. $L_1 = 3D_2$. In this event, the velocity trend is getting packed and tends to the line showing a constant velocity throughout the barrel cross-section with a decrease in the gap between the coaxial cylinders (Fig. 4).

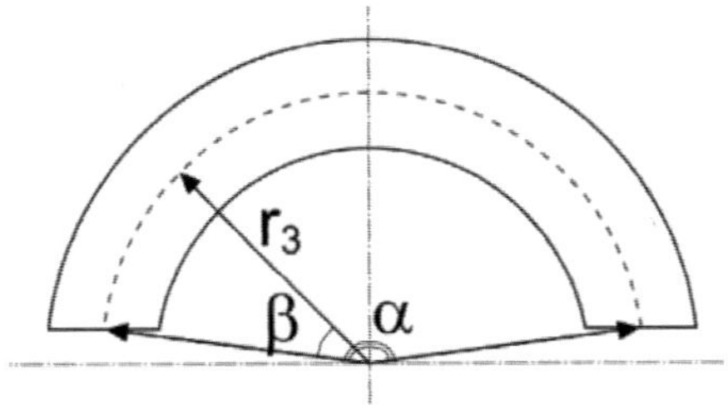


Fig. 3 The fluid flow region in the barrel cross-section: $r_3 = (r_1 + r_2)/2$, $\beta$ is the current angle value, $\alpha$ is the sector size

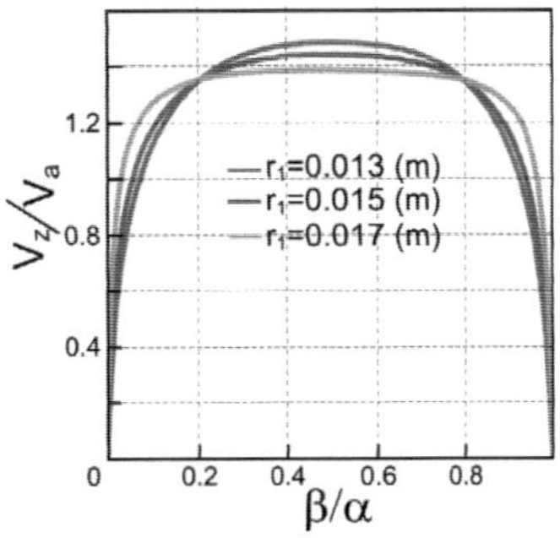


Fig. 4 Trends of the dimensionless axial component of the velocity vector in the cross-section $A_{360}$ around the circumference with radius $r_3$

**Conclusion**

Numerical results show that the initial hydrodynamic portion in the screw insert barrel for the aqueous solution of 0.5% polyacrylamide is limited by the distance $L_1 = 3D_2$. The gap between the outer pipe and the radius of the screw insert core was found to be less in order to ensure the same residence time of the polymer solution flow throughout the barrel cross-section.

This work was possible with the backing of the project No. 19-11-00220 by the Russian Science Foundation.

**References**

[1] Marschik C, Roland W, Löw-Baselli B and Miethlinger J 2017 *A heuristic method for modeling three-dimensional non-Newtonian flows of polymer melts in single-screw extruders.* Journal of Non-Newtonian Fluid Mechanics Vol. 248. pp. 27–39

[2] Li Y and Hsieh F 1996 *Modeling of Flow in a Single Screw Extruder.* Journal of Food Engineering Vol. 27. pp. 353-375

[3] Marschik C, Roland W and Miethlinger J 2018 *A Network-Theory-Based Comparative Study of Melt-Conveying Models in Single-Screw Extrusion: A. Isothermal Flow.* Polymers. Vol. 10, 929

[4] Bessonova M P, Ponomareva M A and Yakutenok V A 2019 *Numerical solution to the problem of polymer melt flow in a single screw extruder.* Chemical Physics and Mesoscopy Vol. 21. No. 2. pp. 198-217

[5] Cruz D O A and Pinho F T 2012 *Analysis of isothermal flow of a Phan-Thien-Tanner fluid in a simplified model of a single-screw extruder.* Journal of Non-Newtonian Fluid Mechanics Vol. 167-168. pp. 95-105

[6] Roland W, Marschik C, Krieger M, Low-Baselli B and Miethlinger J 2019 *Symbolic regression models for predicting viscous dissipation of three-dimensional non-Newtonian flows in single-screw extruders*. Journal of Non-Newtonian Fluid Mechanics Vol. 268. pp. 12-29

[7] Suresh K, Selvam K and Karunanithi B 2019 *CFD Simulation Studies on the Flow Behavior of Power-law Fluids used to extrude the Polymeric Hollow Fiber Membrane through an Angular Spinneret.* AIP Conference Proceedings Vol. 2112, 020160

[8] Xie J and Jin Y C 2016 *Parameter determination for the Cross rheology equation and its application to modeling non-Newtonian flows using the WC-MPS method.* Engineering Applications of Computational Fluid Mechanics Vol 10. Is. 1. pp. 111-129

[9] Cross M M 1965 *Rheology of non-Newtonian fluids: a new flow equation for pseudoplastic systems*. J. Colloid Sci. Vol. 20. Is. 5. pp. 417 – 437

[10] Welahettige P, Lie B and Vaagsaether K 2019 *Computational Fluid Dynamics Study of Shear Thinning Fluid (Drilling Fluid) Viscosity Models in an Open Venturi Channel*. International Journal of Petroleum Science and Technology Vol. 13, No 1. pp. 9-20

[11] Balan C M and Balan C 2010 *µPIV me surement nd numeric l comput tion of the velocity profiles in microchannels*. U.P.B. Sci. Bull. Ser. D. Vol. 72. pp. 121–128

[12] Kashapov LN, Kashapov NF, Kashapov RN 2013 *Research of the impact acidity of electrolytic cathode on the course of the plasma – electrolytic process* Journal of Physics: Conference Series, Vol. 479, Iss. 1, Art. № 012011

IOP Conf. Series: Materials Science and Engineering **915** (2020) 012020 doi:10.1088/1757-899X/915/1/012020

# Formation of by-products during water chlorination as a result of corrosion of elements of the mechanical system.

**O R Karataev [1,3] V F Novikov[2] R I Sayakhov[1] and E S Karataeva[4]**

[1] Kazan National Research Technological University, Kazan city, Karl Marx street, bld.68, Russia
[2] Kazan State Energy University, Kazan city, Krasnoselskaya str., bld. 51, Russia
[3] Volga Region State Academy of Physical Culture, Sport and Tourism, 33, Universiade Village, Kazan, 420138, Russia
[4] Kazan Federal University, engineering institute, Kazan city, Saidashev street, bld.12, Russia

e.s_karataeva@mail.ru

**Annotation** The work considers corrosion of the metal elements of the filtration system as an electrochemical process, and also presents the results of the analysis of the swimming pool water using the modern gas chromatographic equipment with a flame ionization detector and highly efficient quartz capillary columns. It was found that the swimming pool water is contaminated with the low-boiling organochlorine compounds and by-products, what is associated with the formation of the hydroxyl radical FeOOH, which occurs as a result of corrosion of elements of the mechanical system.

The primary stage of water purification is its mechanical purification from suspended impurities. Mechanical contaminants (suspensions) from large streams of water, as a rule, are removed in bulk filters with a combined (zeolite + filtrate) loading (periodic mechanical filters) of batch action.

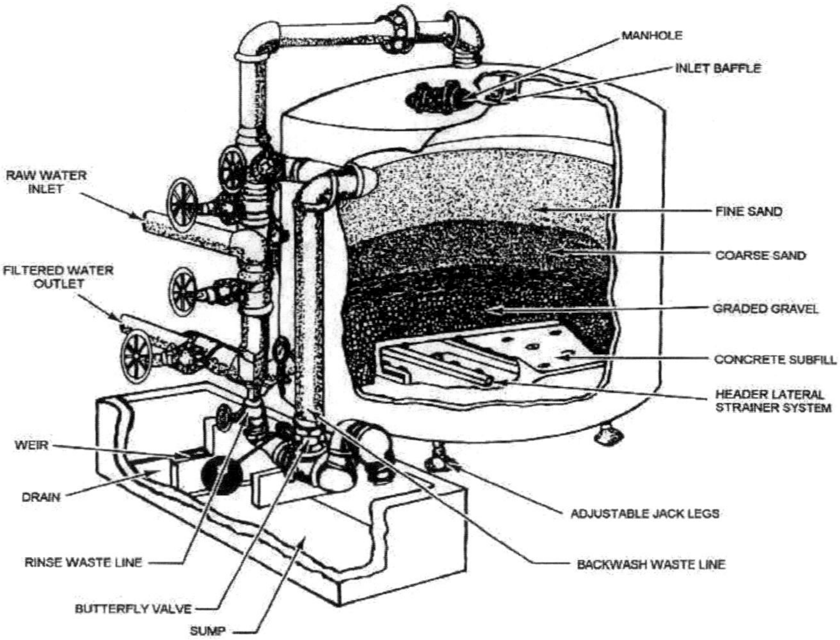


**Figure 1.** Conceptual design of the bulk filter (©Shivam Water Treaters Pvt)

The mechanical bed pressure filter «Shivam Water Treaters Pvt» (Fig.1) is a vertical metal or plastic casing with drainage distribution systems filled with as a rule, it is quartz sand, hydroanthracite, etc. To improve the distribution of the solution over the cross section and reduce the clogging of the holes of the lower drainage device, it is placed into a layer of gravel [1-3].

Filtration of the contaminated water is performed from top to bottom. In this case, large particles are retained in the pores between the granules of loading, and small contaminants, due to some various effects, primarily the electrostatic one, adhere to the particles of loading. The more contaminants are retained by the layer of loading, the narrower the passageways for the liquid are and the higher the depth of the water treatment is. Most of the contaminants are collected at the top of the layer of loading. At a certain point, the layer becomes so dirty that the filtration resistance rises sharply, and productivity falls. An increase in water pressure can lead to a "breakdown", that is, to carrying-out of dirt into the clean water. The filter is stopped and the layer of loading is regenerated.

Regeneration of the granulated loading (loosening) consists in washing it with water from bottom to top at a rate, at which pseudofluidization of loading and its expansion by 30-50% occurs. In this mode, the particles seem to "boil" and the delayed suspended solids are removed from the inter-pore space, and when the particles collide the adhering contaminants are removed from their surface [7-9].

Before being fed to the central networks, the water undergoes mandatory preparation, including cleaning and disinfection, since at the outlet it must comply with the safe drinking water standards. But going through many kilometers of labyrinths of the water supply, the water quality deteriorates – it absorbs corrosion products of the pipes and other mechanical impurities. On the surface of stainless steel equipment, a "point" attack of high concentration chlorides is constantly conducted. Corrosion products may include oxides, chlorides, sulfides:

$$4Fe + 3O_2 \rightarrow 2Fe_2O_3;$$

$$2Al + 3Cl_2 \rightarrow 2AlCl_3;$$

$$4Ag + O_2 + H_2S \rightarrow 2Ag_2S + 2H_2O.$$

Over some time, they "eat out" small holes in the surface. This leads to a premature failure of the heat exchangers, pipelines, equipment.

Table 1 shows the physical and chemical properties of low-boiling organohalogen compounds, being present and formed in the swimming pool water as a result of treatment with a chlorinating reagent.

As we can see from the table, these compounds are characterized by a wide variety of boiling points and solubility. It has also been revealed that the presence of by-products is associated with the formation of free radicals of the hydroxyl group (OH *), which appear during the corrosion of metal elements of the pumping system [4-6].

Table 1

Physical and chemical properties of organohalogen compounds

| No. | Compound | $T_{boil}$, $^0C$ | Solubility in the water, mg/l | Saturated steam pressure at 25$^0$C, mm Hg |
|---|---|---|---|---|
| 1 | Methyl iodide | 42.5 | 14000 | 437 |
| 2 | Chloroform | 61,2 | 10000 | 200 |
| 3 | Ethyl iodide | 72,2 | 4000 | 128 |
| 4 | Carbon tetrachloride | 76,8 | 800 | 113 |
| 5 | Bromodichloromethane | 87,0 | 4500 | 65 |
| 6 | Trichlorethylene | 87,2 | 1000 | 72 |
| 7 | Tetrachlorethylene | 121,2 | 400 | 19 |

| | | | | |
|---|---|---|---|---|
| 8 | Bromoform | 149,5 | 3000 | 5 |

Figure 2 shows a chromatogram of the separation of the swimming pool water, obtained by concentrating a sample with a statistical vapor-phase sampling unit.

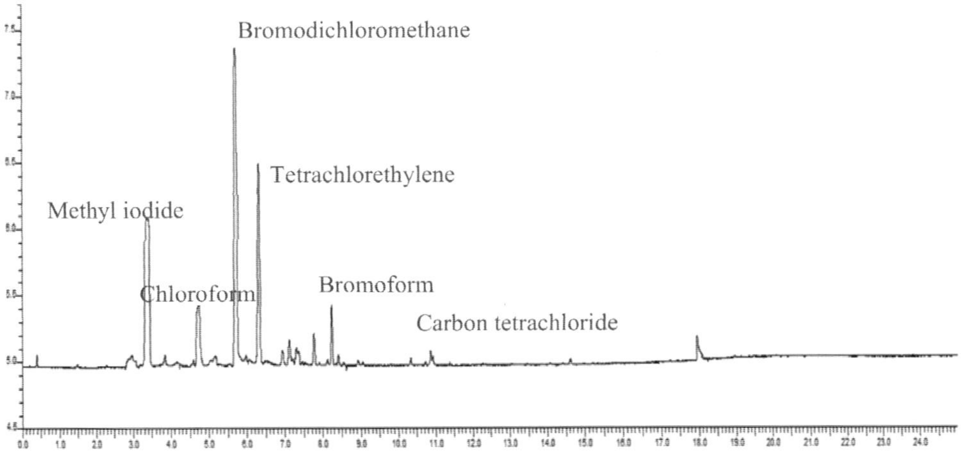


**Figure.2** Results of a chromatographic analysis of the swimming pool water. Chromatograph PerkinElmer Clarus 680.

Corrosion was considered in the work as an electrochemical process occurring due to interaction of the water and oxygen. To protect the metal elements of the filter equipment from corrosion we use the various inhibitors, which, when introduced into the water, significantly reduce the rate of the corrosion process and do not reduce the quality of the water in accordance with the sanitary standards.

**References**

[1] Kuznetsov A. A., Brukhova A. S. Determination of the quantitative content of corrosion products on the surface of reinforced concrete products // Omsk Scientific Bulletin. 2018. No. 6 (162). P. 160–164. DOI: 10.25206/1813-8225-2018-162-160-164.

[2] Karataev O.R, Karataeva E.S., Mechanical filtration, based on elective concentration of particles, as an innovative method of water treatment//IOP Conference Series: Materials Science and Engineering. - 2018. - Vol.412, Is.1. - Art. No. 012030.

[3] Khafizov I I, Nurullin I G and Sadykov Z B. Problems of development of electrochemical production of Russia and possibility of their decision. IOP Conf. Series: Materials Science and Engineering 412 (2018) 012042 doi:10.1088/1757-899X/412/1/012042

[4] Karataev O.R, Novikov V.F, Karataeva E.S, Use of modified molecular sieves in mechanical filters//IOP Conference Series: Materials Science and Engineering. - 2019. - Vol.570, Is.1. - Art. No. 012035.

[5] Kashapov N.F., Semochkin V.N., Fathutdinov R.Kh. Effect of nanostructure on the properties of filtering carbon-filled cellulose materials. Bulletin of Kazan Technological University. 2008. No. 1. P. 68-73.

[6] Galyautdinov, R.T., Kasparov, N.F., Luchkin, G.S. Technology of sputtering high-reflection coatings on ABS-plastic articles (2002) Inzhenerno-Fizicheskii Zhurnal, 75 (5), pp. 170-173.

[7] Karataev O.R., Karataeva E.S. Peculiarities of calculating the longitudinal mixing of adsorbent in a fixed layer with account of surface friction// IOP Conference Series: Materials Science and Engineering. - 2018. - Vol.412, Is.1. - Art. No. 012031.

[8] Kashapova R.S., Kashapov R.N., Kashapov L.N., Kashapov N.F., Lyadov N.M., Faizrakhmanov I.A Investigation of a nickel coating deposition processes from solid nickel electrolyte.// IOP Conference Series: Materials Science and Engineering. - 2018. - Vol.412, Is.1. - Art. No. 012008.

IOP Conf. Series: Materials Science and Engineering **915** (2020) 012020 doi:10.1088/1757-899X/915/1/012020

[9] Karataev O R, Ryzhkov D V, Golubeva G N 2015 Evaluation of the technical condition of walls of the building by means of thermal imaging control Bulletin of the Technological University No 3 T 18 pp 182-184.

IOP Conf. Series: Materials Science and Engineering **915** (2020) 012021 doi:10.1088/1757-899X/915/1/012021

# Influence of various types of fertilizers and equipment on the physical and microbiological properties of soil

**N F Kashapov[1], M M Nafikov[1,2], A R Nigmatzyanov[2], V V Medvedev[3], V A Chernov[3], and M D Nigmatullin[3]**

[1]Kazan (Volga Region) Federal University. Kazan, Russia
[2]Tatar Institute of Agribusiness Personnel Retraining. Kazan, Russia
[3]LLC "center of agricultural Chemistry" Kazan, Russia

arnig76@yandex.ru

Annotation. The cultivation of corn hybrids on silage, corresponding to the terms of ripeness of the district, and their timely harvesting allow to harvest silage containing 6.09-6.29 MJ per 1 kg of feed. High-quality silage should contain about 30 % SV, more than 10.8 MJ of exchange energy per 1 kg SV, at least 32% starch, no more than 4.5% crude ash, about 20% crude fiber and have a coefficient of digestibility of organic mass of at least 75 %. Therefore, it became necessary to improve the technology of its cultivation in the fields of crop rotations in order to increase the concentration of metabolic energy in the feed. It should be noted that an important role in the technology of growing corn should be assigned to the variety, fertilizers, biological products, growth stimulants and biological fungicides.

The main processing of arable land in agriculture is given great importance, since it largely depends on the water-physical, biological and chemical properties of the soil, which, in combination with other techniques, ultimately determines the amount of crop yield [1,2,3].
During the main tillage, the structure of the arable layer improves, the overall porosity increases and the volume of capillary pores decreases, which reduces capillary and increases convection- diffuse evaporation of moisture, and also improves the warming of the soil. When compacting the soil, the total porosity of the soil decreases, and capillarity increases, which creates favorable conditions for the flow of moisture from the lower layers to the upper layers and reduces aeration, which in turn affects the intensity of aerobic processes [4,5,6].

Thus, by regulating the degree of compaction of the soil through processing, it is possible to influence the conservation and accumulation of moisture, as well as the living conditions of the soil microflora [7,8].

The density of the arable layer is the most important indicator of the physical state of the soil [9,10,11,12], which changes significantly during the main processing, the optimal interval corresponds to 1.10-1.30 g / $cm^3$.

In our studies, the density of the arable soil layer during the growing season of maize changed to a greater extent depending on the methods of basic processing, and to a lesser extent – on the level of nutrition. Thus, the density of the soil before sowing and before harvesting was higher in the variant with no-tillage treatment for all layers of the arable soil layer, but it was in the aisles of the optimal level necessary for the growth and development of corn (table 1).

Table 1 - density of soil composition in the years of experiments, 2014-2016, g / cm$^3$

| Processing method (factor A) | food background (factor B) | Before sowing | | | Before harvesting | | |
|---|---|---|---|---|---|---|---|
| | | 0-10 cm | 10-20 cm | 20-30 cm | 0-10 cm | 10-20 cm | 20-30 cm |
| Plowing (control) | Control (without fertilizers) | 1,06 | 1,11 | 1,17 | 1,16 | 1,22 | 1,29 |
| | NPK per 40 t / ha of green mass | 1,07 | 1,12 | 1,18 | 1,17 | 1,23 | 1,31 |
| | PK Background | 1,07 | 1,12 | 1,18 | 1,18 | 1,26 | 1,37 |
| | Background + N40 (anhydrous ammonia) | 1,07 | 1,11 | 1,17 | 1,17 | 1,23 | 1,30 |
| | Background + N60 (anhydrous ammonia) | 1,07 | 1,10 | 1,7 | 1,18 | 1,24 | 1,31 |
| | Background + N80 (anhydrous ammonia) | 1,06 | 1,10 | 1,17 | 1,18 | 1,26 | 1,30 |
| | Background + N100 (anhydrous ammonia) | 1,06 | 1,10 | 1,18 | 1,19 | 1,26 | 1,33 |
| | Background + N120 (anhydrous ammonia) | 1,06 | 1,10 | 1,18 | 1,19 | 1,26 | 1,34 |
| Waste-free treatment | Control (without fertilizers) | 1,07 | 1,15 | 1,20 | 1,18 | 1,25 | 1,32 |
| | NPK per 40 t / ha of green mass | 1,08 | 1,14 | 1,21 | 1,19 | 1,27 | 1,34 |
| | PK Background | 1,08 | 1,14 | 1,21 | 1,19 | 1,27 | 1,34 |
| | Background + N40 (anhydrous ammonia) | 1,07 | 1,13 | 1,22 | 1,19 | 1,29 | 1,35 |
| | Background + N60 (anhydrous ammonia) | 1,07 | 1,15 | 1,22 | 1,20 | 1,29 | 1,35 |
| | Background + N80 (anhydrous ammonia) | 1,07 | 1,15 | 1,22 | 1,20 | 1,30 | 1,36 |
| | Background + N100 (anhydrous ammonia) | 1,07 | 1,15 | 1,23 | 1,21 | 1,30 | 1,36 |
| | Background + N120 (anhydrous ammonia) | 1,07 | 1,15 | 1,23 | 1,21 | 1,33 | 1,36 |

If before sowing, the density of soil addition in a layer of 0-10 cm with non-soil tillage on a non-fertilized background was 1.07 g/cm$^3$, in a layer of 10-20 cm-1.15 and in a layer of 20-30 cm-1.20 g / cm$^3$.

The excess relative to plowing options for soil layers was 0.01; 0.04; 0.03 g / cm$^3$, respectively. Before harvesting corn, the density of the arable layer was 1.18; 1.25; 1.32 g / cm$^3$, which exceeded the control by 0.02; 0.03 and 0.03 g/cm$^3$, respectively. With increasing depth of tillage, the soil density increased with all methods of tillage. If the soil density in the 0-10 cm layer was equal to 1.06 g/cm$^3$ when plowing at the control, then in the 10-20 cm layer it was 1.11 g/cm$^3$ and in the 20-30 cm layer it was 1.17 g/cm$^3$. The background of nutrition did not have much effect on the density of the soil.

In direct dependence on the density, there was also the hardness of the soil. An increase in soil hardness for all layers of the arable layer was observed in all years of research (table 2).

On a non-fertilized background at a depth of 0-10 cm, the soil hardness for plowing (on average for 3 years) in the germination phase was 8.6 kg/cm$^2$, for non – tillage-12.4 kg/cm$^2$. At a depth of 10-20 cm, these indicators were 22.4 and 24.3 kg/cm$^2$, respectively. With the deepening of the arable layer, its indicators increased, and the difference between treatment options was more significant. So at a depth of 30 cm, these indicators were 26.3 and 28.3 kg/cm$^2$, respectively.

Table 2-soil Hardness depending on the studied agricultural techniques, 2014-2016, kg / cm$^2$

| Processing method (factor A) | food background (factor B) | Before sowing | | | Before harvesting | | |
|---|---|---|---|---|---|---|---|
| | | 0-10 cm | 10-20 cm | 20-30 cm | 0-10 cm | 10-20 cm | 20-30 cm |
| Plowing (control) | Control (without fertilizers) | 8,6 | 22,4 | 26,3 | 15,1 | 24,9 | 36,7 |
| | NPK per 40 t / ha of green mass | 8,8 | 23,1 | 26,8 | 15,4 | 25,2 | 37,5 |
| | PK Background | 8,7 | 23,8 | 27,4 | 15,9 | 26,5 | 37,9 |
| | Background + N40 (anhydrous ammonia) | 8,5 | 23,5 | 26,7 | 15,2 | 26,2 | 38,4 |
| | Background + N60 (anhydrous ammonia) | 8,4 | 23,2 | 26,4 | 15,4 | 26,8 | 38,6 |
| | Background + N80 (anhydrous ammonia) | 8,4 | 23,4 | 27,6 | 15,8 | 27,1 | 38,5 |
| | Background + N100 (anhydrous ammonia) | 8,3 | 23,4 | 27,1 | 15,3 | 27,6 | 38,9 |
| | Background + N120 (anhydrous ammonia) | 8,0 | 23,9 | 27,4 | 15,5 | 27,6 | 38,8 |
| Waste-free treatment | Control (without fertilizers) | 12,4 | 24,3 | 28,3 | 17,3 | 28,3 | 39,3 |
| | NPK per 40 t / ha of green mass | 12,6 | 24,9 | 29,5 | 17,9 | 28,6 | 39,6 |
| | PK Background | 12,9 | 24,9 | 29,3 | 18,5 | 28,2 | 39,7 |
| | Background + N40 (anhydrous ammonia) | 12,6 | 25,2 | 28,4 | 18,7 | 29,3 | 39,4 |
| | Background + N60 (anhydrous ammonia) | 12,9 | 25,6 | 28,8 | 18,9 | 28,6 | 40,5 |
| | Background + N80 (anhydrous ammonia) | 13,3 | 26,1 | 28,1 | 18,8 | 29,1 | 40,7 |
| | Background + N100 (anhydrous ammonia) | 13,0 | 26,4 | 29,3 | 18,9 | 29,7 | 41,3 |
| | Background + N120 (anhydrous ammonia) | 13,6 | 26,9 | 29,6 | 18,9 | 29,9 | 41,8 |

By harvest, the General trend of soil hardness in the variants of the experiment remained. For plowing on a windless background in the 0-10 cm layer, it was 15.1 kg/cm$^2$, 10-20 cm-24.9 and the 20-30 cm layer-36.7 kg / cm$^2$. On the variants of non-shaft processing on similar variants, it was equal to 17.3, 28.3 and 39.3 kg/cm$^2$, respectively. The excess was 2.2, 3.40 and 2.60 kg/cm$^2$ compared to plowing.

The highest soil hardness was observed when plowing in the Phon + N100 variant (anhydrous ammonia) and amounted to 38.9 kg/cm$^2$, and for non – tillage in the Phon + N120 variant it was 41.8 kg/cm$^2$.

The total indicator that characterizes the activity of soil micro-organisms in the variants of soil treatment and fertilization is the intensity of decomposition of linen fabric (table 3).

Table 3-Influence of main tillage methods and fertilizers on the rate of linen fabric decay,%, 2014-2016

| Processing method | food background | 30 days | | | 60 days | | |
|---|---|---|---|---|---|---|---|
| | | 0-10 | 10-20 | 20-30 cm | 0-10 | 10-20 cm | 20-30 cm |

| (factor A) | (factor B) | cm | cm | | cm | | |
|---|---|---|---|---|---|---|---|
| Plowing (control) | Control (without fertilizers) | 26,3 | 21,0 | 19,1 | 41,2 | 37,1 | 32,6 |
| | NPK per 40 t / ha of green mass | 29,7 | 25,3 | 21,7 | 45,6 | 39,5 | 35,1 |
| | PK Background | 28,1 | 23,6 | 19,5 | 42,2 | 38,6 | 33,4 |
| | Background + N40 (anhydrous ammonia) | 29,1 | 25,8 | 22,3 | 45,2 | 39,3 | 35,0 |
| | Background + N60 (anhydrous ammonia) | 28,9 | 25,7 | 22,1 | 45,0 | 38,8 | 34,7 |
| | Background + N80 (anhydrous ammonia) | 28,5 | 25,2 | 21,8 | 44,5 | 37,6 | 33,1 |
| | Background + N100 (anhydrous ammonia) | 27,3 | 24,6 | 20,4 | 43,3 | 35,3 | 31,6 |
| | Background + N120 (anhydrous ammonia) | 26,9 | 23,4 | 19,2 | 42,0 | 35,7 | 31,2 |
| Waste-free treatment | Control (without fertilizers) | 36,4 | 29,7 | 25,8 | 44,8 | 38,4 | 32,3 |
| | NPK per 40 t / ha of green mass | 38,1 | 31,4 | 26,3 | 48,2 | 42,6 | 34,0 |
| | PK Background | 37,5 | 30,1 | 26,0 | 46,4 | 41,1 | 32,5 |
| | Background + N40 (anhydrous ammonia) | 38,8 | 32,6 | 26,4 | 48,9 | 43,8 | 34,2 |
| | Background + N60 (anhydrous ammonia) | 38,2 | 32,0 | 26,8 | 48,6 | 43,1 | 32,7 |
| | Background + N80 (anhydrous ammonia) | 38,1 | 31,8 | 26,2 | 47,2 | 42,9 | 32,0 |
| | Background + N100 (anhydrous ammonia) | 37,3 | 30,9 | 26,6 | 47,0 | 41,5 | 31,9 |
| | Background + N120 (anhydrous ammonia) | 36,6 | 29,3 | 25,9 | 46,9 | 41,0 | 31,4 |

For 30 days in a layer of 0-30 cm on the background without fertilizing by plowing, it was in the layer of 0-10 cm 26.3%, in the layer of 10-20 cm-21.0 and in the layer of 20-30 cm-19.1 %.

For non-waste processing, this indicator was 36.4, 29.7 and 25.8%, respectively.

When applying NPK at the rate of 40 t / ha of green mass, the decomposition of linen for ploughing in similar versions was 29.7; 25.3 and 21.7%, and for non – ploughing-38.1; 31.4 and 26.3 %.

With an increase in the level of nutrition during plowing in the Fon + N120 variant, the decomposition of linen fabric was 26.9 %; 23.4 and 19.2%, respectively, and 36.6 %; 29.3 and 25.9% for non – tilling loosening.

With an increase in exposure to 60 days, the decomposition of linen fabric on these variants increased. At the control (without fertilizers) for ploughing it was in the layer 0-10 cm – 41,2%, in the layer 10-20 cm – 37,1 and in the layer 20-30 cm – 32,6%. The decomposition of linen fabric in the same variants was higher in non-fallow processing than in ploughing and was 44.8 %, 38.4% and 32.3%, respectively.

Against the background of the Republic of Kazakhstan, the decomposition of linen by plowing in similar variants was 42.2 %, 38.6 and 33.4 %, and by non – plowing-46.4 %; 41.1 and 32.5 %.

IOP Conf. Series: Materials Science and Engineering **915** (2020) 012021 doi:10.1088/1757-899X/915/1/012021

With an increase in the level of nutrition during plowing in the Fon + N100 variant, the decomposition of linen fabric was 43.3%, 35.3% and 31.6 %, respectively, and 47.0%, 41.5% and 31.9% for non – tilling processing.

**References**

1. Mikhailov, V. V., Bogomazov S. V. soil Treatment in the Penza region / / Зemledelie, 2005. - #4. - Pp. 12-13.
2. Moiseev, A. A. Efficiency of fertilizers for corn for grain in the forest-steppe of the middle Volga region/ A. A. Moiseev, P. N. Vlasov, A.V. Ivoilov / / Bulletin of the Altai state agrarian University. - 2016. - No. 4 (No. 138). - Pp. 28-33.
3. Nafikov, M. M. Cultivation of single-species and mixed crops of sorghum crops/ M. M. Nafikov, N. M. Yakushkin, V. N. Fomin, I. p. Talanov. Moscow, 2015. - 248 p.
4. Nafikov, M. M. Evaluation of productivity of forage crops in single-species and binary crops in the Republic of Tatarstan / M. M. Nafikov, A. R. Nigmatzyanov / / Combined crops of field crops in the crop rotation of the agro-landscape: mezhdunar. science. Ecol. Conf. / ed. by I. S. belyuchenko. – Krasnodar. - Kubgau, 2016. - Pp. 166-170.
5. Nafikov, M. M. cultivation of forage crops in single-species and mixed crops in the Western Zakamye of the Republic of Tatarstan. M. M. Nafikov., By A. R. Hafizov. // Bulletin of the Kazan state agrarian University. - 2010. Vol. 5. - No. 2 (16). - Pp. 138-142.
6. Usanova, Z. I., Influence of calculated doses of fertilizers and the density of standing on the productivity of corn, removal and economic balance of the main elements of nutrition/ Z. I. Usanova, I. V. Shalnov, A. S. Vasiliev / / Agriculture. - 2016. - No. 3. - S. .23-26.
7. Fomin D. V., Nigmatzyanov A. R., Chekmarev P. A., Nafikov M. M. Influence of precursors and nutrition level on soil contamination, agrophysical properties and productivity of sugar sorghum / / Agriculture. 2016. # 5. Pp. 26-28.
8. Fomin V. N., Nafikov M. M., Medvedev V. V. Productivity of corn depending on the methods of soil treatment and fertilization / / Points of growth of agricultural efficiency in an unstable market. Collection of scientific articles. Issue 12. - Kazan: Brig publishing house, 2018.- Pp. 210-215.
9. Kashapov N. F., Nafikov M. M., Gilmanshin I. R., Gazetdinov M.Kh., Nafikova M. M., Nigmatzyanov A. R. epegdu-saving technologies of cultivation of sugar sorghum. In the collection: IOP Conference Series: Materials Science and Engineering Ser. "International Scientific-Technical Conference on Innovative Engineering Technologies, Equipment and Materials 2016, ISTC-IETEM 2016" 2017. P. 012032.
10. Nafikov M. M., Yakushkin N. M., Nigmatzyanov A. R., Petrova V. Ya., Sayfutdinov R. F., Mingazov R. A. Productivity of sugar sorghum depending on the method of sowing and seeding rates // Achievements of science and technology of the agro-industrial complex. 2017. Vol. 31. # 12. Pp. 36-39.
11. Khramtsev, I. F. efficiency of fertilizers in the cultivation of corn for grain on Chernozem soils of the forest-steppe of Western Siberia / I. F. Khramtsev, N. A. Punda / / Achievements of science and technology of the agro-industrial complex. - 2012. - No. 3. - Pp. 24-25.
12. Khramtsov, I. F. Improving resource-saving technologies in Siberian agriculture / resource Conservation and diversification as a new stage in the development of A. I. Baraev's ideas about soil-protecting agriculture. - Astana, Shortandy, 2008. - Pp. 21-26.

ISTC-IETEM-2019 IOP Publishing
IOP Conf. Series: Materials Science and Engineering **915** (2020) 012022 doi:10.1088/1757-899X/915/1/012022

# Features of robotization of agricultural processes

**N F Kashapov[1], M M Nafikov[1,2], A R Nigmatzyanov[2], R F Sayfutdinov[2] and R A Mingazov[2]**

[1]Kazan (Volga Region) Federal University. Kazan, Russia
[2]Tatar Institute of Agribusiness Personnel Retraining. Kazan, Russia

arnig76@yandex.ru

Annotation. The agro-industrial complex, as an industry, currently faces the problems of climate change, limited water resources, soil degradation and biodiversity conservation. It is necessary to significantly increase the production of agricultural products, while it should follow the trend of sustainable development. Resources must be used more efficiently. Intelligent solutions are needed to address conflicts of interest and meet the need for healthy and nutrient-rich foods and feeds. Therefore, it is necessary to assign an increasing role to the digitalization and robotization of agriculture in achieving these goals.

At the 2nd International conference of Ministers of agriculture within the framework of the exhibition "Green week" held in Berlin on January 19, 2019, digitalization is described as a field of activity that covers digital technologies, innovations, information and communication technologies, as well as artificial intelligence. Large corporations, such as Cargill or Du Pont, invest in this area, supporting startups that make it possible to monitor the health and well-being of farm animals. For this purpose, visualizations and custom algorithms are used, which have become instructions for use by specialists of the agro-industrial complex. The estimated positive benefit is from 100 to 200 dollars per cow for lactation. Therefore, many participants of the conference stated that soon the computer and Excel tables will not be enough - you will need a separate control element for the necessary data on livestock farms and crop rotation fields [5].

A huge problem at the moment is that in many cases it is difficult to explain to the layman how modern agriculture functions. the "Greens" criticize the activation of digitalization and robotization in the processes of agricultural production. Therefore, it is important to constantly emphasize that technological progress, including in food production, cannot be stopped. A realistic view of modern agriculture should be included in school textbooks and in the curricula of all agricultural Universities. The idyllic image of green meadows, cows eating flowers, and pitchforks in a haystack does not correspond to the current state of the industry. What is agriculture 4.0? Precision farming and animal husbandry, automation and robotics, remote access-all these are components of agriculture of the previous generation-agriculture 3.0. But only linking these technologies together is the fourth generation of digital technologies [2].

What is happening in modern conditions is quickly becoming clear that the data collected by satellites and their assessment have great potential for technical and crop innovations in agriculture. But the collection of this data in the present requires clear rules - you need to prevent their misuse by third parties. The state should play an active role here . first of all, it should develop legislation on data management, i.e. decentralized data storage and regional networking. Centralized data storage in the cloud poses significant risks, including from the point of view of protection from failure and disruption of primary processes on livestock farms and crop rotation fields in regions with intensive agriculture [2,3,4].

One of the most important conditions is that qualified personnel are required to work in the new format. The technical infrastructure of schools, public institutions, professional development systems, as well as the training of teaching staff is significantly behind the needs of the time. It is necessary to develop educational platforms and support agricultural enterprises in the process of their transition to digitalization and robotics.
The diagram of the transition to the "figure" in the agriculture of the Republic of Tatarstan is shown in the figure (figure-1).

All this requires certain costs and, most importantly, the need to speed up the implementation processes in production. Moreover, significant cost savings will motivate such development. The possibilities of digitalization of animal husbandry largely depend on the type of animals and form of maintenance. While in pig and poultry farming technologies are primarily focused on livestock management and indoor equipment, in dairy farming the role of "front - runners" is played by sensors and individual systems - a variety of biological parameters and evaluate them-body temperature, blood pressure, heart rate or activity (mobility). A special advantage here is the constant documentation of these parameters, which makes it easier to observe animals. Large amounts of data and indicators calculated on their basis contribute to maintaining the health of the herd, and at the same time to the welfare of animals, since early detection of an emerging problem makes it possible to timely intervention [2,3,4,5, 7,8,9,10].

One of the necessary elements is the need for standardization. The question arises whether all the collected data can be used in the production of products, and how significant they are for implementation in practice. After all, the potential for their further use is certainly high, but many indicators detected by sensors require standardization, which is difficult in production conditions. Very often, data are not sufficiently linked, so it will also be necessary to bring them to common standards in the international aspect. An example is genomic selection, which began to be introduced in dairy cattle breeding in 2010 [5].

You can give examples of successful practical use of data in accurate animal husbandry, such as individual feeding, regular control milking, measurement of animal activity, electrical conductivity of milk, automatic detection of hunting, determination in space, respiratory rate, etc.But the most important thing to remember is that everyone - programmers, biostatistics, ethologists, economists, breeders, feeders, zooengineers - sit in the same boat, which certainly causes the need to take into account complex data.

IOP Conf. Series: Materials Science and Engineering **915** (2020) 012022 doi:10.1088/1757-899X/915/1/012022

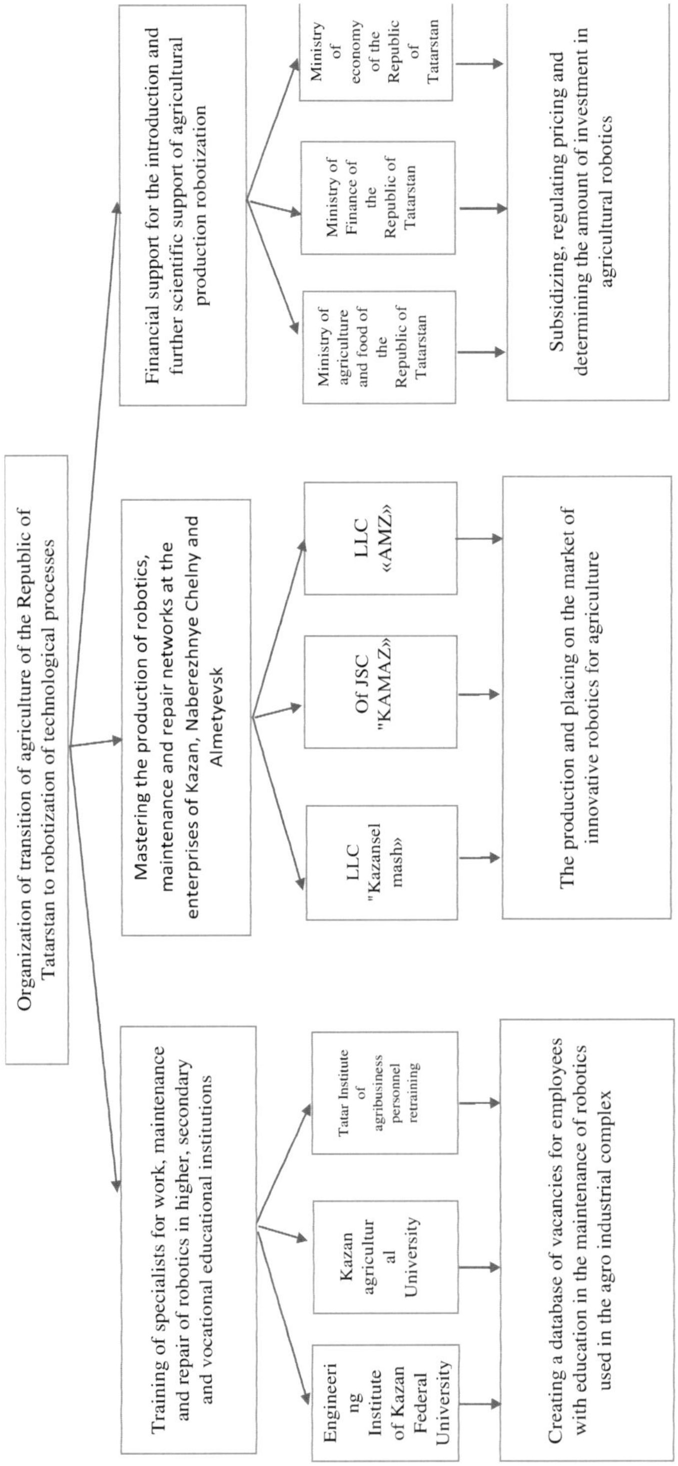


Figure 1. Organization of transition of agriculture of the Republic of Tatarstan to robotics

**References**

[1] Zakharenko V. A. Progressive technologies of application of chemical plant protection products for the purpose of establishment and elimination of harmful organisms that cause emergencies. - Moscow, 2009, 69 p.
[2] Zakharenko V. A. Assessment of the economic feasibility of protective measures in the conditions of the market economy of Russia / / Protection and quarantine of plants, 2019 no. 5 P. 12-15
[3] Kazakhselmash. Technique for plant protection, 18 p. www. Kazansm.ru
[4] Machinery and equipment for applying fertilizers and plant protection products. - Moscow Federal state budgetary institution "Rosinformagrotech" 2012, 96 p.
[5] Digitalization for the future of the agricultural sector // New agriculture, 2019, no. 2. P. 30
[6] Skvortsov E. A., Skvortsova E. G., Sandu I. S., iovlev G. A. Transition of agriculture to digital, intelligent and robotic technologies // Regional economy. - 2018. - Vol. 14, issue 3. - P. 1014-1028
[7] Skvortsov E. A. Improving the efficiency of robotization of agriculture: author. dis. ... Cand. Econ. science / Usau-Yekaterinburg, 2018. - 24 p.
[8] Kashapov, N. F. Innovative technologies and digital solutions in the processes of robotization and mechanization of the agro-industrial complex on the example of soybean cultivation: textbook / N. F. Kashapov, M. M. Nafikov, A. R. Nigmatzyanov. - Kazan: Kazan University Press, 2019. - 142 p.
[9] Kashapov N. F., Nafikov M. M., Nafikov M. M., Nigmatzyanov A. R. Development of digital and robotic agricultural production in the Russian Federation in the collection: Innovative engineering technologies, equipment and materials-2018 (ISTC "IMTOM-2018") Proceedings of the IX International scientific and technical conference. 2018. Pp. 301-305.
[10] Kashapov N. F., Nafikov M. M., Nigmatzyanov A. R., Gainutdinov I. G. (2019). State of the technical level of domestic agricultural machinery. IOP conference series: materials Science and engineering. 570. 012047. 10.1088/1757-899X/570/1/012047

IOP Conf. Series: Materials Science and Engineering **915** (2020) 012023 doi:10.1088/1757-899X/915/1/012023

# Using digital agricultural production based on navigation and information systems

**N F Kashapov[1], M M Nafikov[1,2], M M Nafikov[3] , A R Nigmatzyanov[2], O I Makarova[4] ,V V Medvedev[5], V A Chernov[5], and M D Nigmatullin[5]**

[1]Kazan (Volga Region) Federal University. Kazan, Russia
[2]Tatar Institute of Agribusiness Personnel Retraining. Kazan, Russia
[3]Kazan national research university Kazan, Russia
[4]Kazan state agrarian University Kazan, Russia
[5]LLC "center of agricultural Chemistry" Kazan, Russia

arnig76@yandex.ru

**Annotation**. Due to the peculiarities of the agro-industrial complex of the Russian Federation, the digitalization of agriculture has a number of significant features. At its core, these technologies are innovative, one of the most important functions of which is intellectual. There is a need to clarify the concepts of"transfer of enterprises of all forms of ownership of agro-industrial complex to the figure": is the process of determining the feasibility of implementation of digitalization and robotics, the purchase of innovative technologies in agribusiness companies, adaptation of the technology and production system, operation and determination of the effectiveness of its use in a specific production, taking into account soil-climatic conditions in the area of implementation.

Enterprises of all forms of ownership in agriculture operate in conditions of acute shortage of qualified personnel, low attractiveness of labor in the crop and livestock industries, a decrease in the life expectancy of the rural population and the migration outflow of residents of rural settlements to large cities. Statistics show that the population of rural settlements has decreased by 2.3 million people since 2000. and amounted to 37.1 million as of January 1, 2014. Due to the optimization of rural schools and the lack of qualified medical care, it is assumed that these trends will continue and even strengthen. These circumstances call for the use of labor-saving technologies, including digital, intelligent and robotic [1,2,3].

When implementing digitalization in the organization of agriculture, it is necessary to take into account regional characteristics and specifics of specific branches of agricultural production [4].

Regional features of digitalization and robotization may be related to the peculiarities of the region's development. These features include the level and conditions of socio-economic development of the region, the level of urbanization, infrastructure development, demographic trends, the competitiveness of the agricultural sector in attracting labor in comparison with other industries, etc.

The industry features that cause the need for digitalization and robotization may be related to the need to perform monotonous repetitive processes, the presence of heavy, dangerous and harmful types of activities for the creation of agricultural products in the agricultural sector. An important condition

IOP Conf. Series: Materials Science and Engineering **915** (2020) 012023 doi:10.1088/1757-899X/915/1/012023

for the robotization of processes in agriculture is the ability to compile certain algorithms that underlie the functioning of robotics.

Robotics can be used in animal husbandry for milking animals, cleaning manure, shearing sheep, etc., in crop production for sowing crops, spraying plants with pesticides and fertilizers, weeding weeds, controlling the germination of soil under crops, etc., in auxiliary production for monitoring agricultural land, sorting and packaging products, etc. [4].

The transition of enterprises of all forms of ownership of agriculture to digital, intelligent systems and robotics is caused by the need for a number of changes:

- optimization of labor productivity in agricultural enterprises. In the case of correctly selected robotic systems, labor productivity increases by several times or by an order of magnitude in comparison with manual production [5,6];

- increase of safety and improvement of working conditions at the enterprises of the agro-industrial complex. The use of robotics is very effective in harmful production. In agriculture, this includes working with pesticides, fertilizers, fuels, lubricants, and animal and poultry waste.

- bringing the quality of agricultural products to high, competitive standards [7,8,9,10].

- creation of additional jobs in agriculture-related industries.

The Republic of Tatarstan has accumulated a lot of experience in using digitalization in LLC "agro-industrial complex "FOOD PROGRAM", especially in the agrosignal system – an online service for accounting and monitoring the progress of agricultural work in the crop production industry.

Implementation goals and objectives

1. Interactive map of fields with work history for each field

2. Quality control processing field, the flaws, the speed of movement of vehicles

3. Accounting for field and transport work

4. Data from computers on equipment: seeding rate on seeders, liquid consumption in sprayers, engine hours, fuel, etc.

5. Accounting for the full cycle of movement of fuel and lubricants (columns, fuel trucks, tanks in equipment)

6. Control of the complete chain of movement of finished products (combines, bunkers, transporters, trucks, weight rooms, warehouses)

The transition of agriculture to digital technologies should be targeted. The following goals are pursued:

— the most economical consumption of all necessary resources for production, compared to traditional technologies;

- reduction of harmful emissions to the atmosphere and soil, significant reduction of soil compaction in comparison with traditional technologies due to spot and precise tillage in crop rotations;

- reducing the need for labor resources, in order to save on the remuneration of permanent and employees;

- reducing the impact of the human factor on technological processes and production results, including non-compliance with technology, theft, etc.;

- optimization of costs for labor protection and safety, prevention of occupational diseases;

- maximum obtaining of objective information for making competent, operational management decisions;

- economical use of seeds, fertilizers, and pesticides through the use of precision and spot farming technologies;

- improving the quality of products through the introduction of innovative technologies, as well as quality control of technological operations.

**References**

[1] Skvortsov E. A., Skvortsova E. G., Sandu I. S., iovlev G. A. Transition of agriculture to digital, intelligent and robotic technologies // Regional economy. - 2018. - Vol. 14, issue 3. - P. 1014-1028

[2] Blinova T. V., Bylina S. G. Scenario forecast of the rural population of Russia in the medium term // Regional economy. - 2014. - № 4. - P. 298-308.

[3] Akimov A. Demographic explosion, population aging and labor-saving technologies. In the Interaction in the twenty-first century. // World economy and international relations. - 2016. - Vol. 60. - Issue 5. - P. 50-60.

[4] Skvortsov E. A. Improving the efficiency of robotization of agriculture: author. dis. ... Cand. Econ. science / Usau-Yekaterinburg, 2018. - 24 p.

[5] Kashapov N. F., Nafikov M. M., Yakushkin N. M., Nigmatzyanov A. R. State of technical level of domestic agricultural machines. Innovative engineering technologies, equipment and materials-2018 (ISTC "IMTOM-2018") Materials of the IX scientific and technical International conference. 2018. Pp. 305-309.

[6] Kashapov n. f., Nafikov M. M., Gazetdinov M. K., nafikova M. M., Nigmatzyanov A. R. comparative evaluation of various machines for seeding sorghum VGD conference series: materials Science and technology Ser. "International scientific and technical conference on innovative engineering technologies, equipment and materials 2016, ISTC-IETEM 2016" 2017. P. 012033

[7] Kashapov N. F., Nafikov M. M., Gilmanshin I. R., Gazetdinov M. H., nafikova M. M., Nigmatzyanov A. R. energy-Saving technologies of sugar sorghum cultivation. Innovative engineering technologies, equipment and materials-2016 (ISTC " IMTOM-2016") Materials of the International scientific and technical conference. 2016. Pp. 316-321.

[8] Kashapov N. F., Nafikov M. M., Gilmanshin I. R., Gazetdinov M.Kh., Nafikova M. M., Nigmatzyanov A. R.
Energy-saving technologies of cultivation of sugar sorghum. IOP conference series: materials Science and engineering Ser. "International scientific and technical conference on innovative engineering technologies, equipment and materials 2016, ISTC-IETEM 2016" 2017. P. 012032.

[9] Kashapov N. F., Nafikov M. M., Nigmatzyanov A. R., Gainutdinov I. G. (2019). State of the technical level of domestic agricultural machinery. IOP conference series: materials Science and engineering. 570. 012047. 10.1088/1757-899X/570/1/012047.

[10] Kashapov N. f, Nafikov M M, Gazetdinov m x, Gazetdinov sh M, Nigmatzyanov a R. Modern problems of digitalization of agricultural production / IOP conference series: materials Science and engineering.(2019) 012048

IOP Conf. Series: Materials Science and Engineering **915** (2020) 012024 doi:10.1088/1757-899X/915/1/012024

# Modeling the processes of forming the organizational structure of management in itegrated formations

**N F Kashapov[1] M M Nafikov[1,3] M X Gazetdinov[2] Sh M Gazetdinov[2] and A R Nigmatzyanov[3]**

[1]Kazan (Volga region ) Federal University, Kazan, Russia
[2]Kazan state agrarian University, Kazan, Russia
[3]Tatar Institute of agribusiness retraining, Kazan, Russia

arnig76@yandex.ru

Annotation. The article discusses an approach to the procedure for building a rational organizational structure of management of integrated agricultural formations in modern conditions of functioning. In the modern theory of areas have developed quite separately. Thus, research on optimal long-term planning focuses mainly on a certain frozen organizational structure of management without taking into account its dynamics in changing conditions. In the works on the rationalization of the production structure, which is generally, the alternative character of the recommendations, mapping projects not done a proper analysis of effectiveness and expected outcomes further improve the organizational structure of management. This implies the need for a unified conceptual approach to planning and creating an organizational management structure in the agricultural sector of the economy.

Introduction. The agricultural sector of the economy has accumulated a lot of theoretical and practical experience in long-term industry planning. However, the allocation of new, extremely complex objects of the industry's goal-setting activities naturally entails qualitatively new methodological and methodological aspects of the planning decision-making process. In the agricultural sector, one of these aspects is taking into account the high dynamism of the organizational structure of management and the need to implement digital technologies [1].

The main regularities that cause significant structural changes in the agricultural sector in modern conditions are that the majority of agricultural formations irrationally use the existing resource and production potential and develop disproportionately, while increasing specialization and co-operation of production, forming vertically integrated territorial and production subsystems.

The complexity and growth of interrelations between branch divisions of the agricultural sector of the economy leads to the fact that effective decision-making requires close coordination of their production programs [2]. At all levels of the economic hierarchy, it becomes necessary to implement integrated planning of interconnected high-tech industries.

The study of the experience of forming integrated agricultural units in the Republic of Tatarstan has shown that since 2005, on average, there have been at least 3-5 significant changes in the organizational structure of management (OSU), and the number of changes and additions to the legislative and regulatory acts regulating economic activity has been estimated at several dozen. The prerequisites for this were not only the rapid pace of intensive development of the production and technological structure of the relevant enterprises, but also the integration modifications of intersectoral relations in the agro-industrial complex arising under the influence of scientific and technological progress, the organizational strengthening of which forms an urgent direction in improving the efficiency of the agricultural sector of the economy, the orientation of all its divisions to achieve high final results.

A close relationship with OSU production technology farm structure and economic mechanism in the economic system, require a constant search for new and better forms of management of integrated groups, and search are not isolated from the tasks of prospective planning, and organically embedded in the routine system development plan socio-economic development of rural areas.

Method of research. In long-term planning, such a pattern of OSU means that there is a certain discrepancy between the existing organizational structure of the agro-industrial complex at the time of drawing up the plan and the level of its future economic and social development, especially with large planning horizons. The lag of the agribusiness management system, since it has a significant impact on the adoption and implementation of planning decisions, leads to the fact that the implementation of calculations based on the current OSU does not allow to fully take into account the trends in production development, open all available reserves and ensure the development and implementation of an effective plan for this object. This circumstance, in our opinion, is the initial methodological prerequisite that determines the feasibility of embedding the tasks of searching for a rational OSU in the direct process of long-term planning in the regional agro-industrial complex. In its formulation should take into account the boundaries of the possible changes in TSR during the planning period and its return influence on the choice and implementation of planning decisions, time and resources to conduct relevant activities as well as time costs for adaptation of production units and management bodies to changes in the economic mechanism, etc.

At the same time, fixing the number of levels in the organization's hierarchy allows you to significantly "unload" the original task. In particular, remove the question of the appropriate degree of compression of information at various levels of management, which sharply limits the Arsenal of possible methods of formalized analysis. Therefore, in the future we will focus on the production system, the OSU of which is built on the principle currently used in most regional agro – industrial associations of the country: "center – territorial formations-enterprises". Thus, the initial statement is somewhat narrowed and directly for this object, the task is reduced to the complex development of an optimal long-term plan and a rational territorial management structure, the composition of its territorial divisions.

Results. The problem statement uses an optimization approach to the construction of organizational structures, and as a specific criterion for the preference of options – an indicator of the minimum cost of creating and operating a management system and negative effects in the organization of economic relationships when introducing lower and upper restrictions on the production indicators of their functioning.

The initial information received in the block "formation of OSU" based on the results of calculations of the optimal plan includes: the main indicators of the work program of their constituent agricultural, processing enterprises and infrastructure units, as well as data on the planned intensity of relations between individual economic units, including external suppliers.

Since a significant part of the management costs are so-called fixed costs, we took into account the non-linear nature of the relationship between their size and the volume of activity of production units. For this purpose, several possible types of organization were previously developed for territorial formations, which differ in the scale of production. The latter were set as ranges with lower and upper borders, within each of which the corresponding functional relations were determined. Experimental research in this area was carried out for JSC "AGROSILA" [3]. AGROSILA is a vertically integrated agricultural holding that includes 25 agricultural enterprises located on the territory of the Republic of Tatarstan, in Tukayevsky, Aznakayevsky, Zainsky, Sarmanovsky, Menzelinsky, Nizhnekamsk and

Muslyumovsky districts. The size of land use is more than 330 thousand hectares, the total number of agricultural machinery is more than 11,000 units. The holding includes the poultry complex "Chelny-broiler", the plant "Zainsky sugar", seed and feed mills. In 2019, the net profit of 600 million rubles.

The conducted research has shown that the real dependence of management costs on production volumes can be considered as a function of the following type in the first approximation

$Z(V)=qkV + pk$ ,

where V is the volume of intermediate and final products of an integrated agricultural formation;

qk – unit variable costs of managing a k-m type of organization;

pk is a constant part of expenses of management under the k-m type of organization;

Z(V) – cost management at k-m type of organization.

One of the important components of assessing the quality of economic activity in the agro-industrial sector of the economy is the indicators of loss of agricultural products during its procurement, transportation, storage, processing and sale [4]. The need for rapid and flexible response to rapidly changing business conditions, close coordination of the work of various separate organizational units is manifested here with special force, and, conversely, any delays in the development and implementation of management decisions, manifestations of inconsistency and territoriality significantly reduce the final results of the entire production process [5]. Therefore, when we talk about improving the management style and methods, developing market relations in the economy, etc. at the same time, the negative effects of the existence of territorial borders that occur in practice cannot be discounted when designing an OSU.

Statistical analysis of the supply of crop and livestock products to processing enterprises of JSC "AGROSILA", in particular, showed that the actual loss of raw materials (based on the same delivery radius) during its transportation between enterprises of different territorial formations is 2-3 times higher, compared with similar transportation within the same enterprise. Territorial disunity is also manifested in uneven capacity utilization of processing industry enterprises in the Republic in cases when this does not meet the General interests of production. In setting the task, such factors were taken into account in two ways. First, through cost indicators of expenditures on economic relationships in the system, which were differentiated depending on whether they are "intra-territorial" within the framework of the OSU project under consideration or are established between enterprises of different territorial formations. Secondly, the introduction of special restrictions on a set of technologically acceptable links that regulate the real possibilities of their organization, depending on the specific version of the territorial management structure. These additional conditions were included in the formalized requirements for the OSU for mandatory establishment of all necessary production links between the system elements.

Another important requirement for the construction of the OSU is a restriction on the organization of the formed control objects, which characterizes the possible load on the governing bodies. It can be taken into account in two main ways in relation to the problem under consideration: 1) setting the number of territorial formations; 2) setting the permissible size of each of them.

In experimental studies, joint-stock company "AGROSILA" we have chosen the second method. The model, as already noted above, provided for several possible options for organizing agricultural formations, for which the restrictions included lower and upper limits on the annual volume of total production activity of the incoming enterprises.

In addition, the problem statement takes into account the conditions for each business unit to become part of only one territorial formation, for creating a certain type of organization for each of them, as well as the requirements for non-negativity and integer variables.

In economic-mathematical model of formation of territorial structure of management in a vertically integrated agricultural groups the following notation: j is the index of economic units (the technological process), $j=1 \div N$; i - the index of territorial units, $i=1 \div I$; ki - index of the type of organization i-th territorial formation, ki = 1÷K; Qj - intensity of the j-th technological process (the output of the j-th economic division) at the optimal production plan; Lki , Lki - upper and lower limits on the sizes of the i-th territorial formation in k-m type of organization; pj1 j2 – unit cost of communication between units j1 and j2 during their occurrence in different territorial formation, pj1 j2 = 0 when entering units j1 and j2 in a territorial formation; rj - index of the type of technology associated with technology j, rj = 1÷R, Mj rj - many technological processes, referring to the rj that the organizational environment can be related to technology j; fj, fj1 j2 - coefficients of reduction to the conditional power and conversion of technology intensities; xj1 j2 - the desired intensity of relations between divisions (technologies) j1 and j2; yi j - an additional variable that takes the value 1, if the j-th division is part of the i-th territorial formation and the value 0-otherwise; yiki - an additional variable that takes the value 1, if the I-th territorial formation adopted the variant of the organization ki , and the value 0-otherwise; Si ki (Vi) - the cost of managing the I-th territorial formation at the ki-th type of its organization, depending on the volume of economic activity, where

$$V_i = \sum_{j \in N} f_j Q_j y_{ij}, \quad i = 1 \div I$$

The goal of solving the problem is to determine the variant j = 1 of the territorial management structure in a regional agrarian integrated formation that minimizes the cost of organizing all the necessary production links in the system, i.e.

$$Z = \min \left[ \sum_{i \in I} \sum_{k_i \in K} S_{ik_i} \left( \sum_{j \in N} f_i Q_j y_{ij} \right) + \sum_{i \in I} \sum_{j_1 \in N} \sum_{j_2 \in N} P_{j_1 j_2} x_{j_1 j_2} y_{ij_1} y_{ij_2} + \sum_{i \in I} \sum_{j_1 \in N} \sum_{j_2 \in N} P_{j_1 j_2} x_{j_1 j_2} \left(1 - y_{ij_1}\right) y_{ij_2} \right]$$

with the following restrictions:

1) on the permissible size of individual territorial agricultural formations

$$\sum_{k_i \in K} \underline{L_{k_i}} \; y_i^{k_i} \le \sum_{j \in N} f_i Q_j y_{ij} \le \sum_{k_i \in K} \overline{L_{k_i}} y_i^{k_i}, \quad i = 1 \div I$$

2) the obligation to establish the necessary production links between business units

$$\sum_{j_1 \in M_j r_j} Q_{j_1} - \sum_{j_2 \in M_j r_j} f_{j_1 j_2} x_{j_1 j_2} Q_{j_2} = 0, \; j_1 = 1 \div N, \; r_j = 1 \div R$$

3) on the mandatory entry of each economic division into the composition of one territorial agricultural formation

$$\sum_{i \in I} y_{ij} = 1, \; j = 1 \div N$$

4) to create a certain type of organization for each territorial agrarian formation

$$\sum_{k_i \in K} y_i^{k_i} = 1, \; i = 1 \div I$$

5) on nonnegativity and integer values of variables

$$x_{j_1,j_2} \geq 0; \;\; y_{ij} \div 0, \; 1; \;\; y_i^{k_i} \div 0, \; 1; \;\; j \;, j_1, j_2 = 1 \div N; \;\; i = 1 \div I; \;\; k_i = 1 \div K$$

The model under consideration is uniquely integer in nature, which is why calculations based on it encounter certain algorithmic difficulties. Therefore, when conducting experimental studies, considerable emphasis was placed on a preliminary analysis of the development of the production and technological structure of agricultural formations and related areas of optimization of their OSU. In particular, taking into account the development strategy, scenarios were developed for the most likely situations of changing the territorial management structure in an integrated formation. These scenarios, among other characteristics, differed in the number of separate territorial subsystems. Then, within each scenario, based on the optimal intensity of relations between economic divisions, "Central" divisions were allocated, which were clearly assigned to specific territorial formations. This made it possible to significantly reduce the number of compared OS projects, reduce the dimensions and time to solve the corresponding tasks.

Conclusion. Thus, the results of solving the considered problem allow us to specify a number of technological parameters (coefficients of annual capacity utilization at processing enterprises, characteristics of functioning of infrastructure divisions, expected amounts of product losses, etc.), a number of cost parameters (production overheads for the maintenance of the management apparatus, the effectiveness of various uses of agricultural products, including for livestock products, for industrial processing and sale in retail chains, etc.).

As a result, for JSC "AGROSILA" practical calculations provided material for analyzing the interaction of production, technological and organizational structures of territorial agricultural formations, the interdependence of the long-term plan and the organizational structure of management, in other words, the most important task of modernizing the economy is not to increase the technical equipment of production, but to increase their balance with labor and material resources [6].

**References**

[1] Akmarov P. B., Voitovich V. Yu., Tretyakova E. S. Improving the efficiency of public administration in the conditions of information society development / Science of Udmurtia. 2019. No. 1 (87). Pp. 11-18.

[2] Akmarov p. b, Gorbushina N. V., Knyazeva O. P. features of digital transformation in the agricultural sector of the economy / / digital agriculture-development strategy materials of the International scientific and practical conference (ispc 2019). Ser. "Achievements in the field of intelligent systems research" 2019. Pp. 22-26.

[3] Gazetdinov sh. M., Gazetdinov M. Kh., Semicheva O. S., Gatina F. F., Nigmetzinov A. A. reserves for improving the efficiency of integrated entities / international scientific and practical conference "agriculture and food security: technologies, innovations, markets, human resources" (Kazan, Russia, November 13-14, 2019) / / bio web conferences". Volume 17. 2020. FIES 2019. https://doi.org/10.1051/bioconf/20201700026

[4] Kazakovtseva M. V., Gumarova F. Z., Tsaregorodtsev E. I. formation of competitive advantages of the regional agro-industrial complex as a mechanism for ensuring economic security / / Bulletin of social Sciences. 2015. Vol. 6. No. 3. Pp. 213-220.

[5] Kashapov NF, Nafikov M M, Gazetdinov M CH, Gazetdinov Sh. M and Nigmatzyanov A R Modern problems of digitalization of agricultural production
IOP Conference Series Material Science and Engineering 570:012044 · August 2019 with 11 readings

[6] Kashapov NF, Nafikov M M, Gazetdinov M Ch, Gazetdinov Sh. M and Nigmatzyanov A R On one approach to assessing the technical equipment of agricultural enterprises in the context of economic modernization / series IOP Conference Material Science and Engineering 412(1): 012038 · October 2018 with 15 readings

# Technological preparation of production for product quality management according to IATF 16949:16 requirements

**S V Kasyanov**

Kazan Federal University Naberezhnye Chelny Institute, 423812 Naberezhnye Chelny Mira Avenue 68/19

dndemyanov@kpfu.ru

**Annotation**. The article analyzes the content of the EU CCI in the conditions of Soviet enterprises. The tasks and additional procedures in the project of preparation of production and release of a new product that allow for the production of machine-building products in modern conditions of competition between domestic enterprises and international machine-building corporations are highlighted.

**Introduction.** In our country, the development of the automotive industry began with the purchase of factories from G. Ford for the production of its own models of cars. The state, as a total owner, set the task for enterprises to set up production in as large volumes as possible. All other tasks, such as environmental friendliness of production, were secondary. In order to achieve this goal, the industry people's Commissariat has standardized the activities of all plants in the industry in terms of standard organizational structures, maintenance and repair of equipment, unification of technical documentation, and normalization of labor intensity and resource consumption; conducting end-to-end quality control. Eventually, by the 70th codes of the twentieth century, a well-developed, well-documented system of production activities had developed.

It was focused on the production and delivery of products that meet the established requirements within the specified time frame. Accordingly, a standardized system of preparation of production prescribes requirements for the execution of the engineering works. Their main characteristics are shown in table 1. The high degree of standardization significantly simplified the tasks of engineers.

Table 1 – Tasks and content of technical preparation of production at Soviet enterprises

| № | Planned features | End result | Planned CCI procedures |
|---|---|---|---|
| 1. | The purpose of the production preparation | Ensure the planned volume of output the calendar period | Development of a infeasibility study |
| 2. | Set of customer requirements | Standardized | Approval of the technical task. Preparation of the delivery contract |

| | | | |
|---|---|---|---|
| 3. | Degree of impact on the natural environment | Compliance with current legislation | Activities on a separate plan |
| 4. | Personnel security level | Compliance with current legislation | Approval of project documents with Supervisory authorities. Development of instructions. |
| 5. | Assessment of competitors' capabilities | ------ There is no competition | Since the 1980s, verification of product characteristics with the world's best samples according to the comprehensive product quality management system (CPQMS) |
| 6. | Consideration of stakeholder requests | ------ Not provided | Standardized maintainability analysis. |
| 7. | Production preparation time | Meet the standard | Development and correction of the work schedule. Monitoring the progress of work. |
| 8. | Pre-production verification | Signing the acceptance certificate | According to a standardized procedure. |
| 9. | Metrological support | To ensure the objectivity of control of products | Purpose of measurement and control tools according to the standard. Metrological examination of technical documentation. |
| 10. | Information support for quality management | Standardized sets of design and process documentation | Development and standard control of sets of documents. |

**Theoretical part**

Specialists had to develop and approve a set of technological documentation, which was used as a set of criteria for product quality control and control of compliance with technological discipline. There were no specific requirements for the depth of presentation in the documentation of the content of operations. The documented requirements remained unchanged. They could not be used for flexible management, for regulating processes.

Unfortunately, this system has been in effect for many decades, while since the mid-twentieth century, all the world's leading corporations are constantly improving the management of their pre-production projects. Long-term experience in a constant system of requirements prevents the transition to effective management of modern projects for pre-production and new product release (table 2).

Today's commercial product is a product of much higher quality, and its production technologies should be many times more stable. In addition, it is necessary to constantly minimize the impact on the natural environment, as well as the consumption of energy carriers and other types of natural resources [1-3], to increase the information content of the measurement process in order to use the data obtained for diagnosing the state of technological systems [4-5]. Apply statistical methods to improve product quality [6-8]. To increase the degree of identification and tracking of the causes of defects of a commodity product by stages of its life cycle [9]. Thus, the list of works is significantly expanded (table 2).

There are also completely new procedures for specialists to approve production by the consumer and assess the acceptability of measurement processes. Accordingly, the requirements for the effectiveness of management processes have increased.

Much more information is needed. To get current data, it is necessary to plan processes for monitoring and regulating characteristics and production operations, associated product lifecycle operations, and equipment lifecycle. Therefore, the old scheme of work of specialists in specialized services is no longer suitable for project management.

Table 2-Tasks and additional procedures in the pre-production and new product release project

| № | Planned features | End result | Planned CCI procedures |
|---|---|---|---|
| 1. | The purpose of the production preparation | Ensure that the planned profit is received | Analysis of legal restrictions, development and monitoring of business plans, sales monitoring, project team management. |
| 2. | Set of requirements for the customer | Ensure that all requirements are met | Identification and ranking of the full set of requirements for deliveries. Creating a register of special characteristics |
| 3. | Degree of impact on the natural environment | To minimize the impact | Analysis of legal restrictions. Planning and managing the lifecycle of associated products |
| 4. | Personnel security level | Prevention of industrial accidents | Monitoring of safe working conditions |
| 5. | Assessment of competitors' capabilities | Minimizing the risk of loss of income | Monitoring the actions of competitors, development of improvements. |
| 6. | Consideration of stakeholder requests | Avoid negative perception of the organization | Monitoring the perception of the organization and developing improvements. |
| 7. | Production preparation time | Perform in a reasonably short time | Organization of effective work of an inter-functional team of specialists. |
| 8. | Pre-production verification | Ensure the customer's confidence in the quality and stability of deliveries | Passing the RRAR procedure. |

| | | | |
|---|---|---|---|
| 9. | Metrological support | Ensure that measurement and control processes are acceptable | Development of measurement methods. Assessment of the acceptability of measurement processes. |
| 10. | Information support for quality management | Provide information support in the project lifecycle | Development of flow maps, management plans, work instructions, monitoring of equipment, product quality. |

All these tasks should ultimately ensure that the project receives the planned amount of profit. The new organization of pre-production work requires the creation of an inter-functional team that provides management throughout the project lifecycle. It is also important that each significant decision of the specialists pass a certain approval procedure (by the customer, the company's management, or the team). It is impossible to achieve quality without the depth of study of the content of world-class technologies that is unusual for our specialists. The necessary level of understanding of the content of managed technologies implies a high level of employee motivation, and in the current economic situation, this is an extremely difficult task for the company's director. In [11-13] methods have been developed to ensure the final results of additional CCI procedures in accordance with the IATF 16949:16 requirements.

**Conclusions**

In general, this approach to production preparation in Russia implies a new approach to training specialists:

- it is necessary to master the skills of working in a team. Developing a focus on achieving the planned result;
- it becomes mandatory to develop skills to identify key specific process characteristics in operations that affect the management of product characteristics;
- be sure to understand the need for careful planning of measurement processes as the main method of obtaining information, as well as the development of modern measurement methods;
- it is important to constantly improve the information culture in their activities in terms of forming requirements and obtaining and analyzing data on the values of product characteristics, as well as the willingness to work in accordance with the requirements of the documentation.

The current educational standards of Russian higher education still direct specialists to solve the problems of the last century. Of course, the task of changing standards should be set at the state level. But the state must also do everything to ensure that enterprises have the opportunity to sell modern products in large volumes-and to stimulate exports, and to ease the tax burden, and to establish a system for training a reserve of managers and chief specialists.

**Reference**

[1] Safarova L R, Kasyanov S V and Safarov D T 2019. Planning of wastes generation and development of life cycle technology in preproduction of a new product. *Paper presented at the IOP Conference Series: Earth and Environmental Science*, , 272(3) doi:10.1088/1755-1315/272/3/032076

[2] Safarov D T, Kondrashov A G, Safarova L R and Glinina G F 2017. Energy planning in production shops with numerically controlled machine tools. *Russian Engineering Research*, 37(9), pp. 827-834 doi:10.3103/S1068798X17090209

[3] Safarov D T, Kondrashov A G, Glinina G F and Safarova L R 2017 Algorithm of calculation of energy consumption on the basis of differential model of the production task performed on

machines with computer numeric control (CNC). *Paper presented at the IOP Conference Series: Materials Science and Engineering*, 240(1) doi:10.1088/1757-899X/240/1/012060

[4] Safarov D T, Kondrashov A G and Khafizov I I 2019 Four-point technique development for measuring spherical surfaces of machine parts. *Paper presented at the IOP Conference Series: Materials Science and Engineering*, , 570(1) doi:10.1088/1757-899X/570/1/012084

[5] Safarov D T, Kasyanov S V and Kondrashov A G 2019 Informative value of measurements for quality management of auto parts doi:10.1007/978-3-319-95630-5_177

[6] Safarov D T and Kondrashov A G 2019 Bevel gears monitoring methods development in the total contact patch terms. Paper presented at the *IOP Conference Series: Materials Science and Engineering*, , 570(1) doi:10.1088/1757-899X/570/1/012083

[8] Safarov D T and Kondrashov A G 2020 Methods of quality control manufacturing gears of the differential satellites doi:10.1007/978-3-030-22063-1_28

[9] Safarov D T, and Kondrashov A G 2019 Improving the quality of manufacture of the differential pinion gears through the integrated application of statistical methods of quality management. *Paper presented at the Journal of Physics: Conference Series*, , 1260(3) doi:10.1088/1742-6596/1260/3/032034

[10] Safarov D T and Gimazetdinov A I 2019 Application of blockchain computer technology to trace the causes of defects in the supply chains of engineering products. *Paper presented at the 2019 International Mılti-Conference on Industrial Engineering and Modern Technologies, FarEastCon 2019*, doi:10.1109/FarEastCon.2019.8934435

[11] Safarov D T, Fedorov, K A and Ilyasova A I (2016). Algorithms development of making special techniques in APQP manufacturing process of automotive components. *Paper presented at the IOP Conference Series: Materials Science and Engineering*, , 134(1) doi:10.1088/1757-899X/134/1/012036

[12] Kasjanov S V and Safarov D T 2018 Mathematical modeling of the technological accuracy index deviation structure of the automobile parts. *Paper presented at the IOP Conference Series: Materials Science and Engineering*, 412(1) doi:10.1088/1757-899X/412/1/012039

[13] Kasyanov S V, Biktimirova G. F. Technological documentation as a basis for the competitiveness of the manufacturer of automotive components in the world market / / *Automobile industry No. 6*, 3013-p. 30-33 2.

[14] Kasyanov S V, Biktimirova G. F. formation of rrar according to ISO / TS 16949: 2009 in the process of technological design / / *Certification #4, 2013*-pp. 31-38

IOP Conf. Series: Materials Science and Engineering **915** (2020) 012026 doi:10.1088/1757-899X/915/1/012026

# The increasing of the electric vehicle maneuverability, excluding disconnecting of the survey from the contact network

**A A Khafizov, R S Akhmetshin and R I Valiev**[1]

Kazan Federal University, Naberezhnye Chelny Institute, 68/19 Mira Street, Naberezhnye Chelny, 423810, Russia

almazok75@yandex.ru, zoia2007@mail.ru, rivaliev87@mail.ru

**Abstract**. A technical solution is proposed that allows to increase the maneuverability of an electric vehicle without reducing the speed of movement, excluding the arbitrary disengagement of the current collector from the contact network at road junctions and turns.

**1. Introduction** The proposed technical solution relates to the field of the electrical engineering, in particular to the electric vehicles that can be used to transport passengers along the city route.
It is known [1-4] an electric vehicle, in particular a trolleybus for transporting passengers around the city. This electric vehicle contains from an electric drive with a speed control system and two rods with current collectors.

The disadvantage of [1-4] is that this vehicle is limited in maneuverability, has a lower speed at the road junctions and the intersections. In addition, the rod with the current collector, as a rule, is randomly disconnected from the contact network. These vehicle flaws can be the cause of the congestion on the roads and slow down the total flow along the route [5].

The closest technical solution in essence is [6]. The power supply system of the trolleybus has a traction device and a contact network, which carries a positive potential and is connected to an electric drive. The device for removing of the electric current is made in the form of a single rod current collector.

The disadvantage of the prototype [6] is that the rod collector can unauthorized disconnect from the contact network at transport interchanges, turns. The trolleybus slows down at the above-mentioned places of the route, which causes traffic jams.

## 2.Problem Statement

The aim of this work is to increase the maneuverability of an electric vehicle, without reducing the speed of the movement and the exclusion of the arbitrary disengagement of the current collector from the contact network.

## 3. Technical Solution

The current collector in the form of a bar (1) is mounted perpendicular to the contact network pivotally on the first and the second current conductors (2 and 3) in the form of two rods (Fig. 2). This current collector provides a sliding connection to the contact network (4), and is attached to the insulating platform (5) from below (Fig. 1a). The first and the second current conductors also have springs (11) expanding the rods from the platform.

The mentioned bars are interconnected by an insulator (12), and both of the bars and the insulator do not have protrusions, and the free ends of the beam are bent down. A video camera (6) is installed on the roof of the electric vehicle aimed at the upper mark of the rhombuses. The picture from the video camera is transmitted online to the monitor (7) installed in the cockpit in front of the driver (Fig. 1a). On the bottom, in the outer front of the electric vehicle, there is an induction sensor for metal (8) that monitors the metal linear conductor (9) provided under the road surface in the axis of movement (Fig. 1a, b). It is assumed that the metal linear conductor will be laid at the entrance and exit of the intersection, or the road junction. The length of the metal linear conductor with a length of at least 50 meters is laid on some complex sections of the traffic path (Fig. 3). The electric vehicle contains a high-capacity Li-ion maintenance-free battery in a sealed design.

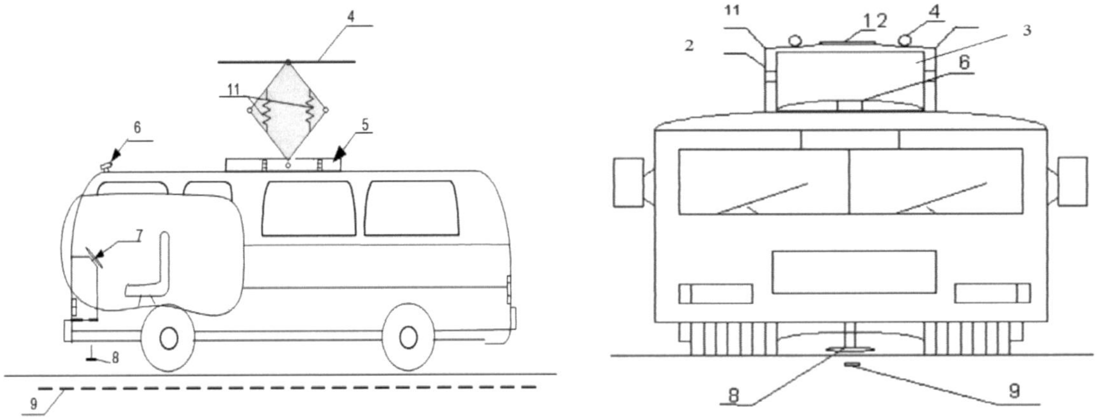


a b

**Figure 1 a and b.** The scheme and devices of an electric vehicle a - side view; b- front view.
2, 3 - conductors in the form of a rhombus of rods; 4 - contact network; 5 - an isolated platform; 6 - video surveillance camera; 7 - video surveillance screen; 8 - induction sensor for metal; 9 - metal linear conductor; 11 - expanding spring; 12 - insulator.

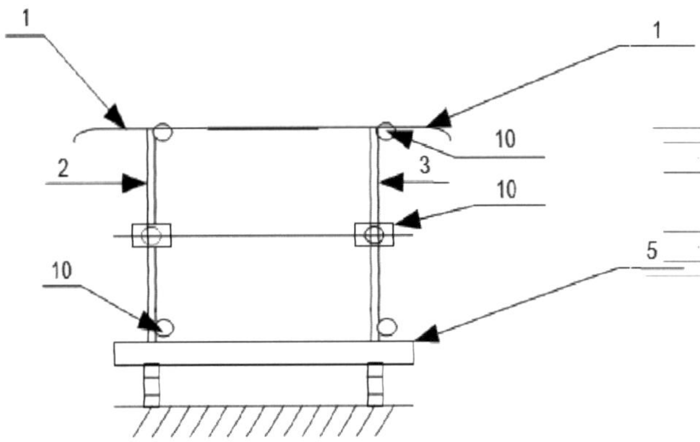


**Figure 2.** The connection diagram of the current collector and the conductors on the roof of the electric vehicle. 1 - current collector in the form of a bar; 2,3 - conductors in the form of a

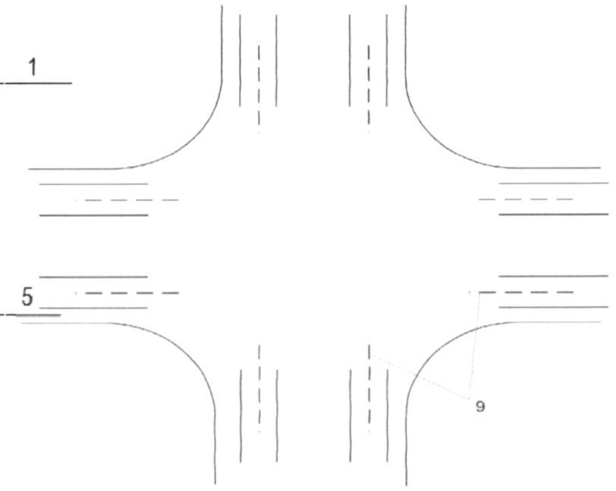


**Figure 3.** Layout of metal linear conductors provided under the road surface in the axis of movement.

rhombus of rods; 5 - an isolated platform; 10 - hinges, fastening the faces of the rhombus of conductors and current collector.

**4.The principle of operation of an electric vehicle** is illustrated in figures 1, 2 and 3, which depicts circuits and devices of an electric vehicle.

The electric vehicle, approaching the intersection or the road junction with the absence of the contact network in these sections, continues to move due to the energy of pre-charged batteries.

Driving out of the intersection or the road junction, a traffic route is automatically selected using an induction sensor on the metal to search for a metal conductor in the road surface. In this case, the poles of the current collector are again in contact with the conductors of the contact network and the driver is made sure using video surveillance.

Shorting the two wires of the contact network by the current collector is impossible, because the length of one pole of the current collector is less than the distance between the conductors of the contact network.

When moving on the straight sections of the route, the batteries of the electric vehicle are charged from the contact network.

In case of unauthorized disconnection of the current collector from the contact network when approaching the intersection or the road junction and when leaving them, the driver does not need to leave the cab to connect the current collector pole to the contact network. This becomes possible due to tensile springs and the properties of hinged fastening (10) of the faces of the current collector rhombuses at its vertices (Fig. 2), and due to the fact that the ends of the contact network in these areas have a rigid mount and are bent down.

## 5. Conclusions

This technical solution ensures the implementation of the goal to increase the maneuverability of the electric vehicle, without reducing the speed and eliminating the arbitrary disengagement of the current collector from the contact network.

This technical solution is protected by the utility model patent [7] and can be implemented in the electric transport industry.

## References

[1] Koryagina Ye Ye, Kos'kin O A 1982 Elektrooborudovaniye tramvayev i trolleybusov. Uchebnik dlya tekhnikumov gorodskogo transporta. - M.: Transport, p 296

[2] Sergey Korol'kov 2017 Elektrobus - tekhnicheskiye osobennosti variantov. Mosgortrans: Nauchno-tekhnicheskiy sovet. ZAO "Tekhnicheskiy tsentr Elektrotransservis"

[3] Maksimov A N Gorodskoy elektrotransport. Trolleybus. Nachal'noye professional'noye obrazovaniye. - Akademiya, 2006. — ISBN 5769523719

[4] Goroshkov Yu I, Bondarev N A 1990 Kontaktnaya set'. [Ucheb. dlya tekhnikumov zh.-d. transp.] /3-ye izd., pererab. i dop. M. Transport p 399

[5] SNiP 2.05.09-90 Tramvaynyye i trolleybusnyye linii

[6] Sistema elektrosnabzheniya trolleybusa RU 140654 U1 Byul. № 14 ot 20.05.2014

[7] Akhmetshin R S, Shakirov Yu I, Il'in V I, Akhmetshin R R 2017 «Elektrotransportnoye sredstvo» Patent poleznoy modeli RU №1712 69 ot 26.05.2017

IOP Conf. Series: Materials Science and Engineering **915** (2020) 012027 doi:10.1088/1757-899X/915/1/012027

# Improving the quality of surfaces of products obtained by electroerosion treatment

**I I Khafizov[1] and I G Nurullin[1]**

[1]Kazan Federal University, 18 Kremlyovskaya street, Kazan, 420008, Russian Federation

khafizov@kpfu.ru, opros16@yandex.ru

**Abstract:** The use of electroerosive processing in production helps to achieve the desired results: high processing accuracy, low roughness of the treated surface and low dependence of productivity on the mechanical properties of the processed materials. Despite this, the need for effective processing of parts and improving the technological process of electroerosion treatment is an urgent topic for research.The trouble-free and long-lasting operation time of machine parts largely depends on the quality of the surface layer. Therefore, it has a high demand, since it is all the destruction begins with the surface.

To improve the quality of the surface layer, you can use both conventional methods under certain conditions, and special hardening methods of processing. The use of hardening methods helps to increase the durability of the machine, reduces the need for spare parts and materials, and reduces the cost of manufacturing and operating machines. According to some experts, when the reliability of equipment increases by 5%, the company's profitability increases by more than 30%.

Many years of experience in research on improving the quality of operational properties of parts shows that it is necessary to carry out processing with a combination of traditional and new technological methods, taking into account the subsequent operating conditions of the part in the future. With this approach, the surface layers of parts can be given the desired combination of performance properties, which are ineffective or impracticable under individual technological influences [1].

Therefore, for the effective use of combined treatments, it is necessary to clearly know what the shaping process is due to. Also, knowing the impact indicators of each individual method, you can assess the feasibility of using them in each specific case, based on the company's capabilities and technical requirements for the part being processed [2].

Accordingly, the need for effective processing of complex parts made of difficult-to-process materials has predetermined the emergence of a number of new methods. The use of electroerosion treatment (EDM) helps us achieve the results we need. EDM occurs at a relatively low pulse energy, the volume of metal removed for each pulse is small, and the depth of the hole is insignificant. This mode, with low productivity, allows you to get surfaces with high accuracy and low roughness.

In our opinion, the main advantages of EDM in comparison with other treatment methods include

1. The processing process does not depend on the characteristics of the processed material: hardness, strength, viscosity, etc.

2. Processes do not exert force on the part, allowing you to process thin-walled elements of various parts without deforming them, cut out of a thin material elements of complex shape with small jumpers, as well as process parts made of brittle materials.

3. No burrs on the parts after processing, which reduces labor intensity and eliminates manual labor

[3] As is already known, the properties of the surface layer change significantly after EDM. The resulting surface layer in thickness can be divided into 5 zones (Fig. 1):

1 - zone of saturation of the working fluid elements;

2 - zone of deposition of the EI material;

3-white layer formed from the molten material of the workpiece;

4 - zone of thermal influence;

5 - zone of plastic deformation.

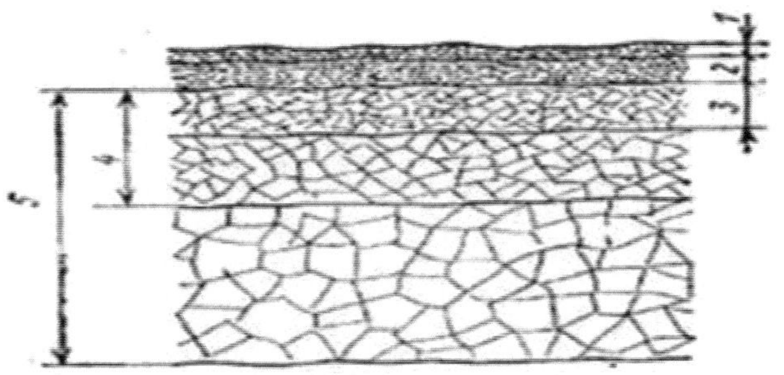

Figure 1-Schematic representation of the surface layer

The number and sequence of formation of zones, their structure and properties depend significantly on the material, the electrode tool, the working fluid used and the conditions of the process. The borders between the zones are almost not clear, and in most cases they overlap each other. But each of the zones has its own characteristics that affect certain properties of the surface and in many cases, determine the possibility of successful operation of the part. At the same time, each zone, even of insignificant thickness, plays a very important role [4].

Modern electroerosive machines allow you to get the roughness of the treated surface RA =

0,1...0.4 μm. These indicators are achieved by connecting the electrode part to the negative pole of the pulse generator and assigning modes with minimal pulse energy. However, it should be noted that the processing of products in finishing cutting modes is characterized by low performance of the EDM process.

In one of the works analyzed by us, it is shown that when processing 38x2n2ma steel on a copying and stitching machine at the finishing modes, the processing performance is 0.05 mm/h.This performance of the EDM process leads to an increase in the technological time and cost of manufacturing the product. Therefore, when composing the technological process, the EDM assigns more rough indicators for the quality of the treated surfaces, which do not allow to fully implement the performance characteristics inherent in the product [5].

Therefore, the development of a technology that allows for finishing the surfaces of parts made of hard-to-process materials obtained by the EDM method is an urgent task.

The use of mechanical processing methods for this purpose is not always effective or feasible, and the use of electrochemical methods of exposure due to the high toxicity of electrolytes requires a large cost to ensure the environmental safety of people and the environment, as well as waste disposal.

The experiment conducted with employees of the Perm NIPU shows that the best solution for this task can be the use of the technology of electrolyte-plasma polishing of complex-profile surfaces (EPP). The method of electrolytic-plasma polishing is based on electric discharge phenomena in the metal-electrolyte system, while the processed part is an anode. In micro-discharges, significant energy is released and there is an intensive process of reducing the height of the surface's micronerities, which leads to its polishing. Also, this technology is environmentally safe and does not require the use of acids, alkalis and other harmful substances in dangerous concentrations.

They experimentally determined that the EPP for 5 minutes of working time can reduce the roughness of the surface treated by the EDM method by an average of 5 times. If the processing of the workpiece by the EDM method to obtain a surface roughness of 1.6 microns, it is necessary to use a mode with a minimum pulse energy (with a current equal to 3 amperes), then the use of finishing technology by the EPP method allows you to obtain a similar roughness value with a more productive EDM mode (with a current equal to 50 Amperes) [6].

It should be noted that the effectiveness of combined treatments depends not only on the combination of effects, but also on the choice of the basic treatment option, which is improved by other methods with known properties.

EDM does not exclude mechanical processing, but in addition to it, it takes its own specific place, corresponding to its features, namely, the ability to process conductive materials with any physical and mechanical properties [7].

We can conclude that in order to improve the quality of product surfaces with minimal costs, we need to correctly select the optimal combination of useful pulse energy, its power, and the frequency of working pulses, and Supplement this with a suitable method of finishing, depending on the required surface quality.

**References**

[1] Grigoriev S N, Safonov S V and Smolentsev V P 2016 Design of combined processes for surface layer modification of standard parts *Vestnik VSTU* №4

[2] Khafizov I I 2013 Ways to reduce the consumption of materials when they are separated by combined methods *Bulletin of Kazan technological University (Kazan: Kazan State Technical University)* № 1 p 208-11

[3] 2019 Electroerosive processing of materials *Modern production technologies: extxe.com [site]* Mode of access: https://extxe.com/2096/jelektrojerozionnaja-obrabotka-materialov Date:30.09.2019

[4] 2019 Brief description of the zones of the surface layer after electrical discharge machining (EDM) *Company "Northern trading company" is [website]* Mode of access: http://sev-torg.com/articles-ip-harakteristika-sloya-posle-EDM.html Date:27.09.2019

[5] Foteev N K 1997 Quality of surface after electroerosion treatment *STIM* p 43-48 № 8

[6] Ablyaz T R, Muratov K R, Kochergin E Yu and Shakirzyanov T V 2018 Improving the quality of surfaces of products obtained by the method of electroerosion treatment, by applying the technology of electrolytic-plasma polishing *Bulletin of PNRPU. Mechanical engineering, materials science* №2

[7] Khafizov I I, Nurullin I G and Sadykov Z B 2018 Application of the method of electroerosion treatment to improve the quality of the treated surface *Materials of IX conference «IMCOM-2018»* vol. 432

[8] Khafizov I I 2015 Processing methods with imposing of electric field at low- waste division of materials *IOP Conference Series: Materials Science and Engineering* vol.86 Art. № 012013

[9] Khafizov I I 2007 Low-waste finishing separation of scarce materials by the combined method. *Abstract of the thesis for the degree of candidate of technical Sciences* (Voronezh: Voronezh State Technical University)

[10] Khafizov I I and Nurullin I G 2017 Product quality as the main factor of increase of competitiveness (on the example of JSC «Kazan helicopter plant») *IOP Conference Series: Materials Science and Engineering* vol.240 Is.1 Art. № 012039

[11] Akhmetov I D, Zakirova A R, Khafizov I I and Sadykov Z B 2017 New electrode-tool for the combined kerf of electrically conductive materials *IOP Conference Series: Materials Science and Engineering* vol.240 Is.1 Art. № 012003

[12] Khafizov I I and Galimov A N 2017 IT-strategy and major aspects of quality management on the market of goods and services *IOP Conference Series: Materials Science and Engineering* vol.240 Is.1 Art. № 012038

[13] Khafizov I I, Karatayev R O and Shamsutdinova R Z 2015 Wastewater treatment by electrochemical methods *Bulletin of technological University* № 22 p 21-3

[14] Khafizov I I, Saveleva T N and Lyubtsov V S 2017 Estimation of parameters of charge carriers in dielectric materials by CELIV method *IOP Conference Series: Materials Science and Engineering* vol.240 Is.1 Art. № 012040

[15] Khafizov I I 2017 Ways of decrease in the material consumption in case of their separation by the combined methods *IOP Conference Series: Materials Science and Engineering* vol.240 Is.1 Art. № 012037

[16] Khafizov I I 2016 Economic efficiency and effectiveness of ways of separating materials electro diamond processing *IOP Conference Series: Materials Science and Engineering* vol.134 №1 Art. № 012014

[17] Kashapov N F, Khafizov I I and Nurullin I G 2018 Influence of introduction of robotics on increase in efficiency of electrochemical production *IOP Conference Series: Materials Science and Engineering* vol.412 Is.1 Art. № 012034

[18] Khafizov I I, Nurullin I G and Sadykov Z B 2018 Problems of development of electrochemical production of Russia and possibility of their decision *IOP Conference Series: Materials Science and Engineering* vol.412 Is.1 Art. № 012042

[19] Gilmanshin I R, Kashapov N F, Gilmanshina S I and Galeeva A I 2016 Landfill energy complex based on the renewable energy installations *IOP Conference Series: Materials Science and Engineering* vol.134

[20] Kashapov N F and Sharifullin S N 2015 Hardening of the surface plasma jet high-frequency induction discharge of low pressure *IOP Conference Series: Materials Science and Engineering* vol. 86 №1 Art. № 012021

[21] Kashapov N F, Gil'manshin I R and Konahina I A 2014 System analysis of the energy complex of engineering enterprise as a basic tool of effective energy management *IOP Conference Series: Materials Science and Engineering* (Bristol-UK) vol. 69

[22] Fazlyyyakhmatov M G and Kashapov N F 2014 Corona discharge in the process of spraying protective powder coatings on piezoceramic materials *High Temperature Material Processes* vol.18 Is.4 p 273-279

[23] Khafizov I I, Nurullin I G and Sadykov Z B 2019 Application of the method of electroerosion treatment to improve the quality of the treated surface *IOP Conference Series: Materials Science and Engineering* vol.570 Is.1 Art № 012050

IOP Conf. Series: Materials Science and Engineering **915** (2020) 012027 doi:10.1088/1757-899X/915/1/012027

[24] Khafizov I I and Nurullin I G 2019 Influence of roughness on operational properties of details *IOP Conference Series: Materials Science and Engineering* vol.570 Is.1 Art № 012048

[25] Khafizov I I, Nurullin I G and Kashapov N F 2019 Influence of application of robotic systems on prospects and directions of development of electroerosive processings *IOP Conference Series: Materials Science and Engineering* vol.570 Is.1 Art № 012049

[26] Kashapov, L.N., Kashapov, N.F., Kashapov, R.N. Research of cobalt chromium alloy surface morphology after sandblasting (2014) *IOP Conference Series: Materials Science and Engineering*, 69 (1), Art № 012017

[27] Kashapov, R.N., Kashapov, L.N., Kashapov, N.F. Research of plasma-electrolyte discharge in the processes of obtaining metallic powders (2017) *Journal of Physics: Conference Series*, 927 (1), Art № 012086.

IOP Conf. Series: Materials Science and Engineering **915** (2020) 012028 doi:10.1088/1757-899X/915/1/012028

# Application of robotic systems in production: new opportunities for development and risks

**I I Khafizov[1], I G Nurullin[1] and N F Kashapov [1]**

[1]Kazan Federal University, 18 Kremlyovskaya street, Kazan, 420008, Russian Federation

khafizov@kpfu.ru, opros16@yandex.ru

**Abstract.** It is impossible to imagine the modern world and the near future without robots. They have already penetrated into absolutely all spheres of our life, including work, are used to perform various functions, which previously only a person could cope with. Automation and robotics of work processes in production, reduces the impact of human factors and leads to improved product quality, while reducing costs.

Today, according to Russian experts, 71% of working time is related to human labor, and the remaining 29% is performed by machines. In 5 years, this indicator is expected to change to 48% and 52%, respectively. The rapid aging of the population of the developed world, the development of new technologies and the cheapening of electronics contribute to the introduction of robots in production, so as not to reduce economic activity and labor productivity.

In recent years, the mass introduction of industrial robots in different countries has given up to 10% of GDP growth. Therefore, the transition to the use of robotic production systems is defined as one of the priority directions of development of the domestic industry. Before the crisis of 2014, the robotization of the industry was taking rapid steps, but the weakening of the ruble slowed this process.

The development of robotics in Russia meets the goals of the national project "Digital economy" and it is expected that by 2024 28.8 billion rubles will be allocated from the budget and 73.8 billion from other sources.

At the moment, the Russian Federation lags behind the world leaders in the number of robots used in industry, and the overall ratio of the number of industrial robots to the number of workers is very low. According to 2017 data, the structure of the Russian market for industrial robots is as follows: 80% of implementations were for robotic complexes (RC) for welding and cutting metals, in second place is the movement and stowage of cargo. Milling, painting, measuring and other types of RC are less in demand [1].

As noted by Eduard Prodikov, a specialist in robotics, due to the strong backlog in robotics, it is almost difficult in Russia to safely switch to a four-day working week, announced recently by the Prime Minister of the Russian Federation.

We have a shortage of qualified workers and specialists in our country. Also, labor productivity is low compared to developed countries in Western Europe, Asia, and the United States. Despite overall competence, effective leadership, and a positive climate in the team, the results are ultimately poor. A

2017 HeadHunter study found that the reasons for poor performance among employees of St. Petersburg companies are the following factors: inconsistency of work between departments (57%), low salaries (46%), incomplete use of personal potential (42%), time and stress (41%), lack of praise and rewards for good work (38%), noisy colleagues and their chatter (33%) [2].

Industrial robots have very high positioning accuracy, and repeatability makes it possible to achieve the desired level of product processing while reducing production defects.

By eliminating the human factor in technological processes, the percentage of operating errors is significantly reduced. This approach has a positive effect on the growth of the company's performance and the overall performance of the industry [3].

We can use modern industrial robots in many processing operations where it was previously impossible to imagine: milling, grinding, polishing, trimming, deburring and cutting. They can be applied to all industries, and they are also available for small and medium-sized businesses, allowing them to develop even with small investments.

Industrial robots are becoming more accurate and robust every year - today it is possible to position the robot with an accuracy of several hundredths of a millimeter. And more advanced controllers make it possible to control several robots at once. This made it possible to use robots in industries where only specialized CNC machines could be used.

To date, the most common area where it is applied to industrial robots is performing repetitive operations on the production lines: welding, moving parts, Assembly, painting, etc. On these lines the robots operate in cycles, and carry out the same operation, replacing routine human work. This allows you to: increase productivity, reduce the factor of human error and injuries, and automate the process as much as possible.

In the process of creating parts, industrial robotic systems (ORS) allow you to save a large amount of materials and raw materials (with a rational organization of the workflow) [4].

But this is not the limit of possibilities: since more expensive robotic systems can make multi-axis movements along the required trajectory, due to the robot's six-power freedom. Thus performing any processing that was previously possible only on specialized machines.

When properly organized, all these operations are managed by a single operator. PRS makes it possible to carry out work where it is difficult and dangerous for a person. They act accurately, accurately, and quickly. They can work 24 hours a day, with the lowest light and temperature of the shop, thus saving on utility costs [5].

Despite all the advantages of using robots, they cannot fully replace humans. Many functions are not yet possible for them, they do not know how to think, analyze, and draw conclusions, and more complex operations require very expensive systems that are not cost-effective to use.

Also, we must remember the axiom: the reliability of any system is equal to the reliability of the weakest link. Using inefficient links and low-cost solutions with unknown quality and manufacturer in the system can lead to huge expenses and downtime if they break down.

The vast majority of working Russians have a negative attitude to robotics, and they can be understood: after all, the introduction of robots into production will lead to the fact that someone will be left without a job.

Robots will leave millions of people without work, blockchain will be able to perform some functions of the state, cryptocurrencies are able to displace banks, artificial intelligence in General threatens to put a fat point on the history of mankind. It is necessary to understand that progress cannot be stopped, it is necessary to find ways to solve the upcoming problem in a rapidly changing world [6].

Robotization will deprive a large number of people of jobs, but only in specific activities. This is not the first time that humanity has faced this problem, and it is necessary to decide where to send people: mechanization in agriculture - the people went to the industry, automation in the industry – the people went to the service.

But the risks are not related to potential layoffs, but to the inability to retrain and create new jobs in a short time. However, there are also some thoughts that the robotization of the industry will take place very slowly, due to the low wages of the labor force in Russia.

But in this case, in the conditions of increasing labor productivity and the introduction of robotics in the developed countries of the world, our domestic goods and services will lose their competitiveness. Progress will eventually lead to the displacement of traditional industries by automatic factories. Therefore, it is necessary not to resist automation processes, but to "swim" in parallel and look for opportunities [7].

Only by keeping up with the times can we become successful. Stabilizing the technical process, improving the accuracy of processed parts and the speed of achieving goals - all these are just a small part of the reasons why companies around the world are implementing robots in production.

**References**

[1] 2019 Industrial Robotics: the threat of unemployment or the basis of the economy of the future? *Truth-teller: www.pravdoryb.info [site]* Mode of access: https://pravdoryb.info/robotizatsiya-promyshlennosti-ugroza-bezrabotitsy-ili-osnova-ekonomiki-buduschego.html Date:30.09.2019

[2] Zhigach A 2017 Interference in operation. Factors that reduce the effectiveness of employees of St. Petersburg companies *Delovoy Peterburg Newspaper: www.dp.ru [site]* Mode of access: https://www.dp.ru/a/2017/02/05/JEffektivnaja_nejeffektivnos Date:27.09.2019

[3] Kashapov N F, Khafizov I I and Nurullin I G 2018 Influence of introduction of robotics on increase in efficiency of electrochemical production IOP Conference Series: Materials Science and Engineering vol.412 Is.1 Art. № 012034

[4] Stone D Robotization of production in the world: scope, examples, pros and cons *Portal FB.ru: www.fb.ru [site]* Mode of access: https://fb.ru/article/406125/robotizatsiya-proizvodstva-v-mire-sfera-primeneniya-primeryi-plyusyi-i-minusyi Date:25.09.2019

[5] Khafizov I I and Nurullin I G 2019 Influence of application of robotic systems on prospects and directions of development of electroerosive processings IOP Conference Series: Materials Science and Engineering vol.570 Is.1 Art. № 012049

[6] Sergeev E G 2017 Our future. Robots are already among us *"Publishing solutions"* ISBN 978-5-44-858013-0

[7] Zemtsov S P Potential robotization and the " economy of ignorance» in Russia's regions *The XIX April conference (HSE)*

[8] Khafizov I I and Nurullin I G 2017 Product quality as the main factor of increase of competitiveness (on the example of JSC «Kazan helicopter plant») *IOP Conference Series: Materials Science and Engineering* vol.240 Is.1 Art. № 012039

[9] Khafizov I I, Nurullin I G and Sadykov Z B 2018 Application of the method of electroerosion treatment to improve the quality of the treated surface *Materials of IX conference «IMCOM-2018»* vol. 432

[10] Khafizov I I, Nurullin I G and Sadykov Z B 2018 Problems of development of electrochemical production of Russia and possibility of their decision *IOP Conference Series: Materials Science and Engineering* vol.412 Is.1 Art. № 012042

[11] Gil'manshin I R and Kashapov N F 2014 Energy service contracts in regional

engineering center for small and medium businesses *IOP Conference Series: Materials Science and Engineering*
[12] Nurullin I G and Khafizov I I 2016 The impact of innovation implementation and certification on the company's activities and reputation Modern technologies: topical issues, achievements and innovations: collection of articles of the winners of the IV international. science.- practical Conf. (Penza: ICSU " Science and education») p 44-6
[13] Akhmetov I D, Zakirova A R, Khafizov I I and Sadykov Z B 2017 New electrode-tool for the combined kerf of electrically conductive materials *IOP Conference Series: Materials Science and Engineering* vol.240 Is.1 Art. № 012003
[14] Khafizov I I and Galimov A N 2017 IT-strategy and major aspects of quality management on the market of goods and services *IOP Conference Series: Materials Science and Engineering* vol.240 Is.1 Art. № 012038
[15] Khafizov I I, Saveleva T N and Lyubtsov V S 2017 Estimation of parameters of charge carriers in dielectric materials by CELIV method *IOP Conference Series: Materials Science and Engineering* vol.240 Is.1 Art. № 012040
[16] Khafizov I I 2017 Ways of decrease in the material consumption in case of their separation by the combined methods *IOP Conference Series: Materials Science and Engineering* vol.240 Is.1 Art. № 012037
[17] Khafizov I I 2016 Economic efficiency and effectiveness of ways of separating materials electro diamond processing *IOP Conference Series: Materials Science and Engineering* vol.134 №1 Art. № 012014
[18] Fazlyyyakhmatov M G and Kashapov N F 2014 Corona discharge in the process of spraying protective powder coatings on piezoceramic materials *High Temperature Material Processes* vol.18 Is.4 p 273-279
[19] Kashapov N F, Gil'manshin I R and Konahina I A 2014 System analysis of the energy complex of engineering enterprise as a basic tool of effective energy management *IOP Conference Series: Materials Science and Engineering* (Bristol-UK) vol. 69
[20] Gubaidullin D A, Kashapov N F and Zaripov R G 2017 Experimental studies for flow velocities of oscillations gas when transiting through resonance in the area of an open end of pipe *IOP Conference Series:Journal of Physics: Conference Series.* vol.789 Is.1 Art. № 012017
[21] Khafizov I I, Nurullin I G and Sadykov Z B 2019 Application of the method of electroerosion treatment to improve the quality of the treated surface *IOP Conference Series: Materials Science and Engineering* vol.570 Is.1 Art № 012050
[22] Khafizov I I and Nurullin I G 2019 Influence of roughness on operational properties of details *IOP Conference Series: Materials Science and Engineering* vol.570 Is.1 Art № 012048
[23] Kashapov, L.N., Kashapov, N.F., Kashapov, R.N., Pashaev, B.Y. The application of additive technologies in creation a medical simulator-trainer of the human head operating field (2016) *IOP Conference Series: Materials Science and Engineering*, 134 (1), Art № 012011
[24] Kashapov, R.N., Kashapov, L.N., Kashapov, N.F. Analysis and development of methods for obtaining metallic powders for selective laser melting (2017) *IOP Conference Series: Materials Science and Engineering*, 240 (1), Art № 012071

IOP Conf. Series: Materials Science and Engineering **915** (2020) 012029 doi:10.1088/1757-899X/915/1/012029

# Application of electric spark doping technology for strengthening the surface of the Part

**I I Khafizov[1] and I G Nurullin[1]**

[1]Kazan Federal University, 18 Kremlyovskaya street, Kazan, 420008, Russian Federation

khafizov@kpfu.ru, opros16@yandex.ru

**Abstract.** To increase the wear resistance and durability of parts, it is necessary to ensure proper roughness of the surface of the part. To achieve this goal, you can use the method of electric spark doping, which is characterized by energy intensity, environmental cleanliness and a number of other advantages. With the help of electric spark doping, both single-layer and multi-layer protective coatings can be applied, which allow you to obtain new surface materials.

Reliable and trouble-free operation of the equipment allows you to reduce operating costs for the production of products and for the maintenance of installations. As you know, the real life of the machine depends on the bearing capacity of the surfaces of parts, which is determined by the quality of the surface layer.

To increase the wear resistance and durability of the machine and mechanism, it is necessary to ensure proper surface roughness of the part. Also, the high surface quality helps to reduce energy losses when working parts and gives them a beautiful appearance [1].

However, under certain operating conditions, the usual physical and mechanical characteristics of materials do not meet the requirements [2]. In such cases, steel is alloyed: other chemical elements are added to the original composition during smelting. As a result, the steel becomes stronger, harder, more resistant to external adverse factors, but loses its plasticity.

To obtain the surface layers of the required quality, various methods of hardening are used:

- quenching with high-frequency currents,
- electroplating,
- nitriding,
- surface plastic deformation, etc.

As an alternative to these methods of hardening and coating, you can use the method of electric spark doping (ESD), which is characterized by energy intensity, environmental cleanliness and a number of other advantages, developed and proposed in the 1930s by B. R. Lazarenko and N. I. Lazarenko. ESD is a phenomenon of electrical erosion and polar transfer of the anode material to the cathode during pulsed discharges in a gas medium. The physical and chemical properties will depend

IOP Conf. Series: Materials Science and Engineering **915** (2020) 012029 doi:10.1088/1757-899X/915/1/012029

on the composition and properties of the electrode materials. Table 1 shows the advantages and disadvantages of using ESD [3].

Table 1. The advantages and disadvantages of electric-spark alloying.

| № | Advantages | Disadvantages |
|---|---|---|
| 1 | no significant heating and deformations of the product during processing | small thickness of the formed layer |
| 2 | the ability to apply coatings of various materials to the surface of the product | high roughness of the treated surface |
| 3 | there is no need for special pre-treatment of the treated surface before applying the coating | relatively low labor productivity |
| 4 | comparative simplicity of the technology of this method of hardening | mainly manual method of surface treatment |
| 5 | ease of maintenance and reliability of equipment that is small and transportable | it is not possible to use non conductive materials |

With ESD, the microhardness of the white layer in carbon steels can be brought up to 230 MPa. And the thickness of the coating layer obtained on some installations can be 0.003-0.2 mm. when reinforced with ferrochrome on the ie-2M installation, a layer depth of up to 0.5-1.6 mm with a microhardness of 50 - 60 MPa can be obtained [4].

For ESD, refractory metals, hard alloys, and less often heat-resistant high-alloy steels and alloys with high mechanical properties are often used. The review of the latest achievements in the field of ESD of metals and alloys has shown that all new technologies of electric spark doping can be divided into 4 groups:

- ESD in heterogeneous media;
- ESD with coated electrodes;
- ESD with powder electrodes;
- Thermosetting electric spark doping [5].

The protective coating on the part or tool in ESD is applied mainly with a compact electrode. There are two methods of coating: the contact method-occurs with direct contact of the hardened product and the electrode, and the non - contact method-the interaction occurs with a certain gap between them.

The process of ESD compact electrodes takes place in several phases: - the formation of electrical contact between the anode and the cathode, after the convergence of the electrodes and breakdown of the interelectrode gap; the electric erosion of the electrodes; polar migration and the formation of the surface layer on the surface of the cathode; - break electrical contact between the anode and cathode, the point of divergence of the electrodes.

Depending on the intensity of the electrode erosion process, there are two doping directions: "finishing" and "rough" processing. The intensity of erosion leads to an increase in surface roughness. In the first case, relatively thin layers are formed on the surface of the cathode, and in the "rough" case, the thickness of the formed layer is greater. In practice, both types are used, but due to the need for precise processing and low roughness, preference is given to "finishing" hardening.

With the help of ESD, you can apply both single-layer and multi-layer protective coatings that allow you to obtain new surface materials. The application of a subsequent layer reduces the roughness value, improves heat dissipation and increases heat resistance.

The method of electric spark doping is used in almost all areas of mechanical engineering and Metalworking. And also because of the possibility of localization of the processing area, the method of electric spark hardening is often used by jewelers and watchmakers [6].

**References**

[1] Khafizov I I, Nurullin I G and Sadykov Z B 2018 Application of the method of electroerosion treatment to improve the quality of the treated surface *Materials of IX conference «IMCOM-2018»* vol. 432

[2] Khafizov I I, Nurullin I G and Sadykov Z B 2018 Problems of development of electrochemical production of Russia and possibility of their decision *IO Conference Series: Materials Science and Engineering* vol.412 Is.1 Art. № 012042

[3] Nazarov D V 2012 Study of the characteristics of the coating made by the method of electric spark doping *dis. ... master's TSU*

[4] 2019 Hardening of the metal surface layer (electroerosive alloying) *Portal Findout.su [site]* Mode of access: https://findout.su/3x32282.html Date:23.11.2019

[5] Yurchenko E V 2016 Technology for obtaining strengthening nanostructured electric spark coatings on aluminum alloys and their use in the restoration and repair of machine parts: *dis. ... doctor of technical Sciences*

[6] Seleznev O V 2019 Metalworking by electric spark alloying *WikiMetall.RU -Portal about metallurgy and Metalworking [site]* Mode of access: https://wikimetall.ru/metalloobrabotka/elektroiskrovoe-legirovanie.html Date:25.11.2019

[7] Khafizov I I and Nurullin I G 2017 Product quality as the main factor of increase of competitiveness (on the example of JSC «Kazan helicopter plant») *IOP Conference Series: Materials Science and Engineering* vol.240 Is.1 Art. № 012039

[8] Khafizov I I 2013 Ways to reduce the consumption of materials when they are separated by combined methods *Bulletin of Kazan technological University* (Kazan: Kazan State Technical University) № 1 p 208-11

[9] Gil'manshin I R and Kashapov N F 2014 Energy service contracts in regional engineering center for small and medium businesses *IOP Conference Series: Materials Science and Engineering*

[10] Khafizov I I and Galimov A N 2017 IT-strategy and major aspects of quality management on the market of goods and services *IOP Conference Series: Materials Science and Engineering* vol.240 Is.1 Art. № 012038

[11] Khafizov I I, Saveleva T N and Lyubtsov V S 2017 Estimation of parameters of charge carriers in dielectric materials by CELIV method *IOP Conference Series: Materials Science and Engineering* vol.240 Is.1 Art. № 012040

[12] Khafizov I I 2017 Ways of decrease in the material consumption in case of their separation by the combined methods *IOP Conference Series: Materials Science and Engineering* vol.240 Is.1 Art. № 012037

[13] Khafizov I I 2016 Economic efficiency and effectiveness of ways of separating materials electro diamond processing *IOP Conference Series: Materials Science and Engineering* vol.134 №1 Art. № 012014

[14] Kashapov N F, Khafizov I I and Nurullin I G 2018 Influence of introduction of robotics on increase in efficiency of electrochemical production *IOP Conference Series: Materials Science and Engineering* vol.412 Is.1 Art. № 012034

[15] Fazlyyyakhmatov M G and Kashapov N F 2014 Corona discharge in the process of spraying protective powder coatings on piezoceramic materials *High Temperature Material Processes* vol.18 Is.4 p 273-279

IOP Conf. Series: Materials Science and Engineering **915** (2020) 012029 doi:10.1088/1757-899X/915/1/012029

[16] Kashapov N F and Sharifullin S N 2015 Hardening of the surface plasma jet high-frequency induction discharge of low pressure *IOP Conference Series: Materials Science and Engineering* vol. 86 №1 Art. № 012021

[17] Kashapov N F, Gil'manshin I R and Konahina I A 2014 System analysis of the energy complex of engineering enterprise as a basic tool of effective energy management *IOP Conference Series: Materials Science and Engineering* (Bristol-UK) vol. 69

[18] Gubaidullin D A, Kashapov N F and Zaripov R G 2017 Experimental studies for flow velocities of oscillations gas when transiting through resonance in the area of an open end of pipe *IOP Conference Series:Journal of Physics: Conference Series.* vol.789 Is.1 Art. № 012017

[19] Gilmanshin I R, Kashapov N F, Gilmanshina S I and Galeeva A I 2016 Landfill energy complex based on the renewable energy installations *IOP Conference Series: Materials Science and Engineering* vol.134

[20] Khafizov I I, Nurullin I G and Sadykov Z B 2019 Application of the method of electroerosion treatment to improve the quality of the treated surface *IOP Conference Series: Materials Science and Engineering* vol.570 Is.1 Art № 012050

[21] Khafizov I I and Nurullin I G 2019 Influence of roughness on operational properties of details *IOP Conference Series: Materials Science and Engineering* vol.570 Is.1 Art № 012048

[22] Khafizov I I, Nurullin I G and Kashapov N F 2019 Influence of application of robotic systems on prospects and directions of development of electroerosive processings *IOP Conference Series: Materials Science and Engineering* vol.570 Is.1 Art № 012049

[23] Kashapov, L.N., Kashapov, N.F., Kashapov, R.N. Research of cobalt chromium alloy surface morphology after sandblasting (2014) *IOP Conference Series: Materials Science and Engineering*, 69 (1), Art № 012017

[24] Kashapov, R.N., Kashapov, L.N., Kashapov, N.F. Research of plasma-electrolyte discharge in the processes of obtaining metallic powders (2017) *Journal of Physics: Conference Series*, 927 (1), Art № 012086.

IOP Conf. Series: Materials Science and Engineering **915** (2020) 012030 doi:10.1088/1757-899X/915/1/012030

# Engineering experiment as an underlying tool for training research engineers

**G R Khamidullina, I A Konakhina, N F Kashapov, I I Khafizov and E M Khusnutdinova**

Kazan (Volga Region) Federal University 18 Kremlyovskaya str., Kazan 420008

rr-088@mail.ru

**Abstract:** The paper analyzes the contribution of the engineering experiment to professional skills of research engineers. The discussion is based on the educational programs and methodology of the disciplines taught at the Engineering Institute of Kazan Federal University.

In a contemporary context, the engineering progress leads not only to an increase in the complexity of test objects, but also it constantly steps up the requirements for the depth of penetration into the essence of the performance of the industrial equipment components, the compatibility with them and their interaction, as well as the measurement accuracy of recorded values. Therefore, the importance of the experiment in modern science, technology and industrial activity is not in doubt. This is the essential and most important method of obtaining new knowledge, innovative technologies and products. Even where the scientific research is theoretical and present in the form of mathematical models, a need always arises to confirm the compliance of the models constructed with real processes, and this is only possible by means of experiments [1, 2].

The training of research engineers has been carried out at the Engineering Institute of Kazan Federal University since 2013 based on the principal undergraduate and master's degree programs in specialized departments «Technical physics and energy», «Quality management», «Biomedical engineering and innovation management».

The training of bachelors-to-be is performed in the following subjects:

Applied physics;

Quality management (Quality Management and Robotics).

The training of masters-to-be is performed in a variety of disciplines:

Measurement and certification;

Biomedical devices, systems and facilities;

Medical and clinical equipment;

Innovation studies: Intellectual property.

In all of the above subjects and focuses of educational programs, the engineering experiment is present either as a basic discipline or as a variable discipline, or is implemented in the form of laboratory studies and independent R&D activity performed under the guidance of experienced teachers.

In obtaining professional competencies, students are aware of scientific research purpose in terms of engineering: identify objective patterns that determine the workflow processes in machines, apparatuses, technological systems and facilities, etc.; study the physical and physico-chemical phenomena that make up these processes; use the obtained scientific results to create innovative designs that are optimal in terms of efficiency, economy, metal consumption, service life, or any other data [1, 3].

They are also aware of the issues facing them as researchers of a fundamentally new equipment, materials, and structures. One of these issues is caused by the fact that often the characteristics of test objects that need to be experimentally determined are increasingly becoming inaccessible for direct measurement. Because of this, the performance indicators used to evaluate the test object or to take organizational and design decisions usually do not coincide with the object parameters determined by the results of a full-scale experiment.

Another issue is the arrangement for testing of objects that have a complex dynamic behavior of performance and are subject to significant influences of changing environmental conditions.

In addition, testing sophisticated facilities enhances the importance of the effects that the test, recording and control equipment has on the running of the test object.

In this regard, the most important challenge of arranging the training of research engineers in the field of design, planning and implementation of the engineering experiment is to master a systematic approach, which involves considering all the tools involved in the experiment as a single system described by the corresponding mathematical model [4]. Thus, mathematical modeling becomes an integral element of research activity applicable to the engineering experiment. Without building a model, today it is impossible to imagine either the design of the experiment, or its performance, nor the processing and analysis of the results [5]. Only the availability of ratios linking the desired technical and economic data of the test object with its parameters allows reasonable judgments about the necessary test measures and their reasonable sequence, the population of the recorded values, the requirements for the accuracy of measurements, the frequency of registration of recorded data, etc. The same ratios serve to determine the estimates of the desired characteristics that satisfy the relevant statistical requirements (bias, minimum dispersion, etc.)

Complex research problems are associated with the need to analyze the interrelation of many factors[6]. For studying processes and phenomena in order to further manage them, it is necessary to find out what purpose each factor individually serves in this process or phenomenon. This is possible only with careful design of a series of experiments and an adequate assessment of the scope of experimental activity. Too small amount of experimental data may not allow to establish a pattern or function, and the researcher may get low accuracy of constant values or not notice a weak effect, which, nevertheless, has great theoretical value. But on the other hand, with an excessively large amount of data, experimental studies take a long time, and data processing can take long, it becomes very time-consuming, complicated, confusing to analyze and obtain calculated engineering dependencies. At the same time, there is a challenge of an obstacle to the detection of relevant effects, the signs of which are literally blurred in giant data sets. This implies the importance of students gaining skills and competencies for self-planning of the experiment.

The methodology of setting the discipline "Design of experiment", which is on the list of basic disciplines of the "Quality management" and "Robotics", includes the development of students' professional skills in the implementation of the following components of the engineering experiment:

1. Statement of the problem of experimental research.
2. Design of experiment

IOP Conf. Series: Materials Science and Engineering **915** (2020) 012030 doi:10.1088/1757-899X/915/1/012030

3. Analysis of experimental data and their conversion into a form convenient for engineering calculations.

The statement of the research problem implies the acquisition by students of skills in choosing independent variables which influence will be studied; substantiation of the assumptions made that facilitate the problem; selection of measuring equipment; assessment of required accuracy; identification of effects that are not capable to direct measurement [7].

The design of the experiment is considered from the point of view of the conventional one-and multi-factor experiment and a randomized experiment. The student acquires skills in determining the required scope and sequence of research work, compiling mathematical models for processing experimental data and describing them.

Analysis of the research results implies the acquisition by students of professional skills in collecting and managing experimental data, maintaining experimental reports, processing (including automated processing) data and their generalization [8,9]. Currently, specialized programs make it possible to display experimental data in the graphical form and even as multidimensional surfaces, nevertheless, the engineering practice urges to obtain simple criteria dependencies that would allow the calculation of similar processes without significant time costs with satisfactory accuracy. In light of this, the «Experiment planning» discipline includes among others: «Similarity theory in engineering research» and «Dimension analysis».

**References**

[1] Konakhin A M and Nazmeyev Yu G 2003 *Fundamentals of the engineering research theory.* Kazan: Kazan state power engineering university p. 283

[2] Vachagina E K, Kadyirov A I, Konakhina I A and Khusnutdinova E M 2019 *Features of hydrodynamic fields in three-way screw insert barrels.* News of the Russian academy of sciences. Mechanics of fluid and gas. No. 3. pp. 28-37

[3] Khamidullina A F, Kuzmina M A and Konakhina I A 2017 *Engineering background: Modern formats and challenges of conceptual engineering.* IOP Conf. Series: Materials Science and Engineering 240 (2017) 012043

[4]Konakhina I A, Khamidullina G R, Kuzmina M A and Khusnutdinova E M 2018 *Target energy monitoring method for the analysis of production companies and facilities of housing and communal services.* IOP Conf. Series: Materials Science and Engineering 412 (2018) 012050

[5]Konakhina I A, Khamidullina G R, Khusnutdinova E M and Kashapov N F 2018 *Engineering education in the system of modern innovative activity.* Proceedings of the IX International scientific and technical conference «Innovative engineering technologies, equipment and materials – 2018» Part 1. pp. 323-326

[6]Khizbullin R N, Khusnutdinov A N, Bazhenov N G and Khusnutdinova E M 2018 *A Joint approach to teaching engineering specialties of higher education and domestic undertakings.* Today's Challenges in updating the content of engineering education: proceedings of VIII All-Russia scientific and methodological conference (Kazan, May 18, 2018). – Kazan: Kazan state power engineering university. Vol. 1. pp. 125-130

[7] Gilmanshina S I, Bendjukevich K G, Gilmanshin I R, Glushkova V A and Gadirova E M *Methods of identifying talent in the field of engineering and natural sciences* // IOP Conference Series: Materials Science and Engineering. - 2019. - Vol.570, Is.1. - Art. № 012028.

[8] Kashapov R N, Kashapov L N, Kashapov N F *Formation of cracks in the selective laser melting of objects from powdered stainless steel 17-4 HP* (2017) IOP Conference Series: Materials Science and Engineering, Vol. 240, Is 1, Art. № 012074

[9] Kashapov LN, Kashapov NF, Kashapov RN, Pashaev B Y 2016 *The application of additive technologies in creation a medical simulator – trainer of the human head operating field* IOP Conference Series: Materials Science and Engineering, Vol. 134, Is 1, Art. № 012011

IOP Conf. Series: Materials Science and Engineering **915** (2020) 012031 doi:10.1088/1757-899X/915/1/012031

# Estimation of precision parameters of milling machines

**R M Khusainov[1], E R Ziangirova[1] and G K Davletshina[1]**

Kazan Federal University, Naberezhnye Chelny Institute, 68/19 Mira Street, Naberezhnye Chelny, 423812, Russia

rmh@inbox.ru

**Abstract**. The paper considers a mathematical model that allows to evaluate the performance parameters of milling machines during the functioning of the working bodies. An example of the task of selecting equipment is given. The mathematical model is based on the mathematical apparatus of coordinate transformation. Using the proposed methodology, it is possible to solve the direct problem - to evaluate the capabilities of an existing machine for manufacturing parts with a given accuracy, or the inverse problem - to choose a machine for manufacturing parts with a given accuracy.

**1. Introduction** One of the indicators of the quality of the operation of machines is the accuracy of the trajectory of their movements. The development of the trajectory with a given accuracy is determined by organized programmed movements of the working bodies that ensure the functioning process, and unorganized, forming a real trajectory of movement. Unorganized movements occur under the influence of systematic and random factors, such as deviations of kinematic and dimensional chains, elastic displacements, temperature changes, wear, measurement errors, programming errors, etc. [1]. Unorganized movements are modeled by introducing component errors into the rules of movement of the coordinate systems of the links of the kinematic scheme of the machine. The accuracy of the machine plays an important role in ensuring the quality of the machined surfaces [2], therefore it is important to have a tool for preliminary assessment of the output accuracy of the machine [3].

The output accuracy of the machine is calculated in the following order [4]:

## 2. Building the coordinate code of a functional system

A functional system can be replaced by an equivalent scheme, as a set of solid bodies, each of which can be replaced by a coordinate system. The origin *O* and *X, Y, Z* axes of the coordinate system are associated with the structural or technological bases of the link.

## 3. Building a functional model

Instead of variables of the coordinate code, matrices of generalized movements (coordinate transformations) are substituted. Each matrix is associated with its own link in the forming system. A tool vector matrix is also added. When machining holes with an axial tool, milling grooves with end and disc mills, and round and internal grinding, the radius of the tool must be taken into account. The coordinate transformation matrices and the tool vector are multiplied in the order determined by the coordinate code.

$$r_0 = A^X \cdot A^Y \cdot A^Z \cdot A^C \cdot r_и \tag{1}$$

where $r_0$ - vector whose elements are the coordinates of the ideal surface of the workpiece; $A^C$, $A^X$, $A^Y$, $A^Z$ – transformation matrices in corresponding coordinates; $r_u$ - vector whose elements are the coordinates of the surface of the cutting tool.

Equation (1) is the basic equation of the functional model, which is a mathematical model of the shaping system and it is a description of the resulting trajectory in the form of a vector.

## 4. Accounting for errors in the links of a functional system

The purpose of this section is to find this error. The required error is the complete variation of the vector of the coordinates of the trajectory:

$$\Delta \boldsymbol{r}_0 = \boldsymbol{\delta A} \cdot \boldsymbol{r}_0 \tag{2}$$

where $\boldsymbol{\delta A}$ – variation matrix.

The variation matrix is a small displacement of the coordinate system associated with the link.

## 5. Estimation of the accuracy of the generated trajectory

To obtain estimates of the accuracy of the size, location and shape of the formed trajectory, it is necessary to build a metrological base [5].

The process of building the base surface is performed in two stages:

*5.1. The first step is to draw up the equation of the base surface.*

Due to the small deviations of the base surface from the nominal, the equation of the base surface can be written as:

$$r_b = r_0 + \Delta r_b \tag{3}$$

where $\Delta r_b$ defined as the sum of the error vectors from position and size:

$$\Delta r_b = \varepsilon_b r_0 + \mathrm{d}r_0 \tag{4}$$

where $\varepsilon_b$ – a matrix whose elements are errors (along the coordinate axes) of the location of the coordinate system associated with the base surface relative to the system in which the equation (1);

$$\varepsilon_b = \begin{pmatrix} 0 & -\delta C_b & \delta B_b & \delta x_b \\ \delta C_b & 0 & -\delta A_b & \delta y_b \\ -\delta B_b & \delta A_b & 0 & \delta z_b \\ 0 & 0 & 0 & 0 \end{pmatrix} \tag{5}$$

$dr_o$ – the full differential of the radius-vector $r_0$, taken for all components of the vector $q_0$ of the dimensional parameters of the surface.

*5.2. The second stage of constructing the base surface is the calculation of the components of the vector $\Delta q$ of position and size errors.*

The components of the vector $\Delta q$ can be found from the matrix equation [4]:

$$H\Delta q = d \tag{6}$$

where $H$ – matrix of order $p \times p$ with elements:

$$h_{ki} = \int_S f_k f_i dS \tag{7}$$

$d$ – vector of order $p$ with elements:

$$d_i = \int_S f_i \Delta r_n dS \tag{8}$$

$f_k$, $f_i$ – $k$-th и $i$-th coordinates of the vector $f$ of normal transfer coefficients:

$$f = G^T n \tag{9}$$

$n$ – unit vector of the normal to the surface $r_0$:

$$\Delta r_n = \Delta r_0 n \tag{10}$$

$\Delta r_n$ – error measured by the normal to the surface, $S$ – the surface of integration is determined by the vector $r_0$.

Elements of the vector $\Delta q$ defined from the expression (6) are used as estimates of errors in the size and location of the treated surface.

To find the values of the point coordinate errors, you must subtract the point coordinate equation from the equation, taking into account the errors (10), the point coordinate equation for an ideal surface (1).

$$\Delta r_0(x, y, z) = rn(x, y, z) - r_0(x, y, z) \tag{11}$$

Based on this equation, you can solve the problem of determining the errors of the treated surface.

## 6. Determination of accuracy indicators of the machine to ensure the formation of the surface with a given accuracy

Let the results of measurements of the size and shape of finished parts are known and you need to find the best estimate of input errors from them. Formally, this problem is reduced to the problem of estimating parameters in the linear regression equation. In fact, equation (2) can be rewritten as:

$$\Delta r_0(x, y, z) = \sum \delta q_i \cdot a_i(x, y, z) \tag{12}$$

where $\delta q_i$ — machine errors; $a_i(x, y, z)$ — known functions, in this case representing transfer coefficients.

You need to evaluate the parameters $\delta q_i$, if you know the result of a multiple measurement $n$ of the value $\Delta r_0(x, y, z)$ at least at $m$ different points.

It is advisable to estimate $\delta q_i$ from the conditions of minimizing the mean square deviations [4] of the observed values $\Delta r_0(x, y, z)$ from those calculated according to equation (10), when their estimates are substituted for $\delta q_i$. The solution of system (12) by the least squares method [4] has the form:

$$\delta = B^{-1} \cdot A^T \cdot D^{-1} \cdot \Delta \tag{13}$$

where $\delta$ — vector of estimated parameters $\delta q_i$; $B$ — matrix of order m × n of a system of normal equations composed by the least square method:

$$B = A^T \cdot D^{-1} \cdot A \tag{14}$$

$A$ — structural matrix of order $n \times m$ of rank $m$ composed of the coefficients of the system (1);
$D$ — a matrix of order $n \times n$ of rank $n$, which is a covariance matrix of measurement results;
$\Delta$ — vector of order $n$ composed of measurement results.

Drawing up the matrix $A$ and the vector $\Delta$ does not cause any special difficulties. The greatest difficulties are caused by composing the covariance matrix $D$, since its elements are usually unknown in advance. However, it is often sufficient to assume that the variances of all values $\Delta$ are constant. Then the formula (14) is simplified:

$$\delta = (A^T \cdot A)^{-1} \cdot A^T \cdot \Delta \tag{15}$$

In this case, you only need to know the matrix $A$, which is obtained from the equations of the variational method for calculating accuracy and measurement results.

Based on the results of the calculation, requirements were formed for the accuracy indicators of the machine for finishing milling of the stamp.

**Table 1.** Requirements for the accuracy of the machine for the finishing milling of the working surfaces of the stamp

| Accuracy indicators | | Requirements for finishing milling |
|---|---|---|
| Radial runout of the conical spindle hole | At the end | 0,001 |
| | At a distance of 150 mm | 0,003 |
| Spindle axial runout | | 0,002 |
| Repeatability of positioning | | 0,001 |
| One-way positioning error at 120 mm length | | 0,008 |

## 7. Conclusions

The given system of mathematical models under the influence of errors of its links allows us to describe the deviations of the trajectory of the working body of the machine. This system allows you to evaluate the accuracy of surface treatment under specific operating conditions and purposefully correct its errors [3].

## References

[1] Balabanov I P, Simonova L A and Balabanova O N 2015 *IOP Conf. Ser.: Mater. Sci. Eng.* **86** 012010
[2] Safarov D T, Kondrashov A G and Khafizov I I 2019 *IOP Conf. Ser.: Mater. Sci. Eng.* **570** 012084
[3] Khisamutdinov R M, Yurasova O I, Yurasov S Yu and Ryabov E A 2018 *IOP Conf. Ser.: Mater. Sci. Eng.* **412** 012044.
[4] Reshetov D N and Portman V T 1986 *Precision of metal cutting machines* (Moscow: Mashinostroyenye) p 336
[5] Markov N N 1995 *Metrological support in mechanical engineering* (Moscow: Stankin) p 468

IOP Conf. Series: Materials Science and Engineering **915** (2020) 012032 doi:10.1088/1757-899X/915/1/012032

# Comprehensive Test Procedure for Digital Instruments and Devices of Automated Versatile Systems

**E M Khusnutdinova[1], P P Pavlov[2], V P Fandeyev[2], R N Khizbullin[2] A N Khusnutdinov[2] and I V CHerepen'kin[2]**

[1]Kazan (Volga Region) Federal University 18 Kremlyovskaya Str., Kazan 420008, Russia
[2]Kazan State Power Engineering University51 Krasnoselskaya str., Kazan 420061, Russia

rr-088@mail.ru

**Abstract.** The authors present a comprehensive test procedure for digital devices, based on the calculation and control of two or more characteristics of a binary signal at a control point. The procedure is aimed at how to increase the reliability of performance monitoring and localization of the failure of digital devices as part of automated versatile systems.

Current automated systems consist of different components and systems (digital controls, monitoring, testing tools, etc.) The results of reliable monitoring and testing provide error-free decisions on safe operation and in-service resource saving. Performance monitoring and troubleshooting is done by hardware and software, selected depending on the principles of operation, design and fault classes of the components. Methods of hardware monitoring and troubleshooting of digital devices subject to logical failure (hereinafter referred to as the “failure”) are usually based on comprehensive testing.

The comprehensive testing of the digital device by comparing a binary signal at a reference point with a valid binary signal ensures failure detection, but requires significant troubleshooting support. Limiting the cost of troubleshooting can be achieved, for example, by choosing the number of transitions, the number of units, the signal signature at the reference point of the digital device as controlled parameters [1]. At the same time, control of the number of transitions, the number of digits, and the signature is inferior to the method of signal comparison by the probability of failure detection. Some of the failures impossible to detect by the signature control can be detected by combining the signature control and the number of transitions or/and digits of binary signals of the digital device [1].

The signal at the reference point of the digital device subjected to testing is usually modeled by a multi-bit binary sequence of digits and zeros, which, in turn, is displayed by a binary polynomial or Boolean function [2, 3, 4]. These models make it possible to calculate the number of transitions, digits, and signal signature.

The number of different binary signal sequences is calculated by the formula

$$M=2^N=2^{2^n}, \qquad (1)$$

IOP Conf. Series: Materials Science and Engineering **915** (2020) 012032 doi:10.1088/1757-899X/915/1/012032

where $N$ is the number of bits of the binary sequence; $n$ is the number of binary input signals (inputs) of the digital device.

A healthy digital device has a single valid binary sequence. The remaining binary sequences correspond to failures of the digital device.

The probability of failing to detect the digital device failure by monitoring the number of transitions or digits with equally probable binary sequences is calculated by the formula

$$q = \frac{1}{M(M-1)} \sum_{k=0}^{N} C_N^k \left(C_N^k - 1\right) \approx \frac{1}{\sqrt{\pi N}}, \tag{2}$$

where $k$ is the number of transitions or digits in binary sequence; $C_N^k$ is the number of $N$ to $k$-combinations; $\pi \approx 3{,}14$.

The binary sequences of the healthy and failed digital device most often differ in only one bit (single failure). The probability of failing to detect the single failure by controlling the number of transitions is calculated by the formula

$$q_1 = \frac{N-1}{2N} \approx \frac{1}{2}. \tag{3}$$

The single failure is detected by controlling the number of digits. Indeed, when changing 1 to 0 or 0 to 1 in one bit, the number of digits in a binary sequence always changes. Thus, controlling the number of digits has an advantage over controlling the number of transitions in terms of detecting single failures.

The binary sequence of the failed digital device may not differ from the binary sequence of the healthy digital device in the number of transitions, but may differ in the number of digits. The corresponding failures of the digital device are not detected by controlling the number of transitions, but are detected by controlling the number of digits [2, 5].

The probability of failing to detect the failure manifesting itself as an $N$-bit binary sequence containing $k$ transitions and $l$ digits is calculated using the theoretically substantiated formula

$$Q_{k,l}^N = \frac{C_N^k \left(C_N^k - 1\right)}{M(M-1)} \sum_{l=1}^{N-1} \frac{M_{k,l}^N - 1}{C_N^k - 1}, \tag{4}$$

where $M_{k,l}^N$ is the number of $N$-bit binary sequences containing $l$ digits among binary sequences containing $k$ transitions. The correctness of formula (4) is proved by the results of a simulation experiment with binary sequences.

The probability of failing to detect the failure when controlling the number of transitions and the number of digits is calculated by the formula

$$Q = \sum_{l=1}^{N-1} Q_{k,l}^N. \tag{5}$$

With an increase in the number of binary bits from four to eight, the probability of failing to detect the failure by controlling the number of transitions and digits approaches the probability of failing to detect the failure by controlling the number of transitions only. Controlling the number of transitions and digits does not provide a significant reduction in the probability of failing to detect the failure.

The probability of failing to detect the digital device failure by the signature control method, is calculated by the formula

$$q_s = \frac{2^{N-m}-1}{2^N-1} \approx \frac{1}{2^m}, \quad (6)$$

where $m$ is the number of bits of the signature analyzer register, many times less than the probability of failing to detect the failure by controlling the number of transitions and digits.

A significant reduction in the probability of failing to detect the digital device failure is achieved by a combination of controlling the signature and the number of digits, since binary sequences of signals of both the healthy and the failed digital device may not differ in signature, but differ in the number of digits [6, 7, 8, 9].

The probability of failing to detect the failure, when combining the methods of controlling the signature and the number of digits of the signal, is calculated using theoretically valid formulas for the signature containing a digit in at least one of the bits (non-zero signature, $s \neq 0$), and the signature containing zeros in all bits (zero signature, $s = 0$).

$$q_{l,s^{1}0} = \frac{1}{M-1}\left(\frac{N-1}{N/2}\sum_{c=1}^{l/2} C_{N/2}^{2c-1} C_{N/2}^{l-(2c-1)} - 1\right), \quad (7)$$

$$q_{l,s=0} = \frac{1}{M-1}\left(C_N^l - \frac{N-1}{N/2}\sum_{c=1}^{l/2} C_{N/2}^{2c-1} C_{N/2}^{l-(2c-1)} - 1\right), \quad (8)$$

where $M$, $N$ are calculated by the formula (1); $l$ is the number of digits in the binary sequence of the signal, $l/2$ is rounded to the nearest integer with an odd number; $C$ stands for the number of combinations; $c$ is the integer variable.

The approximate probability of failing to detect the digital device failure is calculated for both the non-zero and zero signature by the formula

$$q_{l,s} \approx \frac{1}{M-1}\left[\frac{(N-1)!}{l!(N-l)!} - 1\right] \quad (9)$$

Formulas (7) - (9) are obtained for the number of bits of the signature analyzer register equal to the number of binary input signals (inputs) of the digital device. Probabilities (7) - (9) take maximum values at $l = N/2$. The correctness of formulas is proved by the results of a simulation experiment with binary sequences [11, 12, 13, 14]. The maximum probability of failing to detect the failure by a combination of methods of controlling both the signature and the number of digits is significantly less than the probability of failing to detect the failure by signature control only.

The decision on the failure of the digital device is taken if the signal signature does not match the valid signature or the number of digits with the allowable number of digits. The valid signature and the allowable number of digits are calculated or determined experimentally.

The possibility of combining the control of both the signature control and the number of signal digits in order to detect the digital device failure, as justified in the paper, seems to be one of the possibilities of a compromise between conflicting requirements to limit troubleshooting costs and detect failures with high probability. A relatively small increase in the troubleshooting cost in terms of monitoring the number of signal digits results in a significant reduction in the probability of failing to detect the digital device failure due to the joint control of the signature and the number of digits. As a result, operational safety is increased and the system operation costs are reduced.

**References**

[1] Kashapov N F, Sabitov L S, Auhadeev A E, Litvinenko R S and Gatiyatov I Z 2019 Description of a complex technical system of urban electric transport from the standpoint of synergistic

methodology. *IOP Conference Series: Materials Science and Engineering*. Vol. 570, Iss. 1, № 012040

[2] Romanovich Zh. A. 2016 *Troubleshooting, repair and maintenance of control systems for household machines and devices*: textbook / Zh. A. Romanovich, V A Scriabin, V P Fandeyev and B V Tsypin. Dashkov and Co. Publishing Corp. p 316

[3] Filina O A and Tsvetkov A N 2019 Evaluation of the operational life of direct current motors IOP *Conference Series: Materials science and engineering* Vol. 489 № 012016

[4] Fandeyev V P 2001 *Technologies and means of failure analysis of recoverable electronic instruments*: monograph, Penza: Penza State University Press. p 240

[5] Sabitov L S, Kashapov N F, Gilmanshin I R, Gatiyatov I Z, and Kuznetsov I L 2017 Increase of reliability of contact networks of electric transport, due to increase of strength of the joint unit of pipes of different diameters *IOP Conference Series: Materials Science and Engineering*. Vol. 240, № 012058

[6] Kashapov N F, Sabitov L S, Litvinenko R S, Auhadeev A E and Gatiyatov I Z 2019 The approach to the study of the reliability of the electric transport system of the city as a complex technical system. *IOP Conference Series: Materials Science and Engineering*. Vol. 570, Iss. 1, № 012043

[7] Khusnutdinov A N, Rybakov R B and Allamova N M 2014 The use of iterative methods in the design of electric machines *Scientific problems of transport in Siberia and the Far East*. No. 1-2. pp 272-274

[8] Khusnutdinov A N, Khusnutdinova E M and Butakov B M 2019 Operation of brushes with increased service life on traction machines of electric vehicles *IOP Conf. Series: Materials Science and Engineering* Vol. 570, Iss. 1, № 012053

[9] Filina O A and Salnikova O.V. 2020 Construction of verification and diagnostic tests for the functional diagram of the object of diagnosis *IOP Conference Series: Materials Science and Engineering*. Vol. 747, № 012111

[10] Filina O A, Tsvetkov A N, Pavlov P P, Radu D and Butakov V M 2019 Vibration model as a system of coupled oscillators in a direct current electric motor *E3S Web Conf.* Vol. 124, № 02002

[11] Kashapov R N, Kashapov L N, Kashapov N F 2017 *Analysis and development of methods for obtaining methodic powders for selective laser melting* IOP Conference Series: Materials Science and Engineering, Vol. 240, Is 1, Art. № 012071

[12] Auhadeev A E, Idiyatullin R G, Pavlov P P, Butakov V M, Kisneeva L N and Tukhbatullina D I 2019 Improving the theory for calculating the rational modes of traction electrical equipment. *E3S Web of Conferences*. Vol. 124. № 05077

[13] Hizbullin Rob and Hizbullin Rad 2019 Ways to improve safety in the power industry: an automated hardware complex for pre-shift inspection of personnel of power enterprises *E3S Web of Conferences*. Vol. 124. № 05037

[14] Auhadeev A E, Litvinenko R S, Fandeev V P, Pavlov P P, Butakov V M and Litvinenko A R 2019 Research of reliability of urban electric transport system *E3S Web of Conferences*. Vol. 124. № 05078

IOP Conf. Series: Materials Science and Engineering **915** (2020) 012033 doi:10.1088/1757-899X/915/1/012033

# Designing the Fault-Detection and Troubleshooting Tests For the Troubleshooting Target Flowchart

**E M Khusnutdinova[1], O A Filina[2], A N Khusnutdinov[2], A V Yashagina[2], G V Osetinskiy[2], and N K Andreev[2]**

[1]Kazan (Volga Region) Federal University 18 Kremlyovskaya Str., Kazan 420008
[2]Kazan State Power Engineering University 51 Krasnoselskaya str., Kazan 420061, Russia

rr-088@mail.ru

**Abstract.** This paper covers the matter of troubleshooting in urban electric transport and electrical equipment. The solution to this issue requires transferring the qualitative determination of the health of the equipment to some quantitative basis. The formalization of qualitative definitions is a prerequisite for constructing formal (computable) troubleshooting algorithms.

Continuous systems of railway automation, telemetry and communications have features that allow their preference to be given to logic models. The use of logic models is associated with tolerance-based test methods, which are described by the conclusion about the equipment health is based on the evaluation of signal values at reference points (values of test parameters). The parameter test results in this case are reduced to the following estimates: "within tolerance - out of tolerance", "satisfactorily - not satisfactorily", in other words, to estimates of the two-digit type (zero or one) [1-3]. Hence, logic models and various logic methods appear to be useful in such cases.

At the first stage of building the logic model in the system, individual functional elements are distinguished whose inputs and outputs are available for measurement.

At the second stage of building the model, a flowchart of the system as a troubleshooting target is designed to show all selected elements and the connections between them.

The troubleshooting target flowchart (Fig. 1) consists of eight elements: *E1 – E8,* has four external input actions: $x_1 – z_4$, and produces four output responses: $y_5, y_6, y_7, y_8$. Each element produces its own output response $y_i$, while the output responses of the elements *E5, E6, E7, E8* match the output responses of the flowchart. Let us accept that $x_i$=1 and $y_i$=1, if *i*-th input action or output response of the *j*-th element is valid; otherwise $x_i$ = 0 and $y_i$ = 0.

We restrict ourselves to the case of considering only single faults, so the system has nine states:

$S_0$=11111111, $S_1$=01111111, $S_2$=10111111,
$S_3$=11011111, $S_4$=11101111, $S_5$=11110111,
$S_6$=11111011, $S_7$=11111101, $S_8$=11111110.

When handling the logic model, it is assumed that the input of the troubleshooting target receives a single input action, determined by the permissible values of all input signals.

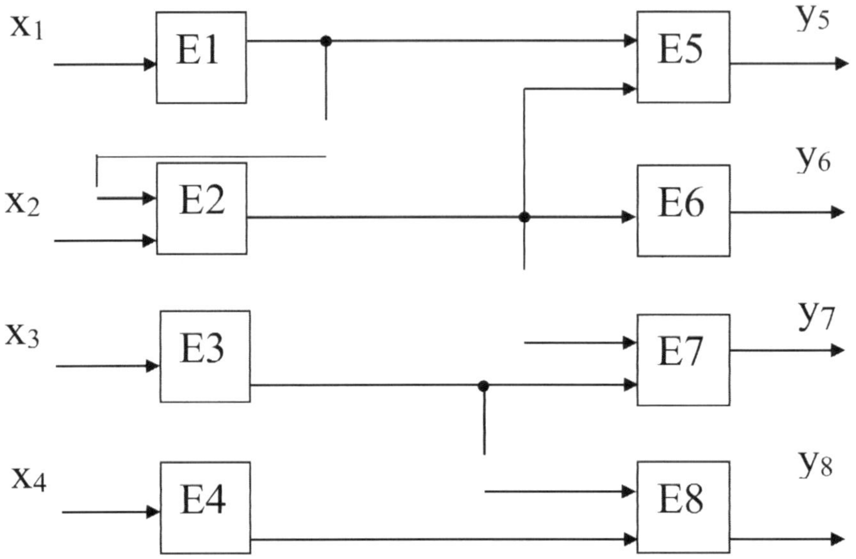


*Fig. 1 The Troubleshooting Target Flowchart*

Therefore, possible simple tests differ only in the reference points to measure. In this event, the purpose of buiding a troubleshooting algorithm is reduced to choosing reference points sufficient to solve a specific troubleshooting problem [4-6]. In practice, a large number of tests cannot be performed, since there is no access to the outputs of some elements; it is impossible to connect directly to the outputs of several elements, etc.

In the case in hand, it is as if only those tests are possible that consist in measuring the response at the output of one of the system elements, and the outputs of all elements are available for measurement. Let us denote the simple test as $\pi_i$ - this is the response control at the output of the i-th element ($i \in \{1, 2, ..., 7\}$).

Table 1 lists fault functions compiled for the system (see Fig. 1).

Table 1

| Test | Test result R for the system in state | | | | | | | | |
|---|---|---|---|---|---|---|---|---|---|
| | $S_0$ | $S_1$ | $S_2$ | $S_3$ | $S_4$ | $S_5$ | $S_6$ | $S_7$ | $S_8$ |
| $\pi_1$ | 1 | 0 | 1 | 1 | 1 | 1 | 1 | 1 | 1 |
| $\pi_2$ | 1 | 0 | 0 | 1 | 1 | 1 | 1 | 1 | 1 |
| $\pi_3$ | 1 | 1 | 1 | 0 | 1 | 1 | 1 | 1 | 1 |
| $\pi_4$ | 1 | 1 | 1 | 1 | 0 | 1 | 1 | 1 | 1 |
| $\pi_5$ | 1 | 0 | 0 | 1 | 1 | 0 | 1 | 1 | 1 |
| $\pi_6$ | 1 | 0 | 0 | 1 | 1 | 1 | 0 | 1 | 1 |
| $\pi_7$ | 1 | 0 | 0 | 0 | 1 | 1 | 1 | 0 | 1 |
| $\pi_8$ | 1 | 1 | 1 | 0 | 0 | 1 | 1 | 1 | 0 |

When the system is in the healthy condition (state $S_0$), the outputs of all elements have valid signal values [7]. Failure of any element causes an invalid signal value at its output and at the outputs of all associated elements.

This fault functions table contains all the data necessary for designing the fault-detection and troubleshooting tests [7,8]. Each table column sets a function defined on a set of tests. The function is equal to one if the test is satisfactory. Let $F$ be the function of the healthy item; $f_i$ is the function of the $i$-th state of the faulty item or the function of the $i$-th fault. For example:

$$F = \pi_1 \vee \pi_2 \vee \pi_3 \vee \pi_4 \vee \pi_5 \vee \pi_6 \vee \pi_7 \vee \pi_8,$$

$$f_1 = \pi_3 \vee \pi_4 \vee \pi_8, \qquad f_2 = \pi_1 \vee \pi_3 \vee \pi_4 \vee \pi_8,$$

$$f_3 = \pi_1 \vee \pi_2 \vee \pi_4 \vee \pi_5 \vee \pi_6, \qquad f_4 = \pi_1 \vee \pi_2 \vee \pi_3 \vee \pi_5 \vee \pi_6 \vee \pi_7,$$

$$f_5 = \pi_1 \vee \pi_2 \vee \pi_3 \vee \pi_6 \vee \pi_7 \vee \pi_8,$$

$$f_6 = \pi_1 \square\square\square_2 \square\square\square_3 \square\square\square_4 \vee \pi_5 \vee \pi_7 \vee \pi_8, \qquad f_7 =$$
$$\pi_1 \square\square\square_2 \square\square\square_3 \square\square\square_4 \square\square\square_5 \square\square\square_6 \vee \pi_8,$$

$$f_8 = \pi_1 \square\square\square_2 \square\square\square_3 \square\square\square_4 \square\square\square_5 \square\square\square_6 \vee \pi_7$$

When designing the test $T_{fd}$ for each fault, the fault-detection function is calculated:

$$\varphi_i = F \oplus f_i. \qquad (1)$$

The function $\varphi_i = 1$ only on those tests where the test results are different for healthy conditions and for a condition containing the i-th fault [9, 10]. In other words, it combines those tests on which the i-th fault is detected.

Fault-detection test

$$T_{fd} = \varphi_1 \varphi_2, \ldots, \varphi_n, \qquad (2)$$

where n is the number of faults.

In this case, n = 8.

In order to find $\varphi_i$, let us sum up the obtained values according to formula (1) to have:

$$\varphi_1 = \pi_1 \vee \pi_2 \vee \pi_5 \vee \pi_6 \vee \pi_7, \quad \varphi_2 = \pi_2 \vee \pi_5 \vee \pi_6 \vee \pi_7,$$

$$\varphi_3 = \pi_3 \vee \pi_7 \vee \pi_8, \qquad \varphi_4 = \pi_4 \vee \pi_8,$$

$$\varphi_5 = \pi_5, \qquad \varphi_6 = \pi_6, \quad \varphi_7 = \pi_7, \quad \varphi_8 = \pi_8$$

Then

$$T_{fd} = \varphi_1 \varphi_2 \varphi_3 \varphi_4 \varphi_5 \varphi_6 \varphi_7 \varphi_8 =$$
$$= (\pi_1 \vee \pi_2 \vee \pi_5 \vee \pi_6 \vee \pi_7)(\pi_2 \vee \pi_5 \vee \pi_6 \vee \pi_7)(\pi_3 \vee \pi_7 \vee \pi_8)(\pi_4 \vee \pi_8)\pi_5\pi_6\pi_7\pi_8 \qquad (3)$$

The equation (2) can be simplified based on the absorption law, which is described by the following equalities:

$$a(a \vee b) = aa \vee ab = a \vee ab = a(1 \vee b) = a \times 1 = a,$$

$$(a \vee b)(a \vee b \vee c) = a \vee b,$$

or in the general case

$$(G_1 \vee G_2)(G_1 \vee G_3) = G_1, \qquad (4)$$

where $G_1$, $G_2$, $G_3$ are any logic functions.

Using the actions:

$$\begin{aligned}
&1)(\pi_6)\times(\pi_6+\pi_7)=\pi_6\times\pi_6+\pi_6\times\pi_7=\pi_6\times(1+\pi_7)=\pi_6\\
&2)(\pi_6)\times(\pi_6+\pi_8)=\pi_6\times\pi_6+\pi_6\times\pi_8=\pi_6\times(1+\pi_8)=\pi_6\\
&3)(\pi_6)\times(\pi_4+\pi_6)=\pi_6\times\pi_4+\pi_6\times\pi_6=\pi_6\times(1+\pi_4)=\pi_6\\
&4)(\pi_1+\pi_5)\times(\pi_1+\pi_5)=\pi_1+\pi_5\\
&5)(\pi_1+\pi_5)\times(\pi_1+\pi_2+\pi_5)=\pi_1\times(\pi_1+\pi_2+\pi_5)+\pi_5\times(\pi_1+\pi_2+\pi_5)=\pi_1\times(1+\pi_2+\pi_5)+\pi_5\times(1+\pi_2+\pi_5)=\\
&=\pi_1+\pi_5
\end{aligned} \quad (5)$$

results in 2 fault-detection tests:

$$\begin{aligned}
T_{\Pi 1}&=\pi_6\times\pi_1\\
T_{\Pi 2}&=\pi_6\times\pi_5
\end{aligned} \quad (6)$$

It follows from the equation that the complete testing of the system requires just to simultaneously apply the permissible signals to the external inputs of elements 1 and 6 or 6 and 5 and measure the response at the output [11, 12, 13]. If the system is healthy, then the output of the element will be a valid signal, if it is faulty, then the output of the element will be an invalid signal.

In general, it is enough to verify all its external outputs to test the healthy condition of the equipent item [14, 15, 16]. However, the logic model and the fault functions table allow one to find such a minimal set of checks that will not include the external outputs of the object, which are also the inputs of the model blocks [17,18].

A troubleshooting test of Tt is designed at the time of solving the problem of detection of a faulty element. For each pair of faults (with numbers i and j), a discriminating function is calculated:

$$\varphi_{i,j} = f_i \oplus f_j \quad (7)$$

The discriminating function obtained by equation (7) is equal to one only in those tests where the test results are different for the conditions with the i-th fault and for the conditions with the j-th fault. In other words, it combines those tests on which the i-th and j-th faults differ from each other.

**References**

[1] Mikhailovsky A E and Yashagina A V 2018 Designing software suites for monitoring and troubleshooting of the condition of power facilities using the scale theory *In the collection: Nanomaterials and nanotechnology: issues and prospects. Proceedings of the VIII International Youth Scientific Conference*. pp 369-372.

[2] Pavlov P P, Fandeev V P, Aukhadeev A E, Litvinenko R S and Butakov V M. 2019 Technique for optimization of diagnostic parameters composition for power systems objects. *IOP Conference Series: Materials Science and Engineering*. Vol. 643, Iss. 1, № 012013

[3] Kashapov N F, Sabitov L S, Litvinenko R S, Auhadeev A E and Gatiyatov I Z 2019 The approach to the study of the reliability of the electric transport system of the city as a complex technical system. *IOP Conference Series: Materials Science and Engineering*. Vol. 570, Iss. 1, № 012043

[4] Khusnutdinov A N, Khusnutdinova E M and Butakov B M 2019 Operation of brushes with increased service life on traction machines of electric vehicles *IOP Conference Series: Materials Science and Engineering* Vol. 570, Iss. 1, № 012053

[5] Kashapov N F, Sabitov L S, Litvinenko R S, Auhadeev A E and Pavlov P P 2019 Consideration of the influence of the reliability of elements of the urban electric transport system on its bandwidth. *IOP Conference Series: Materials Science and Engineering*. Vol. 570, Iss. 1, № 012042

[6] Kashapov N F, Sabitov L S, Auhadeev A E, Litvinenko R S and Pavlov P P 2019 New methodological approach to the development of the theory of electric traction of urban electric transport. *IOP Conference Series: Materials Science and Engineering*. Vol. 570, Iss. 1, № 012041

[7] Kashapov N F, Sabitov L S, Auhadeev A E, Litvinenko R S and Gatiyatov I Z 2019 Description of a complex technical system of urban electric transport from the standpoint of synergistic

methodology. *IOP Conference Series: Materials Science and Engineering*. Vol. 570, Iss. 1, № 012040

[8] Auhadeev A E, Idiyatullin R G, Pavlov P P, Butakov V M, Kisneeva L N and Tukhbatullina D I 2019 Improving the theory for calculating the rational modes of traction electrical equipment *E3S Web Conferences* Vol. 124, № 05077

[9] Auhadeev A E, Litvinenko R S, Fandeev V P, Pavlov P P, Butakov V M and Litvinenko A R 2019 Research of reliability of urban electric transport system. *E3S Web of Conferences* Vol. 124, № 05078

[10] Khizbullin R N and Laryushin A I 2014 Automated medical hardware complex for pre-shift inspection of personnel of energy enterprises *News of higher educational institutions "Problems of Energy"* Iss.1-2 pp 125-133

[11] Khizbullin Rob N, Khizbullin Rad N 2019 Ways to improve safety in the power industry: an automated hardware complex for pre-shift inspection of personnel of power enterprises *E3S Web Conferences* Vol. 124, № 05037

[12] Filina O A and Tsvetkov A N 2019 Evaluation of the operational life of direct current motors IOP *Conference Series: Materials science and engineering* Vol. 489 № 012016

[13] Filina O A and Salnikova O.V. 2020 Construction of verification and diagnostic tests for the functional diagram of the object of diagnosis *IOP Conference Series: Materials Science and Engineering.* Vol. 747, № 012111

[14] Filina O A, Tsvetkov A N, Pavlov P P, Radu D. and Butakov V M 2019 Vibration model as a system of coupled oscillators in a direct current electric motor *E3S Web Conf.* Vol. 124, № 02002

[15] Safarov D T, Kondrashov A G and Khafizov I I 2018 The method of diagnosing machine systems by measuring the accuracy of manufactured parts. *IOP Conf. Series: Materials Science and Engineering* Vol. 412 № 012065

[16] Sabitov L S, Kashapov N F, Gilmanshin I R, Gatiyatov I Z, and Kuznetsov I L 2017 Increase of reliability of contact networks of electric transport, due to increase of strength of the joint unit of pipes of different diameters *IOP Conference Series: Materials Science and Engineering*. Vol. 240, № 012058

[17] Kashapov R N, Kashapov L N, Kashapov N F Formation of cracks in the selective laser melting of objects from powdered stainless steel 17-4 HP (2017) IOP Conference Series: Materials Science and Engineering, Vol. 240, Is 1, Art. № 012074

[18] Kashapov L N, Kashapov N F, Kashapov R N, Semushin V D Changes of thermal balance in a plasma-electrolyte system according to the shape of the applied voltage (2016) Journal of Physics: Conference Series, Vol. 669, Is. 1, Art. № 012028

IOP Conf. Series: Materials Science and Engineering **915** (2020) 012034 doi:10.1088/1757-899X/915/1/012034

# Influence of nonlinear viscosity of waxy crude oils on the thermohydraulic characterization of their flow in helical channels: theory and experiment

**E M Khusnutdinova, I A Konakhina**

Kazan (Volga Region) Federal University 18 Kremlyovskaya Str., Kazan 420008

rr-088@mail.ru

**Abstract:** The authors looked into a matter of how the nonlinear viscosity of waxy crude oils has an effect on the thermohydraulic characterization of their flow in helical symmetry channels. The consideration was given to the energy equation in a helical coordinate system followed by theoretical and experimental values.

The purpose of this paper is theoretical and experimental study of intensified heat transfer during in the flow of a coolant with pseudoplastic rheological characteristics in pipes with a triple-thread screw insert. This work is especially important for fluids characterized by high effective viscosity parameters, since low flow rates cause low thermal efficiency of heat exchangers, where these fluids are used as operating fluids.

The triple-thread screw insert is an intensifier inducing the entire flow in the pipe cross section and can significantly increase the efficiency of the heat transfer process in the laminar flow [1]. The behavior of pseudo-plastic fluids (including waxy crude oils) when flowing in sophisticated channels is different from the behavior of Newtonian viscous fluids. This is expressed by the nonlinear dependence of the effective viscosity on shear rate. Consequently, the study of the heat transfer process requires analysis of how the rheological properties of the liquid influences the hydrodynamic and thermal characteristics of the flow [2, 3].

A mathematical model was built to perform a desktop study of the heat transfer process during the flow of a nonlinearly viscous fluid in screw insert pipes. Mathematical transformations required the tensor and vector calculus to be used. Here, the value of the material function $\varphi_1$ during the flow of a nonlinearly viscous fluid at each point of the area of interest is characterized by a complex dependence on the second invariant of the strain rate tensor. $I_2$The determination of this functional dependence requires an experimental study of the rheological properties of the fluid (waxy crude oil). The most convenient is the form of material functions of $I_2$as a power law [3, 4]:

$$\varphi_1 = \varphi_{10}\,(I_2)^n; \tag{1}$$

where $\varphi_{10}$ – is the value of the material function for $I_2 \to 0$.

To represent the temperature dependence $\varphi_1$, experimental dependences of the main rheological parameter $\varphi_{10}$ are used:

$$\varphi_{10} = A_{10} \exp\left(-\frac{B_{10}}{RT}\right); \tag{2}$$

where $A_{10}$ – is the pre-exponential factor;
R – is the gas constant;
$B_{10}$ – is the activation energy of the flow.

The composition of the energy equation in a helical coordinate system uses a dot product $\overline{V} \cdot \text{grad}\, t$ and an operator $\nabla^2 t$, which is given by:

$$V^1 \frac{\partial t}{\partial q^1} + V^2 \frac{\partial t}{\partial q^2} + V^3 \frac{\partial t}{\partial q^3} = a\left[\frac{1}{q^1}\frac{\partial}{\partial q^1}\left(q^1 \frac{\partial t}{\partial q^1}\right) + \frac{A}{(q^1)^2}\frac{\partial^2 t}{(\partial q^2)^2}\right] + \frac{\Phi}{\rho C_p}, \tag{3}$$

where $\Phi = T^0 : D$ is the dissipative function.

By replacing the velocity component $V^1$ и $V^2$ with functional relationship of the current function ψ, the equation (3) yields as follows:

$$\frac{1}{q^1}\frac{\partial \psi}{\partial q^2}\frac{\partial t}{\partial q^1} - \frac{1}{q^1}\frac{\partial \psi}{\partial q^1}\frac{\partial t}{\partial q^2} + V^3 \frac{\partial t}{\partial q^3} = a\left[\frac{1}{q^1}\frac{\partial}{\partial q^1}\left(q^1 \frac{\partial t}{\partial q^1}\right) + \frac{A}{(q^1)^2}\frac{\partial^2 t}{(\partial q^2)^2}\right] + \frac{\Phi}{\rho C_p}. \tag{4}$$

The single-valued condition for solving the energy equation were taken as:
1. Boundary conditions for temperature:

$$C_1 \lambda \frac{\partial t}{\partial n}|_{\Gamma} + C_2 t|_{\Gamma} = C_3, \tag{5}$$

where the coefficients $C_1, C_2, C_3$ take values:
1) $C_1 = 0; C_2 = 1; C_3 = t_{\Gamma}$ – for thermal boundary conditions of the first kind;
2) $C_1 = -\lambda; C_2 = 0; C_3 = q(q^3)$ – for thermal boundary conditions of the second kind
3) $C_1 = \lambda; C_2 = \alpha; C_3 = \alpha t$ – for thermal boundary conditions of the third kind
2. The initial condition at the channel inlet:

$$t|_{q^3=0} = t_0 = const.$$

The primary characteristic that determines the intensity of heat transfer processes is the heat transfer coefficient $\alpha$ [5]. The temperature distribution as a result of the energy equation allows determination of the local and average heat transfer coefficients throughout the length of the pipe.

The heat transfer coefficient in each section of the channel is determined from the ratio, W/(m²·K):

$$\alpha_M = \frac{\frac{\partial t}{\partial n}|_{\Gamma}}{\bar{t} - t_{\Gamma}}, \tag{6}$$

where $\bar{t}$ – is the average temperature in section, °C;

$$\bar{t} = \frac{1}{F}\iint_{\Omega} t \, d\Omega. \tag{7}$$

The average heat transfer coefficient throughout the length of the pipe is determined from the ratio:

$$\bar{\alpha} = \frac{1}{L}\int_0^L \alpha_л \, dq^3, \tag{8}$$

IOP Conf. Series: Materials Science and Engineering **915** (2020) 012034 doi:10.1088/1757-899X/915/1/012034

where $L$ – is the channel length, m.

The adequacy of the resulted mathematical model was verified by conducting a series of experiments using a trial bench. The thermal boundary conditions of the first kind were provided at a temperature on the wall of channel $t_r = 70$ °C. In the course of the experiments, the constant flow rate of the operating fluid was controlled under the condition $\mathrm{Re} \leq 800$. Processing of experimental data and their subsequent comparison with similar data of the theoretical study showed satisfactory agreement of the results [6,7]. The difference was not more than 15%.

The considered mathematical model of heat transfer in screw insert pipes makes it possible to generalize the results of theoretical and experimental studies of the rheological properties of liquids [8]. In turn, the application of the designed calculation methods of intensified heat transfer allows reducing the metal consumption and dimensions, as well as the temperature of the walls of promising heat exchangers with intensified heat transfer.

**References**

[1] Konakhina I A, Khusnutdinova E M, Kadyirov A I and Vachagina E K 2018 *Hydraulic resistance of channels with triple-thread screw inserts in laminar flows.* The IX International scientific and technical conference «Innovative engineering technologies, equipment and materials – 2018», part 1. pp. 326-330

[2] Khusnutdinova E M, Konahina I A, Khamidullina G R and Vachagina E K 2018 *Comparative analysis of hydraulic losses in long oil pipelines with forming of channel cross-sections.* The IX International scientific and technical conference «Innovative engineering technologies, equipment and materials – 2018», part 1. pp. 408-411

[3] Nazmeyev Yu G 1996 *Flow dynamics and heat transfer of swirling flows of rheologically complex fluids.* (M. Energoatomizdat) p 304

[4] Kadyirov A I, Khalitova G R and Vachagina E K 2016 *The hydrodynamic structure of a viscous flow in ducts with intensifiers as auger inserts.* Papers of Akademenergo. No. 3. pp. 7–16

[5] Bergles A E, Jensen M K, Somerscales E F C and Manglik R M 1991 *Literature review of heat transfer enhancement technology for heat exchanges in gas-fired applications*. Report GRI 91–0146, Chicago, Gas research institute

[6] Kashapov L N, Kashapov N F, Kashapov R N, Semushin V D *Changes of thermal balance in a plasma-electrolyte system according to the shape of the applied voltage* (2016) Journal of Physics: Conference Series, Vol. 669, Is. 1, Art. № 012028

[7] Mamonova A O, Plugatyreva O A, Khusnutdinova E M and Khusnutdinov A N 2019 *Features of very heavy crude oil transportation* IOP Conf. Series: Materials Science and Engineering Vol. 570, Iss. 1, № 012067

[8] Kashapov N F, Kashapov R N, Kashapov L N 2018 *Influence of the electrolytic cathode temperature on the self-sustaining mechanism of plasma – electrolyte discharge.* Journal of Physics D: Applied Physics, Vol. 51, Iss.49, № 494003

IOP Conf. Series: Materials Science and Engineering **915** (2020) 012035 doi:10.1088/1757-899X/915/1/012035

# Improving the performance of cylinder liners by creating thin -layer coatings in the interface

**A T Kulakov[1], E P Baryshnikova[1], S V Snarsky[2], A A Gafiatullin[2], F L Nazarov[2] and A F Malachowski[3]**

[1] Kazan Federal University, Naberezhnye Chelny Institute, Russia, 10A, Syuyumbike avenue, 423800, Russian Federation, Naberezhnye Chelny,
[2] The engine plant of OJSC KAMAZ, 423800, Russian Federation, Naberezhnye Chelny, Avtozavodskiy prospect, 2.
[3] JSC "Remdiesel", 423800, Russian Federation, Naberezhnye Chelny, Menzelinsky tract, 40

alttrak09@mail.ru, 692401@mail.ru

**Abstract**. The paper presents the technological process of brass plating of the cylinder bore. Application of the coating allows to increase the operational reliability of parts of the cylinder-piston group and the technical life of the entire engine

Failures of the internal combustion engine in operation are shown in the form Natarov and scuffing of the cylinder group. The working conditions of the cylinder-piston group are significantly affected by the ovality of the sleeve, due to the design features of the head attachment to the cylinder block and the gas joint [1]. The ovality and non-cylindrical shape of the sleeve is formed during Assembly and is preserved during operation of internal combustion engines.

The cylinder liner is brought to the standard parameters for cylinder shape, ovality, and taper when manufactured at the KAMAZ engine plant, but its shape is distorted after installation in the cylinder block and tightening the cylinder head bolts, deviations reach ±0.02...0.05 mm from the nominal size. The ovality of the sleeve is aggravated during installation (Fig. 1) due to a breach of the bolt torque, and asymmetrical arrangement of the mounting bolts to the cylinder head. When four bolts with the same torque are tightened due to uneven positioning, the larger axis of the oval is located in sectors *A* and *B*, and the smaller axis is located in sectors *C* and *D* [2, 3].

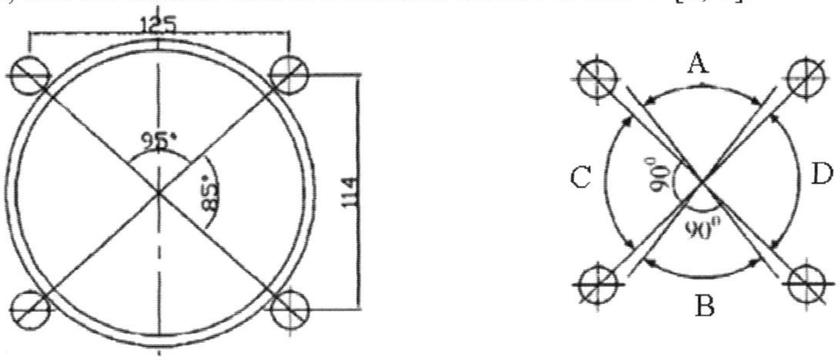

IOP Conf. Series: Materials Science and Engineering **915** (2020) 012035 doi:10.1088/1757-899X/915/1/012035

Fig. 1. Asymmetry of the location of the cylinder head mounting bolts and the shape deviation sectors (maximum (*A* and *B*) and minimum (*C* and *D*)

Physical and chemical, anti-wear and anti-pressure properties of lubricants, properties of the surfaces of sleeves and rings often can not prevent intensive wear and destruction of rubbing surfaces, expressed in the form of rubbing, teasing, cauterization on sleeves and rings (Fig. 2).

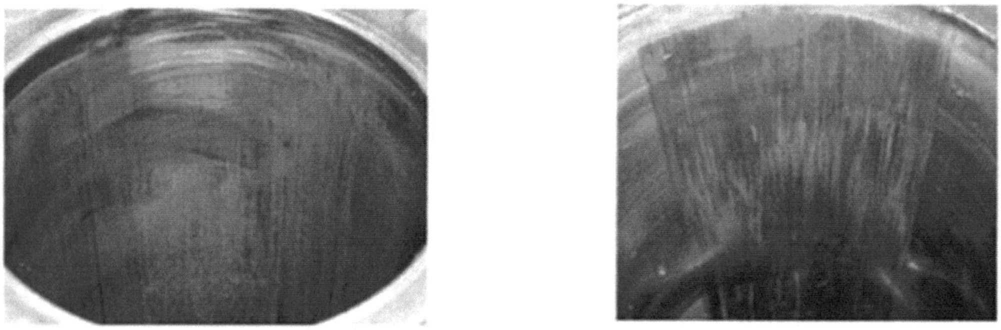

a b

Fig. 2. Rubs (a) and teasers (b) on the working surface of the sleeve

To improve burn-in, the inner surface of the sleeve is lubricated with spray and oil mist, sometimes it can be covered with phosphating or porous chromium.

The effect of selective transfer during friction has a positive effect, when a metal-coating film is spontaneously formed on rubbing surfaces in the contact zone [4].

Thanks to the creation of a metal-coating film that separates the rubbing surfaces even at high loads, the wear of parts under selective transfer conditions can be reduced to zero ( the wear-free effect), and the coefficient of friction can be reduced to values under liquid friction.

Selective transfer conditions for reducing the wear of rubbing surfaces are usually created using simple methods:

- - introduction of copper- containing additives into the oil;
- applying a layer of non-ferrous metal (brass, bronze, copper) to the work surfaces of parts);
- installation of inserts from non- ferrous metals in machine parts.

In KAMAZ engines with an unloaded gas joint on the cylinder liner , it is advisable to apply a self-working coating that adheres well to the cylinder surface (Fig.3), and when working due to the removal of the rings creates a form that ensures complete fit of the entire surface of the rings to the sleeve, and thus provides good burn-in characteristics and work in conjunction with the compression rings and the piston without rubbing and teasing [5].

IOP Conf. Series: Materials Science and Engineering **915** (2020) 012035 doi:10.1088/1757-899X/915/1/012035

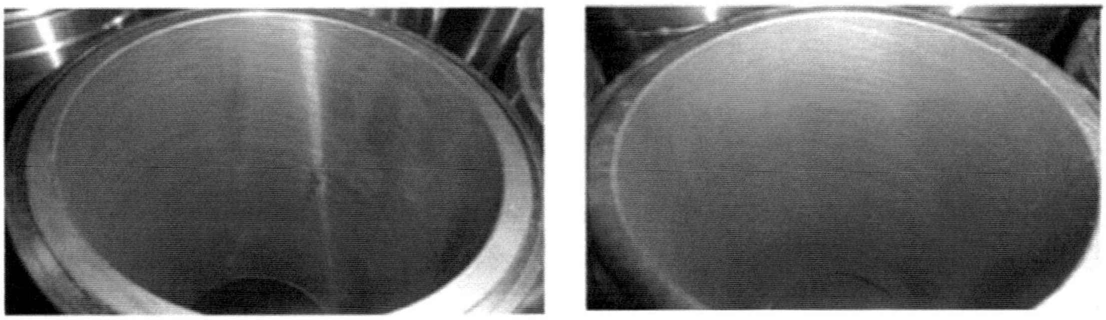

a b

Fig. 3. The Shell before (a) and after (b) brass plating

Joint research by PJSC «KAMAZ», VNPO «Remdetal», NIIAT and KFU obtained results on the impact of finishing antifriction-free treatment with brass in the form of tribolatuning (FABO) and surface-plastic deformation (FABO PPD) on the wear and tear resistance of the cylinder liner - piston ring pair of the KAMAZ-740 engine [6].

The coating was obtained using LMC 58-2-2 brass (Cu 57-60%, Mn 1.5-2.5%, Pb 1.5-2.5%, the rest is zinc) [7]. Rollers are made of brass rods for installation in a modified honing head (Fig. 4).

a b c

Fig. 4. The process of latuning on the honing machine SS6101 , converted for the process of latuning cylinder liners (a), the device for installing the sleeve (b), the honing head with installed rollers for latuning cylinder liners (c)

As an activating solution, a process liquid is used, which is a mixture of glycerol with metal chlorides and in one liter contains the following components (g/l): zinc chloride ($ZnCl_2$) 24 g; copper chloride ($CuCl_2$) 10 g; tin dihloride ($SnCl_2$) 16 g; iron chloride ($F_2Cl_3$) 14 g; urea ($Co(NH_2)_2$) 4 g; technical glycerine ($C_3H_5(OH)_3$) 200 g; drinking water. The materials used are standardized, produced by industry in mass quantities, and are relatively cheap.

The layer of applied metal at FABO (8-10 microns) was very effective in the initial period of operation and allowed to exclude cases of unsatisfactory burn-in of the cylinder-piston group during acceptance tests.

A special advantage of the internal combustion engine with this treatment of the cylinder liner is that the layer does not require final processing and can be carried out during installation in any conditions when assembling a new engine and repairing it.

IOP Conf. Series: Materials Science and Engineering **915** (2020) 012035 doi:10.1088/1757-899X/915/1/012035

During bench tests of engines for failure-free operation for 1000 hours, it was found that the average wear of latuned sleeves in the upper dead center zone is 30% less than that of uncoated sleeves.

The wear intensity of the raw sleeve processed by the PPD-FABO and FABO methods is two orders of magnitude higher ($2 \cdot 10^{-10}$) when tested on a friction machine compared to the sleeve processed by the PPD-FABO and FABO methods ($1 \cdot 10^{-12}$ and $4 \cdot 10^{-12}$) (table. 1).

Table 1 - Ehe wear rate of the liners and chrome -plated rings

| A method of brass plating raw shells | Wear rate | |
|---|---|---|
| | sleeve | ring |
| Crude | $2 \cdot 10^{-10}$ | $8 \cdot 10^{-11}$ |
| PPD-FABO | $1 \cdot 10^{-12}$ | $4 \cdot 10^{-11}$ |
| of FABO | $4 \cdot 10^{-12}$ | $5 \cdot 10^{-11}$ |

Using the liner coating on «Remdiesel», it was possible to eliminate grates and teasers during the burn -in process and get an improvement in operational parameters [8].

The obtained results of samples with brass coatings applied in two different ways (using the FABO and FABO PPD method ) during laboratory and bench tests allow us to recommend processing in the production of KAMAZ engine cylinder liners.

**References**

[1] Internal combustion engines: Design and strength calculation of piston and combined engines / Under the editorship of A. S. Orlin, M. G. Kruglov. - M.: Mashinostroenie, 1984 – 384 p.

[2] Kulakov A. T. Features of construction , operation, maintenance and repair of power units of trucks: studies. manual / A. T. Kulakov, A. S. Denisov, A. A. Makushin. - Moscow: Infra-Engineering, 2013. – 448 p.

[3] Kulakov A. T. Evaluation of performance and residual resource of the piston rings of KAMAZ engines that were in operation / A. T. Kulakov, M. I. Mistrikov / / Automotive industry. - 1995. - No. 12-Pp. 16-20.

[4] Garkunov D. N., Polyakov A. A. Increasing the wear resistance of aircraft structural parts . Moscow: Mashinostroenie, 1978. – 234 p.

[5] Kulakov A. T., M. I. Mastrikov Friction brass liners // Car industry. - 1990. No. 3. - P. 33.

[6] Kulakov A. T., Burumkulov Kh., Andreeva A. G., Mistrikov M. I., Glazova I. V., and Khotenko A. A., Bezmotornye bench tests of the coupling sleeve-ring of the KAMAZ-740 engine / / Vestnik mashinostroeniya. - 1992. No. 3. - Pp. 13-16.

[7] Patent no. 2570683 Russian Federation, IPC S23C 26/00. Method of processing the cylinder liner with the creation of antifriction coating / Kulakov A. T. Yakubovich I. A., Yakubovich A. N., Malakhovetsky A. A., Makushin A. A., Malakhovetskaya A. N., Luzhbin D. S.; patent holders Kulakov A. T. Yakubovich I. A., Yakubovich A. N., Malakhovetsky A. A., Makushin A. A., malakhovetskaya A. N., Luzhbin D. S.; - no. 2013111140/02 ; application. 13.03.2013; publ. 10.12.2015, Byul. No. 34 – 6 sec.

[8] Kulakov A. T. Increase in wear resistance and resource of diesels by creating thin-layer coatings in the coupling "sleeve-ring" during repair / A. T. Kulakov, A. A. Makushin / / Motor transport enterprise. - 2008. - No. 7. - Pp. 40-41.

IOP Conf. Series: Materials Science and Engineering **915** (2020) 012036 doi:10.1088/1757-899X/915/1/012036

# Innovations in the management system of industrial enterprise to improve the efficiency of its activities

**I A Kuzmina[1] and I I Khafizov[2]**

Kazan Federal University, 18 Kremlyovskaya street, Kazan, 420008, Russian Federation

iranina@mail.ru

**Abstract.** The modern competitive advantages of enterprises increasingly depend on the ability of employees to develop and implement innovations, and not on the degree of ownership of material assets and capital resources, which is the basis of economic growth based on scientific and technological progress and leads to an expanded attraction of investment in intellectual capital. Innovation activity is the basis for the development of industrial enterprises, and also makes it possible to increase the competitive advantages.

The activity of industrial enterprises is the basis of innovation processes, the activity of which is expressed in a variety of forms of development and implementation of organizational and managerial innovations.

Enterprise management must be understood as a new management technology. “Innovations” are created by reforming existing management systems, processes, and technologies.

Organizational innovations contribute to improving the efficiency of the enterprise through the development of advanced technological processes and methods of organizing work.

Management innovations represent the development of the functioning methods of the management apparatus, the improvement of management processes and technologies.

Organizational and managerial innovations are aimed at the development of enterprise management system, and their main goal is to increase the efficiency of functioning, improve entrepreneurial activity, and increase the competitiveness of products.

Currently, the successful functioning and economic growth of any enterprise is closely related to the effectiveness of innovative activities, which increases the profitability of the enterprise and provides it with significant advantages [1].

Effective management of the innovation process is carried out on the basis of a combination of financial, economic, managerial, technical, organizational, legal, moral and psychological information factors and their relationship, which contributes to the successful implementation of innovative activities and increase the efficiency of the enterprise.

A prerequisite for the effective implementation and development of innovation activity in industrial enterprises is to provide it with qualified personnel, for this it is necessary to develop and implement a training system that includes a set of courses that provide knowledge of various and depths, requirements for employees and a test system for testing knowledge [2].

To improve the quality and effectiveness of information support of innovative activities, enterprises are recommended to develop and implement the following measures: improving the information culture of personnel; improving the quality of work with information; increasing information security.

Using innovations will help industrial enterprises accelerate their growth, develop new markets, create new jobs, ensure full employment of staff, flexibility in the use of resources and compliance with market requirements [2].

Innovation activity is able to provide increased production efficiency, its competitiveness and the overall development of the structures and enterprises themselves. Thus, all that is associated with innovative activity in the industry is of great importance [4].

A well-organized mechanism for introducing innovative processes is the basis for the success of industrial enterprises.

Ensuring the innovative orientation of industrial enterprises in modern market conditions will help the efficiency and balanced functioning of the markets as a whole and increase the consumption of manufactured products. Therefore, innovation can be considered the key to sustainable economic development.

The types of activities divide management into a general of enterprise, management of the main and service production, its technical training, economic and organizational, innovative and investment management [3].

Management is understood not only as a form (system, hierarchy), but also as content, since forms are created for content (for liquid dishes, for car cargo, for metal working, metalworking machine, for woodworking, etc.).

Properly implemented innovation policy by an industrial enterprise increases its competitiveness in the market.

The application of an integrated approach to innovation at enterprises of the industrial sector will lead to the dynamic development of the enterprise, strengthening competitiveness, and strengthening marketing policy [5].

Managers and specialists of production divisions agree with the fact that design and technological activity should become much more market-oriented and it should be extended to the entire life cycle of a new product, which, as you know, includes, along with its design and production, implementation. Management should also be consistent with this development of innovative activity.

Market-oriented management fully realizes itself only when the process becomes its object, or, in other words, when functional management enters into interaction with process management. These two types of management in unity, and as a whole, acting as innovative management, ensure its quality.

The level of management rises at enterprises that create unified scientific and technical structures in the form of design and technological departments, experimental design bureaus as part of enterprises with expanded powers or experimental design bureaus at enterprises providing them with broad independence. With the transition to such structures in the field of innovation, management mobility increases.

Then the design and technological structures of industrial enterprises due to the development of the market component of their activities will provide themselves with the solution of a number of very important tasks:

– personal and social needs of specialists engaged in new developments are met;

– a higher level of organization of labor of all categories of workers and production is achieved, which is provided as a condition for creating a competitive product;

– the pace of introducing innovations into practice is accelerating, since materializing into a market product, provided with the demand, they find quick and profitable use;

– the possibility of increasing of labor productivity and reduction at this enterprise and at enterprises where innovations and production costs will be implemented increases; this is achieved due to the expansion of types of competitive innovative developments and their introduction into production.

IOP Conf. Series: Materials Science and Engineering **915** (2020) 012036 doi:10.1088/1757-899X/915/1/012036

**References**

[1] Safronov D M 2014 *To the question of evaluating the effectiveness of innovative activities of an enterprise.* Siberian Financial School, No. 2 (103). pp. 83-87

[2] Shevelev V V 2015 *Improving of the management mechanism of an industrial enterprise on the basis of the formation of innovative infrastructure.* Creative Economy, T. 9. No. 5. pp. 587-600

[3] Levshtanova Y Y 2018 *Financial model of investment of an innovative project* / Y Y Levshtanova, I I Khafizov // Siberian financial school, No. 1 (126). pp. 70-75

[4] Kashapov N F, Khafizov I I, Nurullin I G 2018 *Influence of introduction of robotics on increase in efficiency of electrochemical production* IOP Conf. Series: Materials Science and Engineering Vol. 412, Is.1, Art. № 012034

[5] Khafizov I I, Nurullin I G, Sadykov Z B 2018 *Problems of development of electrochemical production of Russia and possibility of their decision* IOP Conf. Series: Materials Science and Engineering Vol. 412, Is.1. Art. № 012042

IOP Conf. Series: Materials Science and Engineering **915** (2020) 012037 doi:10.1088/1757-899X/915/1/012037

# The use of mathematical models to improve the effective functioning of enterprises

**I A Kuzmina**[1]

Kazan Federal University, 18 Kremlyovskaya street, Kazan, 420008, Russian Federation

iranina@mail.ru

**Abstract.** In the process of ensuring the innovative development of machine-building enterprises within the framework of the concept, it is necessary to use mathematical models that correspond to machine vision technologies, robotic technology, automated and intelligent production and control systems at enterprises. Mathematical models of robotic technology allow us to create new equipment, improve technological processes and production organization systems.

Various enterprises must be reliable in terms of production safety and the effectiveness of their functioning. It is necessary to strictly comply with the technological process of production, control and prevention of emergency situations, stable operation of the equipment. The effectiveness of the enterprise involves the stability of the enterprise and minimize risks. All this implies a mandatory preventive repair plan, which allows you to follow the progress of repair works, both scheduled and extraordinary, as well as to ensure the priority of those that are directly related to product safety. In addition, the company needs control of incoming and outgoing flows of raw materials, identification of risks associated with the production of a particular product, the likelihood of this risk occurring, and so on.

All of this can be achieved experimentally, but it will require a long time and serious costs [1]. Time and costs can be significantly reduced if mathematical modeling techniques are used to increase the efficiency of production functioning.

Depending on the type of problem being solved, various topologies (structures) and types of neural networks are used. Machine vision technologies based on neural networks are widely used for diagnostics and quality control in industry, in particular in mechanical engineering, and have a lot of advantages over such classic models as gradient, statistical, local adaptation, template, etc. The key advantages of neural networks over other mathematical models are self-learning, fault tolerance and speed of work.

Used in conjunction with robots, machine vision systems can significantly expand the range of tasks in production [8]. Among the tasks of the "vision + robot" system we can distinguish the movement of products, loading / unloading, classification of objects, rejection of products, inspection of objects from all angles.

The advantages of using of machine vision for robots are: robot control during production, accuracy of work, comprehensive control of products, eliminating of need to buy high-precision equipment, preventing accidental collisions of robots, the ability to process various objects without complex reconfiguration, etc.

Machine (technical) vision systems allow you to automate the management of production processes by analyzing visual information and controlling products. Industrial video cameras are used to create images. The software of machine vision systems analyzes what they see and passes this information to the operator, automated control system, robot or actuators to control the production process. The special effect of machine vision systems is manifested when the complexity, volume or speed of the analyzed information significantly exceed the abilities of the operator.

In the process of developing and implementing such systems, stochastic geometry methods, neural network technologies, regular grid interpolation models, etc. are widely used [5].

To solve various problems in mechanical engineering, robots, mechatronic and robotic systems are more often used. This is due to their growing functionality, due to the use of more advanced control systems, the development of which is based on the well-known achievements of computer technology [9].

At machine-building enterprises, automated systems for the production and assembly of parts and machines, automated process control systems, which, as they develop, take the form of intellectual production systems, are widespread now [2].

The mathematical model is the basis for the functioning of these elements,

and modeling is carried out using computer technology, since it requires a significant amount of computation [5].

The wide range of mathematical models is used to model intelligent manufacturing systems [3].

To solve the problems of quality control, tracking the movement of products, monitoring the availability of objects, measuring their geometric dimensions, comparing with a sample, counting, identifying and classifying objects, rejecting products, inspecting objects from different angles, and highly accurate measurements of elements (for example, blanks or individual parts) etc. in automated systems in production, stochastic geometry models, interpolation models on regular lattices, cluster analysis methods, and artificial neural network models can be used [7].

The results of such innovations can be an improvement of working conditions, an increase in product quality, a reducing the need for labor force and a systematic increase in the profits of engineering enterprises [6].

The most promising models for implementation at engineering enterprises are models of artificial neural networks, which is associated with a wide range of possible tasks, which includes modeling of machine vision, mechatronic and robotic systems, production automation tasks, and intelligent production systems.

It should be noted that the widespread dissemination of high technology has opened access to vast amounts of information. The data stream is constantly growing. But the more information is collected, the more difficult it is to see in it trends and patterns hidden from a superficial view. It is important in these conditions to be able to quickly and timely find useful information and effectively use it. Improving competitiveness is almost impossible to achieve randomly. Therefore, a set of methods and techniques is needed that form an innovative competitiveness management system. The application of methods of economic and mathematical modeling has expanded thanks to modern PC software [4].

Attracting innovative mathematical methods and models is especially important for improving the scheduling of production processes, for example, with the project method of organization of production. This task is associated with the allocation of limited resources for project operations. Often the task of optimal allocation of resources in a project (set of operations) is called the task of scheduling. Note that such a task refers to complex multi-extreme or combinatorial optimization tasks.

Features of mathematical models of industrial production depend on the features of the structure, composition and functioning of the enterprise systems.

IOP Conf. Series: Materials Science and Engineering **915** (2020) 012037 doi:10.1088/1757-899X/915/1/012037

Currently developed models and methods for description and analysis of the main processes at engineering enterprises cover various engineering, economic and software aspects.

**References**

[1] Nikonorov L V 2012 *To the issue of increasing of the efficiency of production activities of an industrial enterprise* / L V Nikonorov // Bulletin of the A.S. Pushkin Leningrad State University. Series Economics, Issue. 3. pp. 145-158

[2] Nikonorov L V 2012 *Modeling of the industrial enterprises management: the monograph* / L V Nikonorov. Norilsk: Norilsk industry. Institute. p 182

[3] Sadovskaya T G, Drogovoz P A, Dadonov V A, Melnikov V I 2009 *The application of mathematical methods and models in the management of organizational and economic factors of the competitiveness of an industrial enterprise.* Audit and financial analysis, No. 3. pp. 364 - 379

[4] Kuzmina I A 2018 *On the Geometric Singularities of Surfaces of Pseudo-Euclidean Space* / I A Kuzmina, P Peška // INTERNATIONAL ELECTRONIC JOURNAL OF GEOMETRY VOLUME 11, NO. 2. pp. 104-110

[5] Kuzmina I A 2016 *The group of invertible elements of the algebra of quaternions* / I A Kuzmina, M Chodorová // Acta Universitatis Palackianae Olomucensis, Facultas rerum naturalium, Mathematica Vol. 55, №1. pp. 53-58

[6] Kuzmina I A 2019 *Innovations in the management system of an industrial enterprise to increase the efficiency of its activities* / I A Kuzmina, I I Khafizov // Materials of the X International Scientific and Technical Conference "Innovative Engineering Technologies, Equipment and Materials - 2019" (ISTC "IMTOM-2019") Kazan Part 1. pp. 258-261

[7] Kashapov R N, Kashapov L N, Kashapov N F 2017 *Analysis and development of methods for obtaining metallic powders for selective laser melting* IOP Conf. Series: Materials Science and Engineering Vol. 240, Art. № 012071

[8] Kashapov L N, Kashapov N F, Kashapov R N 2014 *Research of cobalt chromium alloy surface morphology after sandblasting* IOP Conf. Series: Materials Science and Engineering Vol. 69, Art. № 012017

[9] Kashapov R N, Kashapov L N, Kashapov N F 2017 *Research of plasma–electrolyte discharge in the processes of obtaining metallic powders* // Journal of Physics: Conf. Series. Vol. 927, Art. № 012086

IOP Conf. Series: Materials Science and Engineering **915** (2020) 012038 doi:10.1088/1757-899X/915/1/012038

# Development of apparatuses based on kinetic calculation data of the process of removing liquid from solid materials by reducing pressure of vapor-gas medium

**V A Lashkov[1], S G Kondrasheva[1], D A Khamidullina[1] and E S Karataeva[2]**

[1] Kazan National Research Technological Univesity, Kazan city, Karl Marx str, 68, Russia
[2] Kazan Federal University, engineering institute, Kazan city, Saidashev street, bld.12, Russia

valashkov1959@yandex.ru

**Abstracr.** The possibility of developing a apparatus based on the results of kinetic calculations of the processes, occurring in it, is considered. The studies were carried out with the system solid-liquid-vapor-gas medium with a continuous increase in vacuum above the surface of a layer of the dispersed material. The presence of the liquid phase and the vapor-gas phase in equilibrium with it makes it possible to provide a change in concentration, structure, or cooling in the system due to artificial disturbance of the dynamic equilibrium of the system. To develop a mathematical model of the liquid removal process, a model of ideal mixing in the vapor-gas phase and a layer of the dispersed material was adopted. The calculation results are presented in the form of kinetic dependences of the main process parameters. Calculation of the apparatus is carried out at the highest load in the vapor-gas phase during the period of removal of the free moisture. In this case, the influence of rarefaction on the physical characteristics of the removed fluids and the design features of the apparatus being calculated was taken into account.

**General information.** The technological processes in the systems with a liquid phase can be classified into three groups: liquid-solid-vapor-gas medium; solid-liquid-vapor-gas medium; liquid-gas medium. The presence of the liquid phase and the vapor-gas phase, which is in equilibrium with it, makes it possible to provide a change in a concentration, a structure, or cooling in the systems due to artificial disturbance of the dynamic equilibrium of the system. Any deviation from the equilibrium state leads to a nonuniform distribution of the potential of transfer as well as an occurrence of the flows of transfer. Restoration of the equilibrium will be ensured by evaporation of a part of the moisture under the adiabatic conditions by reducing the accumulated thermal energy of the system [1-10].

Depending on a set of the apparatuses that introduce disturbances into the external environment, reduction of the pressure in the working volume can be carried out using the system of a vacuum pump-condenser. The specificity of the processes implies airtight conditions for their implementation, ensuring the reduction of combined-gas emissions into the environment. The principle of sealing the working volume is the basis for creation of the waste-free technologies, in which the apparatuses for carrying out the process and the vapour and gas purification equipment are technologically connected.

The steam and gas flows along the components from the surface of the wet material or solution are determined from solving a system of the equations that make up the internal task of the processes. The evaporation surface and the volume of the gas-vapor space determine the design parameters of the apparatus.

**Theoretical part.** To develop a mathematical model of the liquid removal process, we adopt a model of the ideal mixing in the vapor-gas phase and a layer of the dispersed material.

A mathematical description of the non-stationary modes of the objects with the lumped parameters in the phases is based on ordinary differential equations, often presented in the form of finite-difference relations between input and output values of the flows of the substances.

Heat and mass transfer in the vapor-gas phase. For non-stationary conditions of the processes, a difference between the inflow and discharge will be accumulation of the mass and energy in the free volume of the apparatus $V$ [1, 4, 8].

When removing the multicomponent liquids in a system solid-liquid-vapor, those which have an interfacial area, the inflow of the mass of $i$ -th component into the free volume of the apparatus and the temperature change of the vapor-gas mixture for the ideal mixing model are determined by the expressions:

$$V d\rho_i = dm_{i\text{вх}} - dm_i ,$$

$$V d\rho_\text{г} = -dm_\text{г} ,$$

$$c_p \rho_{\text{см}} \frac{dT}{d\tau} = \left[ \alpha(T_F - T) + T_F \sum_{i=1}^{n} j_i c_i \right] \frac{\mu_{\text{см}}}{R_{\text{пр}}} - c_p \rho_{\text{см}} T \frac{I_{\text{с.пг}}}{V} ,$$

where $dm_{i\text{вх}} = j_i F d\tau$ , $dm_i = I_{\text{с.пг}} \rho_i d\tau$ .

The equations given above include the geometrical parameters of the apparatus and the volumetric capacities of the vapor and inert gas removal systems. After their transformations, the differential equations for changing the parameters of the gas-vapor mixture $p_i$ , $p_\text{г}$ , $T$ are written in the form

$$\frac{dp_i}{d\tau} = \frac{FR^* T}{VM_i} j_i - p_i \left( \frac{I_{\text{с.п}}}{V} - \frac{1}{T} \frac{dT}{d\tau} \right),$$

$$\frac{dp_\text{г}}{d\tau} = p_\text{г} \left( \frac{1}{T} \frac{dT}{d\tau} - \frac{I_{\text{с.г}}}{V} \right),$$

$$\frac{dT}{d\tau} = \left\{ \frac{\left[ \alpha(T_F - T) + T_F \sum_{i=1}^{n} j_i c_i \right] FR}{(c_{p\text{г}} p_\text{г} + \sum_{i=1}^{n} c_{pi} p_i)} - I_{\text{с.пг}} \right\} \frac{T}{V} .$$

The steam and gas flows along the components from the surface of the wet material $j_i$ , included in the equations, are determined from solving a system of equations that make up the internal task of specific processes.

Heat and mass transfer when removing free moisture from capillary-porous materials. For a one-dimensional problem with the constant thermophysical properties of the material and the absence of a gradient of total pressure (due to a gradual decrease in pressure), the differential heat and mass transfer equation can be reduced to the form [4]:

$$c dT_\text{м} = U \sum_{i=1}^{n} r_i dx_i + dU \sum_{i=1}^{n} r_i x_i ,$$

which is the heat balance equation for the case when the heat accumulated by the wet material goes only to the evaporation of a multicomponent liquid. By analogy with the simple distillation process, changes in the concentrations of the components in a liquid are determined by the ratio

$$dx_i = \frac{x_i}{U}\left[\frac{\exp(A_i - B_i / T_M)}{\sum_{i=1}^{n} x_i \exp(A_i - B_i / T_M)} - 1\right] dU \cdot$$

If the pressure in the drying chamber decreases due to the operation of the system vacuum pump-condenser, then in the absence of inert gas in the drying chamber, the mathematical description of the condensation process is reduced to the heat balance equation. The thermal load of the condenser when the aggregate state of the system changes is written in the form of the equation:

$$dQ = m_0\left\{\sum_{i=1}^{n}(U_н - U)x_i c_{жi} dT_м - U\sum_{i=1}^{n}[r_{0i} + (c_i - c_{жi})\right\}\times(T_м - 273)]dx_i - \sum_{i=1}^{n}\bar{x}_i[r_{0i} + (c_i - c_{жi})(T_м - 273)]dU \cdot$$

Heat and mass transfer when removing bound moisture from capillary-porous materials with a mobile zone of phase transformations. The solution of a system of nonlinear differential equations with a moving zone of phase transformations and a variable criterion for vaporization by coordinate and time, taking into account the considerable complexity of the problem even in relation to an unlimited plate, it is often reduced to consider a more simplified mechanism in the form of a deepening of the evaporation surface As the first approximation, a linear law of distribution of the moisture content and temperature in the evaporation zone is adopted:

$$U_{i1} = \left(U_{iF} - U_{i\xi}\right)\frac{r - R}{\xi} + U_{iF} ,$$

$$T_{M1} = \left(T_{M.F} - T_{M.\xi}\right)\frac{r - R}{\xi} + T_{M.F} .$$

and a parabolic law of distribution of the same parameters in the wet zone

$$U_{i2} = U_{iц} - \left(\frac{r}{R_M - \xi}\right)^2 \left(U_{iц} - U_{i\xi}\right),$$

$$T_{M2} = T_{M.ц} - \left(\frac{r}{R_M - \xi}\right)^2 \left(T_{M.ц} - T_{M.\xi}\right).$$

The boundary conditions for the equations have the form:

the evaporation zone

$$a_{mil}\rho_c \frac{U_{iF} - U_{i\xi}}{\xi} + a_{mil}\rho_c \delta_{Til} \frac{T_{MF} - T_{i\xi}}{\xi} + K_p \frac{p_i - p_{i\xi}}{\xi} - j_{iF}(\tau) = 0 ,$$

$$U_{iF} = b_i \left(\frac{p_i}{p_{iнас}}\right)^{n_i} ,$$

where $b_i$ , $n_i$ are constants.

the wet zone

$$\lambda_{M1}\frac{T_{MF} - T_{M\xi}}{\xi} = -\lambda_{M2}\left[\frac{2}{R - \xi}\left(T_{Mц} - T_{M\xi}\right)\right] - \sum_{i=1}^{n} r_i j_i(\tau),$$

$$a_{mil}\left(\frac{U_{iF} - U_{i\xi}}{\xi} + \delta_{Til}\frac{T_{MF} - T_{M\xi}}{\xi}\right) + \frac{K}{\rho_c}\frac{p_i - p_{i\xi}}{\xi} = 2a_{mi2}\left(\frac{U_{i\xi} - U_{iц}}{R - \xi} + \delta_{Ti2}\frac{T_{M\xi} - T_{Mц}}{R - \xi}\right).$$

The temperature of the liquid is related to the pressure in the capillary by the Thomson dependence, which, taking into account the Antoine equation, can be written this way:

$$p_{i\xi} = \left\{ \exp\left(A_i - B_i / T_{M\xi}\right) + \left[ - \frac{2\sigma_i \rho_i}{\rho_{жi} r_{кап} \exp\left(A_i - B_i / T_{M\xi}\right)} \right] \right\}.$$

The length of the section of the capillary, on which the transfer of the vaporous mixture occurs, is determined from the relation, transformed as applied to an unlimited plate

$$\frac{d\xi}{d\tau} = \frac{d}{\mu_{CM}} \frac{\rho_{CM}}{\rho_c U_H} \frac{dP}{dx},$$

where d is a permeability coefficient.

The amount of evaporating liquid under adiabatic conditions depends on changes in the heat content of the body. To maintain a high rate of moisture removal, the drying process is desirable to combine with the supply of heat in a closed coolant cycle, for example, with the convection.

**Discussion of the results of mathematical modeling.** The system of equations describing heat and mass transfer when removing a multicomponent liquid from a layer of dispersed capillary-porous material was solved by the numerical Runge-Kutta method. The calculation results are presented in the form of kinetic dependences of the main process parameters.

The fig. 1-3 show the results of the calculation of the kinetics of the removal of a three-component liquid mixture of diethyl ether-ethyl alcohol-water from active recovery coal. An analysis of the obtained data makes it possible to evaluate the influence of regime parameters and the state of the material on the processes of moisture removal and condensation of its vapor. It was established that a decrease in pressure in the drying chamber leads to evaporation of moisture, which is carried out by reducing the thermal energy of the wet material, and is accompanied by a decrease in its temperature (Fig.1).

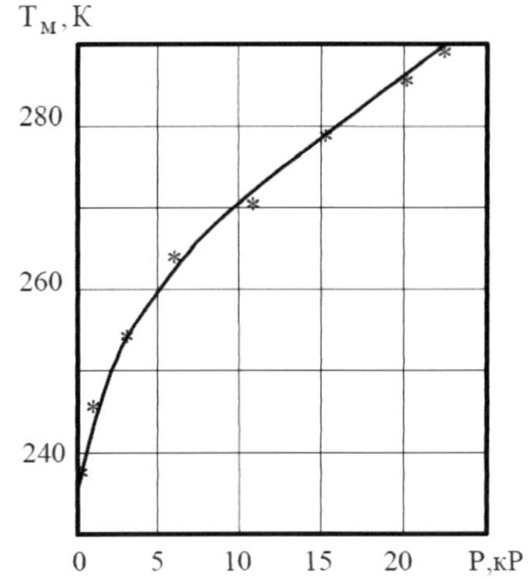


Fig. 1. Pressure dependence of wet material temperature

In the calculation, the temperature of the wet material was set within 293 K, and the initial moisture content was set 0.35 kg/kg.

A joint analysis of the curves of the total moisture content and the moisture content of the components of the liquid mixture allows us to conclude that, when a multicomponent solvent is removed, the latent heat of vaporization of the removed liquid has a significant effect on the evaporation process. As can be seen from the curves (Fig. 2), the moisture content of the low-boiling component - diethyl ether, the evaporation rate of which is greatest, with a decrease in pressure, it changed from 0.166 kg / kg to an equilibrium value. In this case, the change in moisture content in ethyl alcohol and water amounted to 7% and 3.5%, respectively. The analysis of the curves shows that the higher the initial temperature of the material, the greater the moisture removal from a unit mass of the product during the drying process. A decrease in the pressure of the medium above the surface of the material in the first drying period causes a decrease in the temperature of the evaporating liquid, and, therefore, requires a decrease in the temperature of the heat and mass transfer surface to condense the vapors.

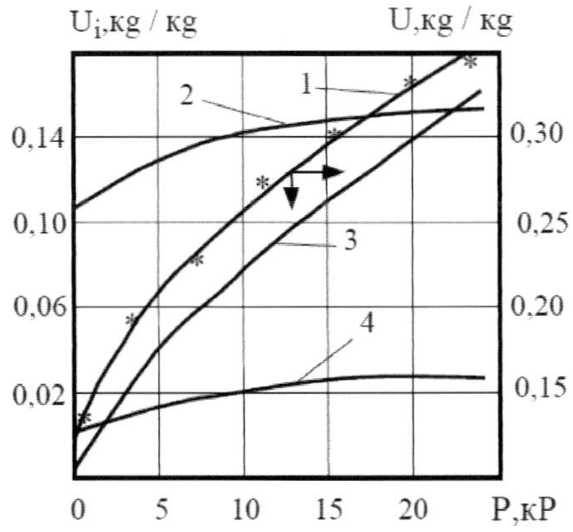


Fig. 2. Dependence on the residual pressure of the total moisture content of activated carbon (1) and moisture content by components: 2 - diethyl ether; 3 - ethyl alcohol; 4 - water (the initial ratio of the components, respectively is 0.5, 0.45: 0.05):

The dependences of temperature and total moisture content on residual pressure with a high degree of certainty ( $R^2 = 0{,}99$ ) were approximated by power and polynomial functions, respectively:

$$T = 235{,}39P^{0{,}119},$$

$$U = -0{,}0004P^3 - 0{,}0015P^2 + 0{,}0582P + 0{,}0467.$$

Figure 3 shows the calculated curve of the change in the specific cold flow rate for vapor condensation. The analysis showed that at a sufficiently low residual pressure in the apparatus, the specific consumption of cold for vapor condensation increases significantly. This requires the use of large-capacity refrigeration units, which is not economically feasible.

**Design part.** The nature of the functional relationships of the equipment used to implement the process of removing moisture from dispersed materials by lowering pressure, largely depends on the type of apparatuses and the features of their work.

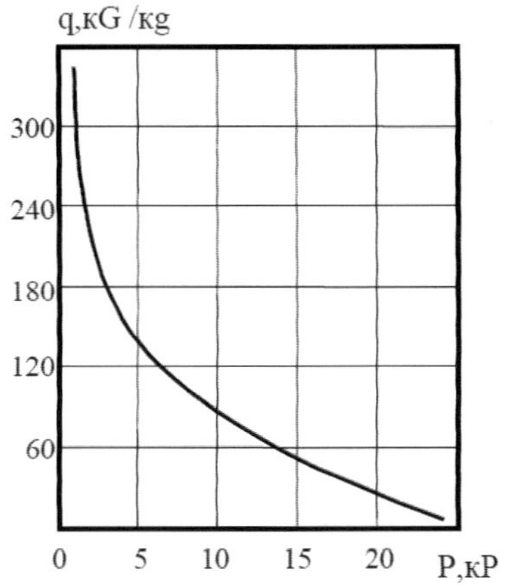

Fig.3. Dependence of the specific flow rate of cold for vapor condensation on the residual pressure in the apparatus

Volumetric capacity of the gas removal system - $I_{с.г} = I_{в.н}$ depends on the design of the vacuum pump and is determined by its operating characteristic. So for a rotary vacuum pump

$$I_{в.н} = Q_0\left(1 - k^{**}\frac{P_{атм}}{P}\right).$$

The volumetric capacity of a condencer $I_к = I_{с.п}$ can be determined from the heat balance equation

$$I_{с.п} = \frac{k_т \Delta T_{ср} F_к}{\sum_{i=1}^{n} \rho_i \left[c_{iпер}(T - T_{iнас}) + c_i T_{iнас} + r_{0i}\right]}.$$

In the formula, the heat transfer coefficient - $k_т$ for the condensation process of a multicomponent vapor-gas mixture is calculated according to the parameters of the cooling agent using the Kolborn-Hougen method or the Ward model.

The main indicator characterizing the process conditions in the local volume of the main apparatus for the system in question is a variation in the moisture content of the product. The magnitude of the variation is a regulated one, and it determines the difference in the pressure across a layer of the material.

The permissible difference in the moisture content of dispersed particles depends on the technological requirements for the product and can be determined on the basis of the equation of the Freundlich isotherm [7]

$$\Delta U_д = 1 - (1 + \Delta P / P)^n .$$

The implementation of the ideal mixing model in a layer of dispersed material is provided by introduction of the perforated distribution elements of the vacuum-pumping system into the layer. The volume of the material is divided into a number of layers of small thickness, which ensure homogeneity of the product in the final moisture content.

The distance between the distribution elements, taking into account the permissible difference in moisture content and the resistance of the layer to mass transfer, is determined by the Ergan formula [8]

$$L = \Delta P / [150 \frac{(1-\varepsilon_0)^2}{\varepsilon_0^3} \frac{w\mu}{d^2} + 1{,}75 \frac{1-\varepsilon_0}{\varepsilon_0^3} \frac{w^2 \rho}{d} ].$$

The calculations found that the distance between the distribution elements at $\Delta U_д = 0{,}001$ kg/kg is $L = 0{,}15$ m.

Calculation of the apparatuses is carried out at the highest load in the gas-vapour phase. This corresponds to the period of free moisture removal (period of constant drying speed). As the residual pressure decreases, the vapor density lowers. With a constant flow rate and a low density of the vapor phase, entering the condenser, a branch pipe diameter connecting the apparatus to the condenser increases significantly. To reduce the hydraulic resistance between the apparatus and the condenser, as well as to create the same process conditions at different points of the dispersed layer, a design of the vacuum apparatus was developed in which the condenser was combined with the working volume of the drying apparatus (Fig.4).

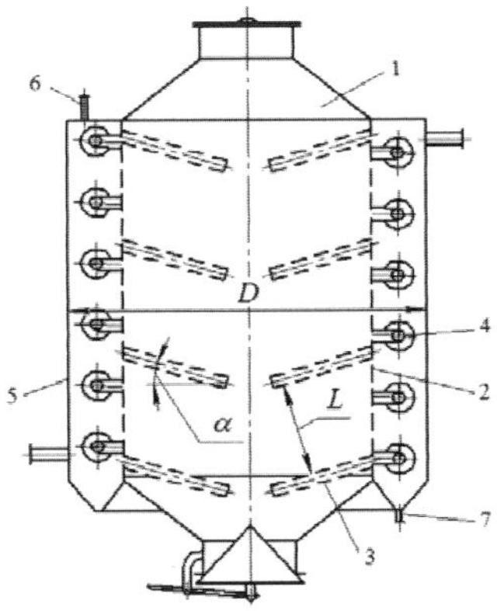


Fig. 4. Apparatus for removing liquid from dispersed materials by lowering the pressure

The apparatus comprises a vertical drying chamber 1 with a perforated shell 2 and a system of perforated distribution elements 3, a surface condenser 4. The apparatus and the condenser are placed in a common housing 5 and communicate with each other through a perforated wall and distribution elements. The apparatus is equipped with the loading units 6 and unloading ones 7.

Distribution elements 3 are installed at an angle of the natural slope ( $\alpha$ ) to the perforated shell 2. For activated carbon, the angle of the natural slope, depending on humidity, varies within $\alpha = 34 - 40^0$ .

The device operates as follows. A wet material is loaded into the apparatus 1 through the loading unit 6, after which the apparatus is sealed and connected to a vacuum system. With a decrease in the residual pressure, the moisture, evaporated from the material through the system of perforated elements 3, enters the condenser, where it contacts the cooled surface and condenses. The absence of the inert gas (air), which is removed from the apparatus by a vacuum pump, increases the efficiency of the phase transition.

To make the strength calculation of the apparatus, working at the external pressure, some various CAD/CAE systems can be used that provide a calculation and an analysis of the results of the static and dynamic [9], structures.

The strength calculation of the apparatus is based on determining the wall thickness at the largest amount of vacuum. In this case, the nominal wall thickness of the housing is taken as the calculated value [10]

$$s' = 1{,}18D\left(\frac{p_{н}}{E^t} - \frac{l'}{D}\right)^{0{,}4},$$

with a check of the results on the admissible pressure

$$p_{нд} = \frac{2\sigma_{сд}(s - C_к)}{D_в\left[1 + 1{,}02\frac{(l')^2 D_в}{(s - C_к)^3}\left(\frac{\sigma_Т^t}{E^t}\right)^2\right]} \ge p_н .$$

The calculations found that the nominal thickness of the outer wall of the apparatus with a diameter of $D = 1000$ mm is $s^{'} = 5{,}4$ mm.

**Legend keys.** $R^*$ - universal gas constant, $P, p$ - full and partial pressure, $j$ - flow of mass, $M$ - molecular mass, $\mu$ - coefficient of dynamic viscosity, $T$ - temperature, K; $c$ - mass heat capacity, $\rho$ - density, $w$ - speed, $\lambda$ - heat conduction coefficient, $\tau$ - time, $\alpha$ - heat-transfer coefficient, $F$ - surface, $V$ - volume, $r, r_0$ - latent heat of vaporization at the current temperature and 0 $^0$C, $A, B$ - empirical coefficients in the Antoine equation; $U$ - moisture content, $\varepsilon$ - free space in the layer, $\xi$ - evaporation zone thickness, $m$ - mass porosity, $a_m$ - mass conductivity coefficient, $\delta_T$ - relative coefficient of thermal diffusion, $K_p$ - molar transfer coefficient, $d, r$ - diameter and radius, $\sigma_T^t$ - material yield strength, $\sigma_{сд}$ - permissible compressive stress of the material, $E^t$ - modulus of elasticity of the material.

## References

[1] Lashkov V.A., Safin R.G., Kondrasheva S.G. Heat and mass transfer in chemically reacting media with decrease in partial pressures of vapor // Inzhenerno-Fizicheskii Zhurnal, volume 73(5), pages 1012-1020 (2000), doi: 10.1007/bf02681591

[2] Lashkov V.A., Khasanov T.G., Savin R.G. Mathematical model of solvent electrodesorption from capillary porous materials in conditions of pressure lowering // Nisha Yanjiu/Journal of Sediment Research, number (4), page 266-270 (1998), ISSN: 0468155X.

[3] Lashkov V.A., Petrova A.V., Safin R.G. Application of the Luikov approximate solutions to mathematical description of the process of self-freezing of materials by decreasing pressure // Inzhenerno-Fizicheskii Zhurnal, volume 75(5), pages 34-37 (2002), doi: 10.1023/A:1021190902719

[4] Lashkov V.A., Safin R.G., Andrianov V.P., Kondrasheva S.G. Heat and mass transfer under the conditions of the external problem for processes occurring with a decrease in the pressure of the medium // Inzhenerno-Fizicheskii Zhurnal, volume 73(3), pages 550-556 (2000), doi: 10.1007/bf02681795

[5] Lashkov V.A., Zajnutdinov E.A., Safin R.G., Buzuev V.V. Crystallization of hexamethylene tetramine in the case of continuous increase in the vacuum // Inzhenerno-Fizicheskii Zhurnal, volume 75(2), pages 157-160 (2002), doi: 10.1023/A:1015618427655

[6] Lashkov V.A., Levashko E.I., Safin R.G. Mathematical simulation of the drying of suspensions and colloidal solutions by their depressurization // Journal of Engineering Physics and Thermophysics, volume 79(3), pages 539-546 (2006), doi: 10.1007 / s10891-006-0132-z

[7] Kashapov N.F., Gilmanshin I.R., Konahina I.A. System analysis of the energy complex of engineering enterprise as a basic tool of effective energy management (2014) // IOP Conference Series: Materials Science and Engineering, 69 (1) 012024, doi=10.1088%2f1757-899X%2f69%2f1%2f012024

[8] Lashkov V.A., Kondrasheva S.G. Low-waste technologies based on the improvement of technological equipment/V.A. Lashkov, S.G. Kondrasheva // Bulletin of Kazan Technological University. 2012. T.15. No.15. P. 204-207

[9] Shelofast V.V. Practical training course. CAD / CAE system APM WinMachine. Tutorial. M.: Publishing House APM, 2013. - 144 p

[10] Karataeva E.S., Lashkov V.A., Brodt A.A., Akhlyamov M.N. Development of turning-filtering mechanism of centrifuges // IOP Conf. Series: Materials Science and Engineering 570 (2019), 012037, doi:10.1088/1757-899X/570/1/012037.

IOP Conf. Series: Materials Science and Engineering **915** (2020) 012039 doi:10.1088/1757-899X/915/1/012039

# Quality management when operating the power equipment

**A O Mamonova[1], E M Khusnutdinova[1] and A N Khusnutdinov[2]**

[1]Kazan (Volga Region) Federal University 18 Kremlyovskaya Str., Kazan 420008
[2]Kazan State Power Engineering University 51 Krasnoselskaya str., Kazan 420061, Russia

rr-088@mail.ru

**Abstract.** This paper looks into the possibilities of designing an integrated quality management system for the operation of power equipment to ensure the fabrication of competitive products, increase the reliability and safety of production.

At the present stage of development, a specific focus is given to the issues of quality assessment and the formation of its management system in all areas of industry, including applications related to the power sector. Except for some regulatory documents, these issues are not fully addressed, although the products of the power industry have a high public demand and a rather specific character [1]. In view of this, there is a need to form a system for assessing and managing the quality of both the products and internal processes of its production [2], in particular, the operation of power equipment.

The power equipment means equipment that provides the company with all types of utilities (for example, boilers, heaters, electrical grids, etc.). Unlike the production equipment, the power equipment is not directly engaged in production, but the trouble-free operation of all plant and machinery that provides each business unit with all types of fuels and utilities relies on its healthy condition.

Thus, the matter of designing the main provisions of the quality management philosophy for the operation of existing power equipment is raised. The solution to this problem, taking into account the regulatory and engineering documentation in a single integrated system, will make it possible to form effective approaches for energy technology complexes to improving the quality of both the products and the entire production cycle.

Each stage of the development of the power system sets a certain level of quality management, which provides for the corresponding organizational impact on production for its further enhancement. At the same time, ensuring and maintaining the level of product quality required at the stages of its design and fabrication, installation and testing, improvement and modernization, operation or disposal, which are performed using systematic quality control, together constitute a continuous process of the system, covering the general provisions and the principles of the process approach in assessing and improving the quality of energy technology equipment [4].

Different methods of theoretical analysis, process approach, statistical forecasting and diagnostics are used to study how the industrial equipment, processes, factors and control actions have an effect on the product quality. The identification of parameters that have a potential impact, as well as the analysis of risks and losses, are possible through the use of various tools for a qualitative analysis of the relationship [3], which, in turn, have characteristic features and properties depending on the study target. Ishikawa diagram gives a clear idea of such impact, which makes it possible to reveal factors determining the quality of products and the degree of significance of individual parameters.

Such a strategy for studying the key characteristics and distinctive features in the energy complex provides the opportunity to take into account the features of the creation of utilities; an integrated approach to the enhancement of product quality, equipment operation and environmental management;

and optimization of the studied processes by impact factors. It is possible by using the methods of generalized variables, target functions, fuzzy sets both in absolute values of parameters and in their dimensionless values [5]. The latter significantly expands the scope of the practical application of the qualitative analysis results, causing a similar approach of assessing the performance of power systems be a one-size-fits-all approach.

Thus, the QMS is becoming a core element in the coordination of engineering, economic, design, technological, environmental, biomedical and other issues regarding agreements and disagreements at all stages of the life cycle of power production: development - design, production - testing, operation - continuous improvement (innovation, modernization, revamp), the solution of which will help to raise the level of competitiveness and operational safety in the workplace.

The decision-making process at different stages of the life cycle is identical, but the specific content of certain clusters (for example, methods for information sourcing and evaluating, approaches to approving and formulating recommendations, etc.) can be diverse [6].

For power production facilities, operating with the continuity and cyclicality of technological processes [7], the following assessment and quality control at different stages from design to revamp (disposal) can be proposed:

- at the stage of product engineering, the required level of quality is achieved in the design and development of technology for its fabrication, for example, by establishing the necessary level of standardization;
- the stage of product testing produces a control system and the corresponding production management subsystem, which defines some estimated indicators of product quality;
- the operation stage makes it possible to obtain data for adjusting the QMS in the first two stages and determine reliable indicators;
- at the stage of introducing innovations, it is necessary to determine comprehensive quality indicators, identify various types of inconsistencies and eliminate their causes, and design and justify further actions to improve the quality of products and the entire production process [8].

Consequently, the integration and consistent use of the four management functions (planning, organization, motivation and control) creates the basis for continuous in-process perfection of the QMS and helps minimize risk factors [9] related to both production processes and their management processes.

The practical application of modern it-technologies for handling databases, software systems in the form of expert systems, elements of artificial intelligence significantly widens opportunities of qualitative analysis and increases the speed and efficiency of specific technical decisions [10,11]. This, in turn, significantly improves the overall progress of matching the power system operating conditions with consumer loads, which is especially important for power installations, since it facilitates to improve the quality of their products and operational reliability [12,13].

**References**

[1] Mamonova A O and Khusnutdinova E M 2018 *Energy Efficiency Management in the Oil Industry.* Final scientific and educational conference of students of Kazan Federal University in 2018: collection of abstracts. (Kazan: Kazan University Press) Vol 1. pp 24-25
[2] Aktsoraeva N G and Sboyev P E 2018 *Quality Management in the Power Sector.* Alley of Science Scientific and Practical Electronic Journal. (Tomsk: Quantum Press) Vol 5. No 4 (20). pp 559-602
[3] Khvastunov R M, Feofanov A N, Korneyeva V M and Nahapetyan E G 2009 *Qualimetry in Mechanical Engineering.* (Moscow: Industrial Engineering) p 285
[4] Gorlenko O A and Mankevich I G 2008 *The Process Approach to Quality Management.* (Bryansk: BSTU) p 168
[5] Totay A V, Kazakov V S and Klimenko T V 2010 *The Process Approach in Assessing the Quality of Operation of Energy Technology Equipment.* Quality Management of Products and Services: proceedings of 3rd International Science and Technology Conference. (Bryansk: BSTU) Vol 2. pp 259-264
[6] Kashapov L N, Kashapov N F, Kashapov R N 2013 *Research of the impact acidity of electrotytic cathode on the course of the plasma – electrolytic process* Journal of Physics: Conference Series, Vol. 479, Iss. 1, Art. № 012011

[7] Khusnutdinova E M, Khusnutdinov A N and A O Mamonova 2018 *Heat Transfer in the Laminar Flow of a Rheologically Complex Viscous Fluid in Three-way Screw Insert Barrels.* The Role of Fundamental and Applied Science in the Socio-Economic Growth of Society: collection of international scientific and practical conference. M.: Impulse. pp 591-593

[8] Khafizov I I and Galimov A N 2017 *IT-strategy and major aspects of quality management on the market of goods and services*. IOP Conference Series: Materials Science and Engineering. Vol.240, Is.1. - Art. № 012038.

[9] Savina N V and Kudryashov A N 2013 *Quality Management in the Electric Power Engineering. The process approach* (Blagoveshchensk: Amur State University Press) p 104

[10] Gulbrandsen T H, Padalko L P and Chervinsky V L 2010 *Energy efficiency and energy management. (*Minsk: BSATU) p 240

[11] Khafizov I I, Nurullin I G and Kashapov N F *Influence of application of robotic systems on prospects and directions of development of electroerosive processings*. IOP Conf. Series: Materials Science and Engineering 570 (2019) 012049

[12] Hajieva S R, Gadirova E M, Sharifova N S, Gilmanshina S I and Gilmanshin I R *Economy of energy conservation* IOP Conference Series: Materials Science and Engineering. - 2019. - Vol.570, Is.1. - Art. № 012031.

[13] Kashapov LN, Kashapov NF, Kashapov RN, Pashaev B Y 2016 *The application of additive technologies in creation a medical simulator – trainer of the human head operating field* IOP Conference Series: Materials Science and Engineering, Vol. 134, Is 1, Art. № 012011

# The current quality management system at Magnitogorsk iron and steel works

**M N Moliboga, E M Khusnutdinova**

Kazan (Volga Region) Federal University 18 Kremlyovskaya str., Kazan 420008

rr-088@mail.ru

**Abstract**: The paper covers the basic principles of the quality management system applicable in Magnitogorsk Iron & Steel Works (hereinafter - MMK PJSC or MMK). The authors refer to various monitoring methods and methods allowing a conclusion on the acceptability of a particular system, describe the types of work performed to improve and facilitate the growth of the company, and apply a new approach to quality management (Group certification) based on a single regulation. Based on the study of the matter described in this paper, a conclusion can be made that this system is necessary for coordinating the further performance of the entire company and its subsidiaries.

Magnitogorsk Iron and Steel Works PJSC is one of the world's largest steel producers and holds a leading position among Russian iron and steel companies. The company's assets in Russia are a large metallurgical complex with a full production cycle from the treatment of iron ore raw materials to the downstream of ferrous metals. MMK produces a wide slate of metal products with a predominant share of products with high added value [1].

Currently, the MMK quality management system (QMS) is certified for compliance with the following international standards: ISO 9001 (since 2002), ISO/TS 16949 (since 2005, IATF 16949 in 2018) and STO Gazprom 9001 industry standards (since 2009). Every year, the compliance of the MMK quality management system is confirmed by TÜV Nord Cert [8].

The current MMK QMS includes the approved “Quality Policy”, annual quality objectives, mandatory documented procedures, quality records, etc. For introduction of one of the key principles of the ISO 9001 standard - process approach - and as dictated by the actual activities of MMK, the company sets up the QMS processes of the first, second and third levels, the sequence of their interaction, the criteria for evaluating the processes of the first level, which are described in the Quality Management System Guide as the main document of the QMS. For the purpose of a deeper assessment of the QMS processes, the owners of the processes of the second and third levels have designed maps of the corresponding processes, which determine the criteria for evaluating the performance and efficiency of the processes of the second and third levels [2].

The QMS package includes standards, instructions, regulations, procedures, etc. Today, MMK PJSC operates an electronic document management system called Nord automated information data system, which facilitates the availability and relevance of regulatory documents at all workplaces [4,5].

Quality improvement is an ongoing process at all levels of the company: from suppliers of resources, equipment and services to the consumer affairs.

Pursuant to the requirements of the IATF 16949 and STO Gazprom 9001 standards, development of suppliers is in progress:

- determining the nomenclature of products, having a high effect on the quality of finished products, suppliers of such nomenclature are included in the audit program by MMK employees;

- digitizing the liaison of all suppliers with MMK in the form of a rating contributing to the further work of suppliers;

- changing the approach to handling non-spec products of the supplier. Upon confirmation of such cases, the supplier shall let MMK PJSC know the reasons for return of such products and corrective action taken.

In the business units engaged in the production of finished products, monthly analyzes of non-spec products, defects, acknowledged claims using various methods and software products. A system of statistical methods for Statistical Process Control (SPC) is in place, Measurement System Analysis (MSA) is in progress [6, 7].

For the timely performance of customer orders, a system of operational scheduling was designed and a software product was introduced that allows building the load of each unit, optimize it, and also enables customers to on-line track the progress of orders. The client can track the order using his/her mobile device: order accepted, order in steel casting, rolling, order shipped.

The company has built a customer support team, a single window system, and a complaints handling system. All this makes it possible to collect feedback from customers about the MMK products, helps to identify bottlenecks, see what the client is dissatisfied with in addition to the standard questionnaire [3].

At MMK PJSC, digital tools are constantly being introduced to help effectively reduce internal costs: for example, the Sniper project, which annually saves significant resources on expensive ferrous alloys in metal production. The commercial service progresses the projects of Industry 4.0 - a special software-based robot for interacting with scrap metal suppliers. In the HR department, a robot recruiter is launched. The key purpose of the project is to make personnel processes transparent, as well as reduce the time spent on the search and selection of qualified personnel [9,10].

In 2017, group certification was conducted at the industrial site of MMK, which is a precedent in the metallurgical industry. Unified group certification of the MMK Group quality management system covered 12 companies of the Group.

The unified group certification at MMK was the result of many years of consistent work to centralize the management of the management systems within the Group [11]. This resulted in uniform requirements for the main processes, which are mandatory for MMK PJSC and all Group companies; a unified approach to conducting system audits; monitoring of the main processes of the functioning of the systems within MMK Group during in-house audits performed by the Group's auditors for the plant's quality system, in which management system control functions are concentrated. Such a single approach - with common and uniform standards - allows to bring the management system applied in the Group to a higher level, reduce the number of documents regulating the basic requirements for the systems, reduce costs and increase the efficiency of the company's business processes [12]. The results of the audit did not reveal any discrepancies. TÜV NORD CERT experts issued group certificates of compliance with ISO 9001-2015.

MMK PJSC, like any modern enterprise, seeks to improve and develop its quality management system aimed at increasing customer satisfaction, including through the introduction of new technologies in its operational processes [13].

## References

[1] Official website of Magnitogorsk Iron and Steel Works [On-line resource]. URL: www.mmk.ru

[2] Information portal of Magnitogorsk Iron and Steel Works [On-line resource]. URL: https://mmk-portal.mmk.ru

[3] Nikolaeva L A, Khusnutdinov A N 2018 A study of the absorption of nitrogen oxides from the boiler flue gases *Thermal Engineering* Vol. 65 Iss. 8 pp 575-579
he official website of *Vedomosti* [On-line resource]. URL: https://m.vedomosti.ru

[4] Khafizov I I and Nurullin I G 2017 *Product quality as the main factor of increase of competitiveness (on the example of JSC Kazan helicopter plant).* IOP Conference Series: Materials Science and Engineering. Vol.240, Is.1. - Art. № 012039.

[5] Soskovets O N *Ferrous metals magazine*, 2018

[6] Nikolaeva L A, Khusnutdinov A N 2018 Purification of gas emissions of chemical industry enterprises by carbonaceous cutting *Ecology and Industry of Russia* Vol. 22 Iss. 8. pp 14-18

[7] Safarov D T, Kondrashov A G and Khafizov I I The method of diagnosing machine systems by measuring the accuracy of manufactured parts. IOP Conf. Series: Materials Science and Engineering 412 (2018) 012065

[8] Gilmanshina S I, Gayfullina A Z and Gilmanshin I R Formation of environmental culture in the conditions of new informational environment (on the example of science disciplines) // Proceedings of INTED 2018. - Valencia, Spain, 2018. - P. 1664-1669.

[9] The official website of *Rossiyskaya Gazeta* [On-line resource]. URL: https://rg.ru/

[10] The news and the articles from industry of metallurgy [On-line resource]. URL: http://metal.rftoday.ru/mmk_i_rashnikov

[11] Kashapov R N, Kashapov L N, Kashapov N F *Formation of cracks in the selective laser melting of objects from powdered stainless steel 17-4 HP* (2017) IOP Conference Series: Materials Science and Engineering, Vol. 240, Is 1, Art. № 012074

[12] Kashapov L N, Kashapov N F, Kashapov R N, Pashaev B Y 2016 *The application of additive technologies in creation a medical simulator – trainer of the human head operating field* IOP Conference Series: Materials Science and Engineering, Vol. 134, Is 1, Art. № 012011

[13] Kashapov L N, Kashapov N F, Kashapov R N *Research of cobalt chromium alloy surface morphology after sandblasting (2014)* IOP Conference Series: Materials Science and Engineering, Vol. 69, Art. № 012017

IOP Conf. Series: Materials Science and Engineering **915** (2020) 012041 doi:10.1088/1757-899X/915/1/012041

# Development of cast dispersion-hardening ferrite-carbide steel

**G F Mukhametzyanova[1], M S Kolesnikov[2], V I Astashchenko[1] and I R Mukhametzyanov[1]**

[1]Kazan Federal University, Naberezhnye Chelny Institute, Department of materials, technologies and quality, 423812, Russia, Naberezhnye Chelny, Syuyumbike av.,10A
[2]Kazan Federal University, Naberezhnye Chelny Institute, Mechanical engineering department, 423812, Russia, Naberezhnye Chelny, Syuyumbike av.,10A

gulnara-ineka@mail.ru

**Abstract**. A method is proposed for producing cast dispersion-hardening ferrite-carbide steel for manufacture of cast molds of hot deformation, molds for die casting under pressure and forging dies for solid-liquid forming, providing production of steel with fine-grain structure and uniform distribution of carbides in a ferrite matrix without formation of casting cracks. In the process of steel production waste carbide ceramic-metal inserts of a cutting tool from alloys are used, they are of the following types: T5K6, T15K6, T5K10, T21K8, TT7K12, TT8K6, TT10K9, TT20K9 and T30K4 containing titanium carbides (TiC), tantalum carbides (TaC), tungsten carbides (WC).

Steel for hot deformation dies and die casting molds should have high heat resistance, wear resistance, thermal stability and resistance to interaction with melts. For the tool group under consideration, martensitic class steels are mainly used: 4Х5МФС, 3Х2В8Ф, 4Х4ВМФС (ДИ – 22), 5Х3В3МФС (ДИ – 23), 5Х2МНФ (ДИ – 32), etc. [1 - 7].

Dies and die casting molds, as a rule, are made of forged blanks, which are associated with high costs in the manufacture of complex engraving of the press tool. The experience of using cast dies from 35Х5НМФСЛ steel for hot deformation at PJSC "KamAZ-Metallurgy" conditions showed that slow cooling during crystallization of die castings in zircon forms leads to the formation of a "rough" cast structure, which leads to low operational performance in comparison with the press – a tool from forged blanks.

A promising direction in the manufacture of a molded press tool for hot deformation is the use of directional crystallization, which provides dispersed structure during accelerated cooling during casting in a cooled chill mold [1]. However, the low plastic characteristics and the high coefficient of thermal expansion of martensitic steels do not allow their use in accelerated cooling due to the formation of casting cracks. This problem can be solved by using types of steel with an isomorphic structure.

In addition, during the operation of a heavily loaded press tool, the temperature of which on the engraving reaches values higher than polymorphic transformations (800 - 900°C) [2], the structural-phase hardening in steels with an isomorphic structure is eliminated due to $\alpha \leftrightarrow \gamma$ transformations during each loading cycle and as a result, the tendency of steels to crack of thermomechanical fatigue decreases.

For the manufacture of cast press tools (hot deformation dies, solid-liquid stamping (SLS), die casting molds of aluminum-based alloys, etc.), cast dispersion hardening ferrite-carbide steel without polymorphic $\alpha \leftrightarrow \gamma$ transformations was developed [8].

To obtain the developed cast dispersion hardening ferrite-carbide steel after the melting of 35Л steel, ferrotitanium and ferronickel are introduced into the melt to ensure the chemical composition of the steel containing, by weight (%): carbon (C) 0.27 - 0.32; nickel (Ni) 0.7-0.9 and titanium (Ti) 5.8-6.2. Then, waste carbide ceramic-metal inserts of a cutting tool from alloys of the following type are introduced: T5K6, T15K6, T5K10, T21K8, TT7K12, TT8K6, TT10K9, TT20K9, and T30K4, containing titanium carbides (TiC), tantalum (TaC), tungsten (WC). They are introduced into the melt at temperature of 1500 - 1520 °C followed by exposure at given temperature for 30 min, which ensures the dissolution of the cobalt binder entering the carbide inserts and uniform distribution in the carbide melt. Then the casting of the melt into a chill mold is carried out, followed by cooling to obtain steel castings, while the steel castings has the following chemical composition, wt. %: carbon (C) 0.27 - 0.32; titanium (Ti) 5.8-6.2; nickel (Ni) 0.7-0.9; cobalt (Co) 0.02; carbides: titanium (TiC), tantalum (TaC), tungsten (WC) 0.5-1.5%; ferrum (Fe) and the rest.

In this case, the cobalt binder dissolves, and the carbides (TiC, TaC, WC) in the carbide compositions are evenly distributed in the melt in the form of inoculators.

Based on the experimental design and simplex studies, the chemical composition of the cast tool steel was optimized for hot deformation dies and die casting molds. Based on the results of resistance to thermomechanical fatigue cracking, the optimum ratios of the components in ferrite-carbide were determined by the content of carbon, titanium, nickel and titanium carbides introduced into the melt during suspension casting [9].

Figure 1 shows the microstructure of the developed 30T6HTiC – 1.5 steel with the optimum content of all components.

Dies for hot deformation and other types of press tooling are obtained by pouring the obtained molten steel into a chill mold. Due to the accelerated cooling of castings when pouring the obtained steel melt into a chill mold, a fine-grained structure is formed with a uniform distribution of carbides in the ferrite matrix without the formation of foundry cracks. The presence of Ni in the alloy allows the use of dispersion hardening regimes with the formation of fine particles of the $Ni_3Ti$ type, which provide an additional increase in hardness and resistance to abrasive wear of the press tool.

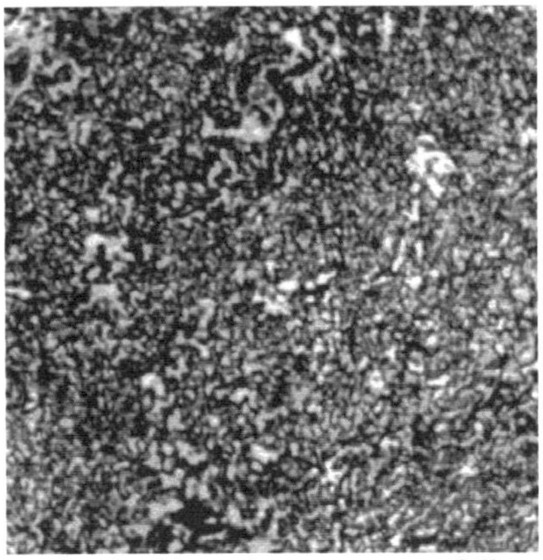

Fig. 1 - The microstructure of dispersion-hardening ferrite-carbide steel 30T6HTiC - 1.5, x500.

The combined effect of all components within the specified limits provides an increase in working capacity, which is confirmed by the test results of the developed steel for thermomechanical fatigue. Physical and mechanical properties of the developed steel (30T6NTiC - 1.5) are presented in table 1.

Table 1. Physical and mechanical properties of steel.

| Steel grades | Thermal conductivity, λ, J/ m·⁰C | Temperature coefficient of linear expansion, $\alpha \cdot 10^{-6}$1/ degree | Mechanical properties | | |
|---|---|---|---|---|---|
| | | | $\sigma_B$, MPa | $\sigma_{0,2}$, MPa | δ, % |
| 4X5MΦC | 41.4 - 42.6 | 12.3 - 13.6 | 1840 - 1900 | 1730 | 6.8 – 7.2 |
| 30T6HTiC-1.5 | 28.6 - 32.8 | 10.4 – 11.8 | 1826 - 1828 | 1716 - 1732 | 8.0 – 8.6 |

The test for thermomechanical fatigue (TMF) was carried out according to the procedure [10]. The number of loading cycles of the samples before the appearance of cracks of TMF with dimensions of 0.1 mm in depth of the sample was estimated. The test mode corresponded to the operational loading of the "Autoforge" dies during brass dying: $T_{max}$ = 780 °C on the surface of the sample, $T_{min}$ + 460 °C, contact time of the sample with the stamped material $\tau$ = 3 with a pause of $\tau_p$ = 2 s. Sample dimensions: height $h$ = 12 mm, thickness $b$ = 2 mm, sample length $l$ = 80 mm. As can be seen from table 1, the proposed ferrite-carbide steel in terms of thermophysical characteristics differs little from the known austenitic-carbide steel. The resistance to crack development of TMF of the proposed steel is 9-10% higher than that of the known austenitic-carbide steel.

The use of a cast press tool made of ferrite-carbide steel, instead of traditionally manufactured by mechanical and electrophysical methods from forged blanks of martensitic steels, can drastically reduce the manufacturing time and production costs by eliminating mechanical operations to produce complex engraving tools.

Replacing complex alloyed austenitic-carbide steel with economically alloyed ferrite-carbide steel simplifies the steelmaking process and also reduces production costs by reducing the content of alloying elements. High hardness of steel castings (46-48HRC) is achieved due to precipitation hardening during aging as a result of precipitation of a dispersed γ phase of the $Ni_3Ti$ type from ferrite, as well as due to the presence of titanium carbides.

The developed steel can be used for the production of hot-formed cast dies, injection molds, forging dies for solid-liquid dying ("Autoforge" dies) of copper-based alloys, etc., while solving the environmental problem of recycling carbide cermet cutting tool inserts.

**References**

[1] Poznyak L A, Skrynchenko Yu M, Tishaev S I 1980 Die steel *M: Metallurgy* p 240

[2].... Trakhtenberg B V 1964 Stability of dies and ways to increase them *Kuibyshev Book Publishing House* p 274

[3] Mukhametzyanova G F, Kolesnikov M S and Mukhametzyanov I R 2018 Mechanism and Kinetics of Formation of Intermetallic Layers on the Surface of Steel Press Molds in Casting of Silumins *Metal Science and Heat Treatment* vol 60 is 3 – 4 pp 190 – 193

[4] Mukhametzyanova G F, Kolesnikov M S and Mukhametzyanov I R 2016 The erosion resistance of tool alloys in foundry melt the Zamak 4 – 1 *IOP Conference Series: Materials Science and Engineering* vol 134 is 1 art № 012033

[5] Mukhametzyanova G F, Kolesnikov M S and Mukhametzyanov I R 2018 Study of the solubility of die-cast steels in silumin AK12 (AL2) *IOP Conference Series: Materials Science and Engineering* vol 412 is 1 art № 012054

IOP Conf. Series: Materials Science and Engineering **915** (2020) 012041 doi:10.1088/1757-899X/915/1/012041

[6] Mukhametzyanova G F, Kolesnikov M S, Mukhametzyanov I R and Astatshenko V I 2017 Process research into metallic pipe wear of hot chamber die casting machines and methods of increasing wear resistance *IOP Conference Series: Materials Science and Engineering* vol 240 is 1 art № 012053
[7] Mukhametzyanov I R, Kolesnikov M S, Mukhametzyanova G F and Astashchenko V I 2019 Cracks Pattern Formation and Thermomechanical Fatigue in Die Steels *HELIX* vol 9 is 5 pp 5590-95
[8] Patent 2605017 RF 2016 MPK C21C 33/04 и C22C 38/14 A method of obtaining cast dispersion hardening ferrite-carbide steel / M S Kolesnikov, G F Mukhametzyanova, D A Gurtovoy, V I Astashchenko, I R Mukhametzyanov, published 20.12.2016 Bulletin № 35
[9] Mukhametzyanova G F 2017 Determination of temperature and power stresses in the "Autoforge" dies to simulate tests of the working capacity of materials *Metallurgy mechanical engineering* vol 3 pp 41 – 44
[10] A s 879400 USSR 1981 A method for studying thermomechanical fatigue of materials / M S Kolesnikov, B L Kuznetsov, V S Kondratenko, A G Shishkin, published 07.11.1981 Bulletin № 41

IOP Conf. Series: Materials Science and Engineering **915** (2020) 012042 doi:10.1088/1757-899X/915/1/012042

# Development of hardening technology for working surfaces of microsurgical instruments made of titanium alloy

**F I Murataev and E R Galimov**

Kazan national research technical university named after A.N. Tupolev, Kazan, Russian Federation

89871726737@mail.ru

**Abstract:** The structure and properties of materials and working surfaces of needle holders were studied to improve the technology of their manufacture according to the characteristics of hardness and functionality of a microsurgical instrument. The types and modes of volumetric and surface hardening of VT6 titanium alloy (Ti-6Al-4V) using laser technologies adapted to the working surfaces of the tool are proposed. The possibility of using the alloy to replace maraging steel in the manufacture of needle holders is evaluated

Using instruments made of heavy metals makes it more difficult to perform microsurgeries and increases doctor fatigue. In addition, working surfaces of needle holders often suffer from wear or local damage due to their insufficient wear resistance caused by unstable parameters of manufacturing processes. That's why, using light and degradation resistant titanium and new technologies for instrument production will improve quality and efficiency of microsurgeries.

We suggest using a new material to manufacture needle holders and treat their working surfaces. It is VT6 titanium alloy (Ti-6Al-4V) with stabilized structure after heat treatment [1, 2]. In order to address the issue of hardness and wear resistance of working surfaces (a significant flaw of titanium alloys), laser machining is used. It increases the degree of treatment localization, accuracy, wear resistance [3-7], and robotic systems ensure stable machining conditions, which is important for microsurgical instruments. The foreign materials of working surfaces for needle holders have been replaced by the Russian materials and technologies that have no equivalents in foreign countries [12-18].

We studied the structure, microhardness of the samples of steel 03Kh11N10M2T2 and VT6 titanium alloy (KMIZ and Kazan National Research Technical University (KNITU-KAI)), metal of needle holder working surfaces and coatings made in Tatarstan and Germany.

The hardness of as-received (KMIZ) steel 03Kh11N10M2T2 is HRC 30, of VT6 titanium alloy is HRC 28. The average metal hardness (longitudinal) of atraumatic microneedle metal is HRC 45 (with a diameter of 0.4 mm), HRC 56 (with a diameter of 0.15 mm) [4]. A hardness of HRC 50 is achieved with heat treatment of steel 03Kh11N10M2T2 under conditions of quenching and age hardening, which have been proposed by KNITU-KAI (it satisfies the requirements for working instrument surfaces) [4]. Hardness of VT6 alloy subjected to harsh quenching and age hardening under different conditions is HRC 36 (Fig. 1). VK10 coating of instrument (Tatarstan) on steel 03Kh11N10M2T2 has a poor adhesion, it is not even and complete in the thickness range of 0 to 0.013 mm [4] (Fig. 2, a).

IOP Conf. Series: Materials Science and Engineering **915** (2020) 012042 doi:10.1088/1757-899X/915/1/012042

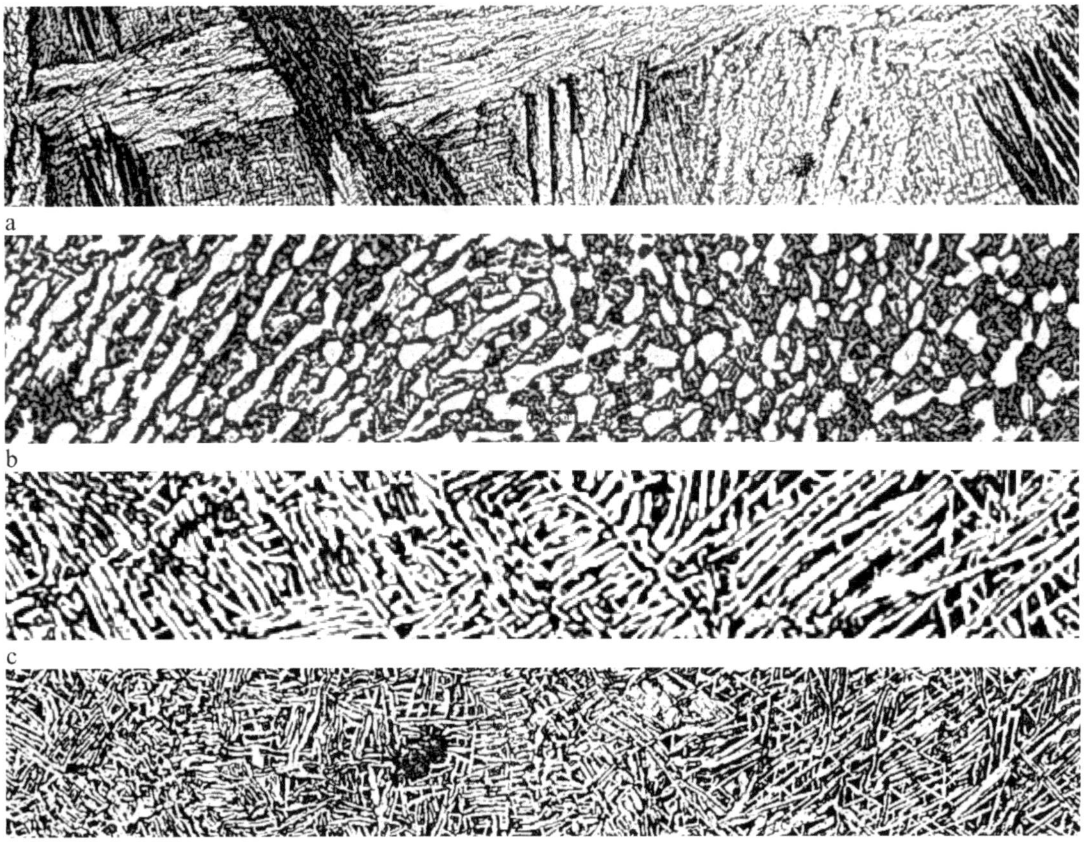


**Figure 1.** The microstructure of VT6 alloy ($\alpha + \alpha^{1} + \alpha^{11} + \beta 1$): (a) "harsh" quenching (from β-region) and age hardening without high temperature thermomechanical treatment; (b) mild quenching (from α+β-region) and age hardening with high temperature thermomechanical treatment; (c) "harsh" quenching (from α + β-region) and age hardening; (d) "harsh" quenching (from α + β-region) and double age hardening. ($^{x}500$ /a/$^{x}1300$)

VK10 coating on the surface of VK6 (Tatarstan) instrument is very uneven, some areas are not covered completely, and its thickness range varies from 0 to 0.036 (Fig. 2, d). Deposited "diamond coatings" of needle holders (Tatarstan) are brittle, and their hardness is low (HRC 42) (Fig. 2, b). The working surfaces of needle holders (Germany) made of steel with the actual composition 20Kh15 ensure a hardness of HRC 51 for a discontinuous coating within the thickness range from 0 to 0.011 mm (Fig. 2, c). Plates made of $W_6C$ alloy welded to the needle holder (Germany) made of steel 20Kh13 (the Russian equivalent) have a hardness of from 73 to 75 HRC. As the used coatings of working surfaces in needle holders do not meet the requirements: they are discontinuous and peel off, welded plates are not easy to manufacture and impractical. Hence, the conventional methods of increasing wear resistance of working surfaces in needle holders are not very effective and not reliable.

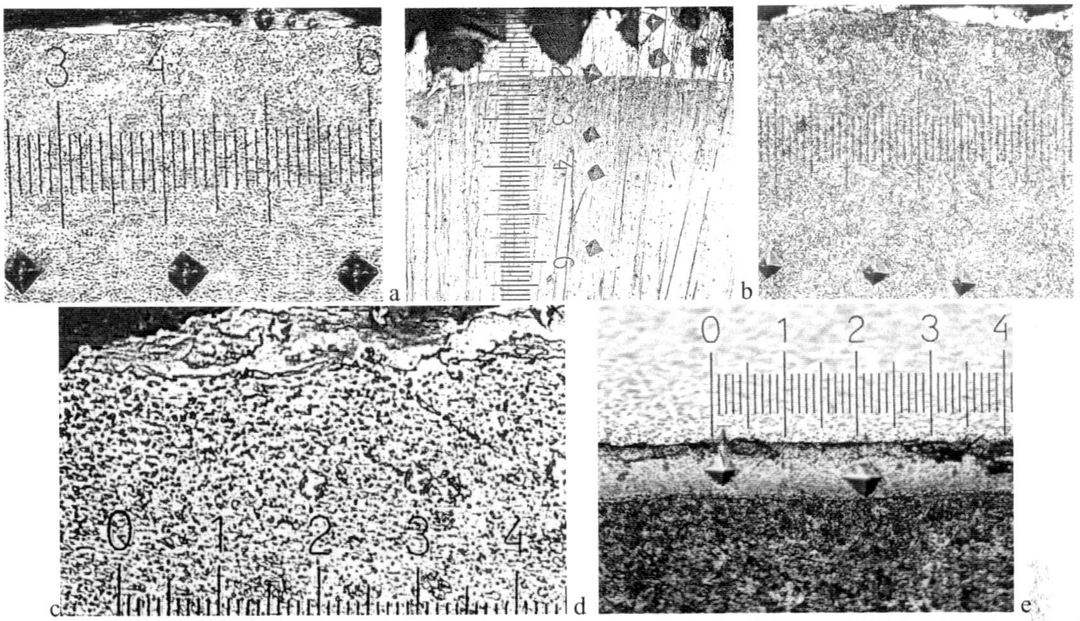

**Figure 2**. Microstructure of working needle holder surfaces with the following coatings: VK10 on steel 03Kh11N10M2T2 (Tatarstan) (a), "diamond" (Tatarstan) (HRC42) on steel 03Kh11N10M2T2 (b), coating (Germany) (HRC51) on steel equivalent to 20Kh15 (c), plates VK10 on VT6 alloy (d), welded plates $W_6C$ (Germany) on steel 20Kh13 (the Russian equivalent) (e)

Laser technologies are the effective way to harden the martensite type of titanium alloys [3-6]. For the purposes of this paper, FMark-20 RL laser station and laser module based on LS-10 fiber oscillator were used as hardening sources [3]. Figure 3 shows the example of the sample for adjustment of laser hardening process, and the microstructure of one area (fusion) in VT6 alloy with as-cast structure. The microstructure was recorded with Auriga CrossBeam scanning electron microscope [8, 9] using SmartSem software.

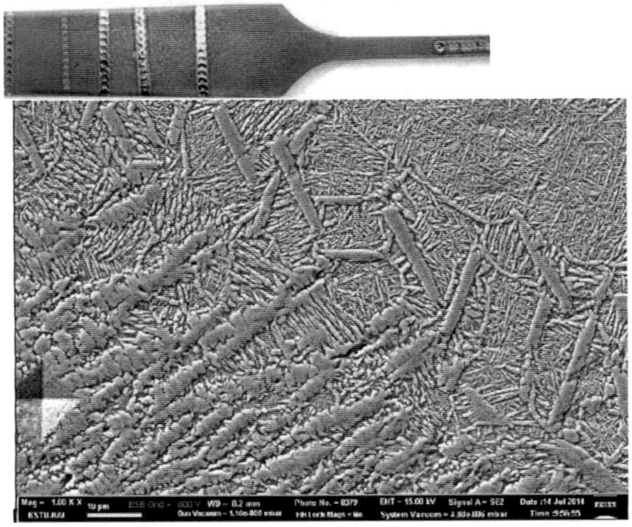

**Figure 3**. Examples with the traces of laser machining: sample of VT6 alloy, structures of the area with partially melted metal and measured $HV^{100}$ hardness.

Figure 4 shows the results of microhardness ($HV^{100}$, MPa) measurement in laser-machined areas (across and inward) of VT6 alloy and degree of laser hardening in relation to the base areas up to 167%, which corresponds to ~ HRC 60 under the following conditions of a laser generator: machining speed of V = 20 mm/s, number of runs of n = 20, average power of N = 20 (100 %), f = 21 500 Hz, laser spot size of d = 40 μm.

The structure analysis of laser-machined areas shows that the structure consists of heat affected zones, fusion zones, and a crater. It has the areas where it interacts with the environment directly, i.e. a crater ("plateau") with melted heterogeneous structure of metal and area of melted structure, which is isolated from interactions, including the partially melted fragment. The hardness increases slightly above the base VT6 alloy in this area. The geometry of heat affected zones and specifically of fusion zones in titanium alloys depends significantly on the conditions of laser machining. In certain instances, laser machining can have negative results (alloy cracking) due to a combination of large thermal input of highly concentrated energy and some thermophysical properties of titanium [3]. It means that, with significant temperature gradients, the pressure of evaporating titanium vapors presses a portion of molten metal out. It increases the melting ability of a laser beam due to its deeper penetration through a small crater at the center of the fusion zone, which is sealed by melting at the last stage of pulse, then solidifies as a "plateau". It was found with INCA X-MAX spectrometer that Ti and V were strongly oxidized and burnt out on the surface of the crater (Fig. 5). Such complex compounds as aluminum spinel, carbonitrides, and SiC are formed. With that, tremendous internal stresses occur and increase the brittleness of hard "plateau".

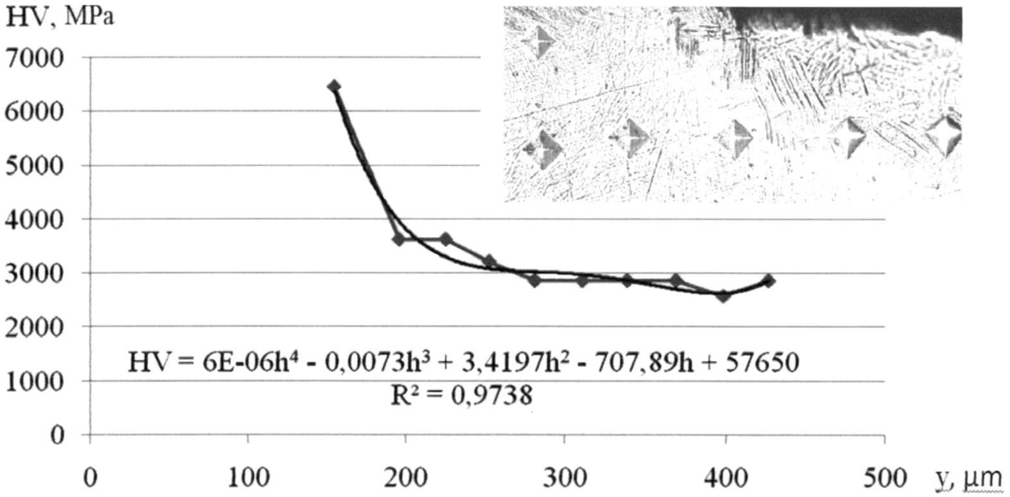

a

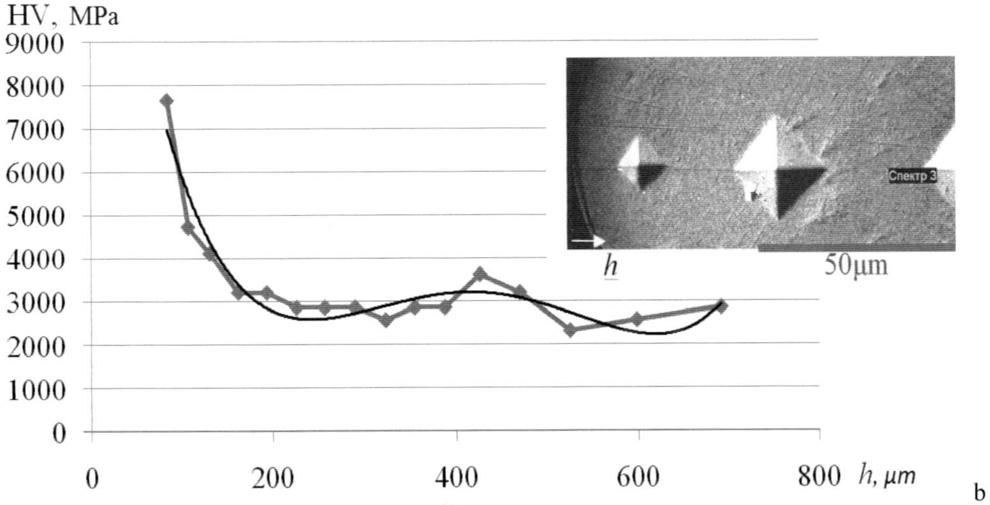


**Figure 4.** Change in HV[100] hardness along the "plateau" in the *y*-direction (a) and inwards - h (b) from the crater to the fusion zone

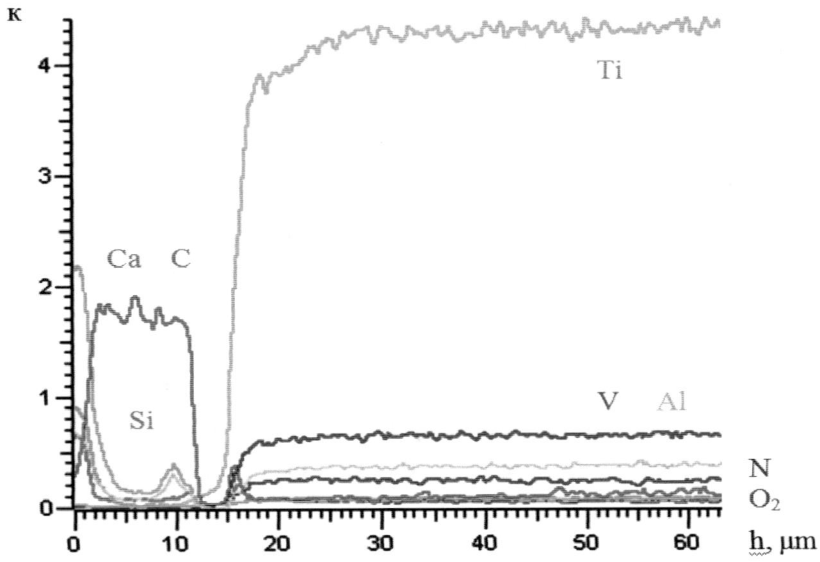

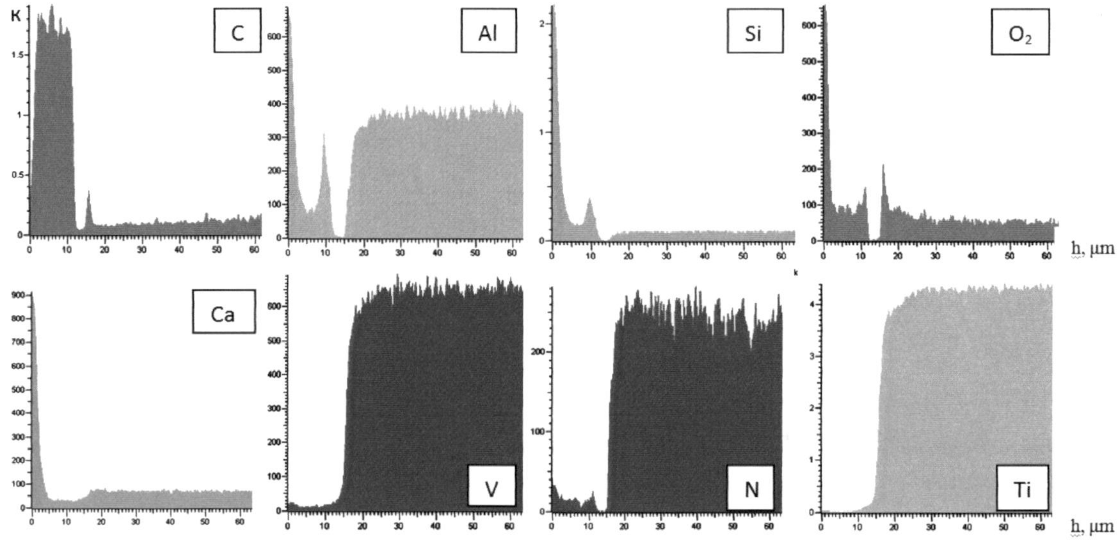


**Figure 5.** Change in element-by-element composition of VT6 alloy inwards the laser-machined areas (from the crater to the fusion zone)

Phase age hardening should be used to provide stable properties, improve hardening of VT6 alloy after laser quenching with a hardness of > HRC 60. In addition to treatment localization and higher hardness, the positive effects are as follows: stable geometry and dimensions of microsurgical instrument. Hardening performance is defined by the increase rate of temperature gradient, nature of alloy interaction with the environment, and concentration of supersaturated phases $\alpha + \alpha_п + \alpha^I + \alpha^{II}$ and $\beta_п$ of the transformed solution.

Meanwhile, performance and service life of working instrument surfaces are related to an increase in structural strength [1, 2, 10, 11], which is defined by the morphology of the phases [2] of the constituents of structure of phases $\alpha + \alpha_п + \alpha^I + \alpha^{II}$ и $\beta_п$ of solid titanium alloy solution. It was analyzed how the key indicators of structural strength ($S_K$, $m^{(0)}$, $\psi$, and $E_K$) were influenced by a dimension ratio of $\alpha/\alpha_п$-phases (see Fig. 1) in the variants of alloy structural conditions. Figures 6-8 and [2] show the examples of conformance of the mentioned characteristics of the structure and the properties ($\psi$, $E_K$, $m^{(0)}x10^2$).

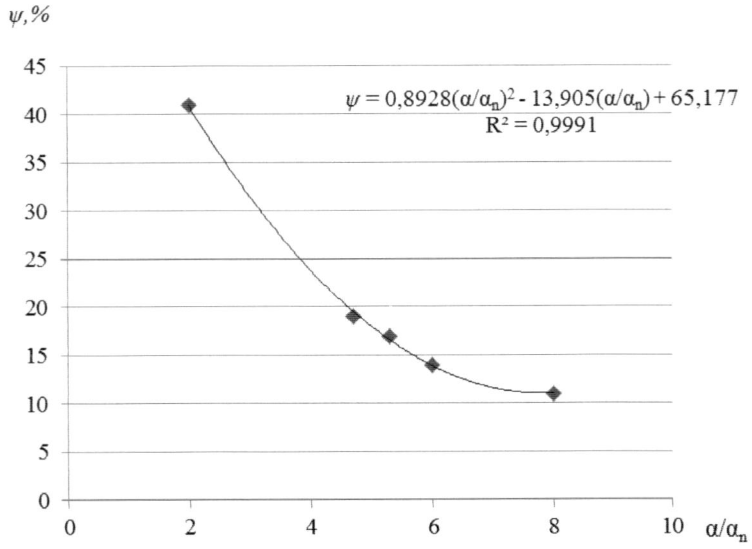


**Figure 6.** Relationship between phase ratio and contraction - $\psi = f\,(\alpha/\alpha_{п})$

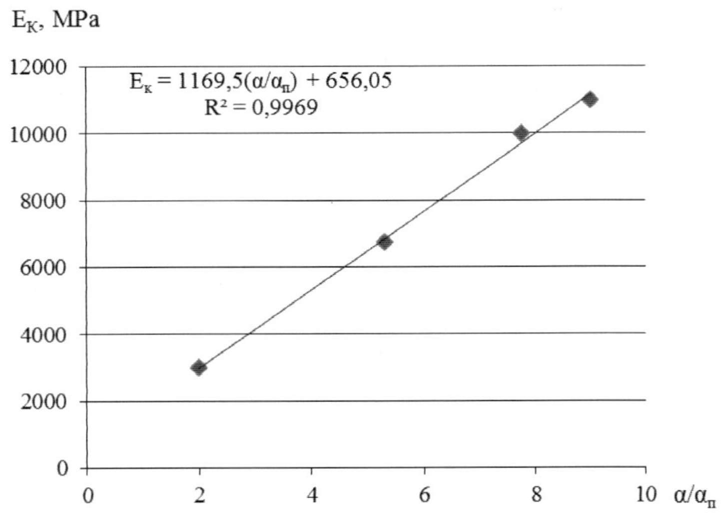


**Figure 7.** Relationship between phase ratio and Ilyushin finite module (limit value of cutting module) [1]) – $E_K = f\,(\alpha/\alpha_{п})$

According to the defined regression coefficients: $R^2_1 = 0.9991$, $R^2_2 = 0.9969$ и $R^2_3 = 0.8542$ (see the curves in Fig. 6 и Fig. 7) the formulas:

$$\psi = 0.899(\alpha/\alpha_n)^2 - 13.91(\alpha/\alpha_n) + 65.18 \quad (1)$$
$$E_K = 1170(\alpha/\alpha_n) + 656.1 \quad (2)$$

- establish more stable connections than the formula (3) with a lower value of $R^2_3$ (according to the curve in Figure 8)

$$m^{(o)} \times 10^2 = -0.713\,(\alpha/\alpha_n) + 10.43 \quad (3)$$

By solving (2) and (3) together with the formulas [7], we get a simple linear expression of true strength from the structure parameters [2]

$$S\kappa = 1700\text{-}41(\alpha/\alpha п) \qquad (4)$$

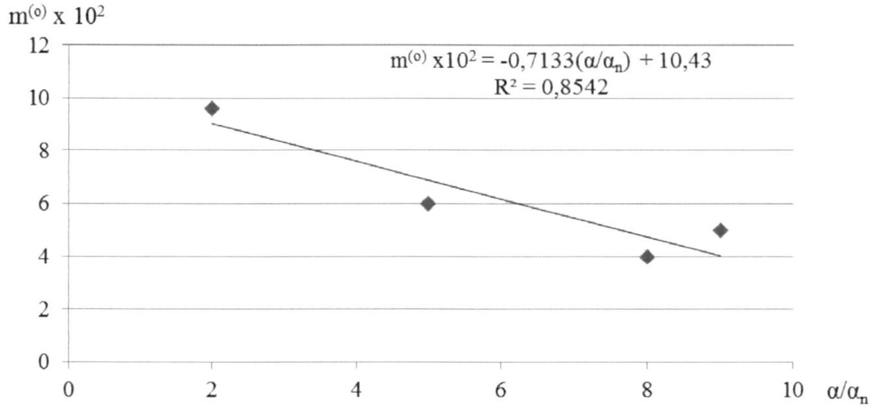


**Figure 8.** Relationship between phase ratio and strain-hardening coefficient $m^{(o)}=f\,(\alpha/\alpha_п)$

The capabilities of the station based on fiber oscillator LS-10 with the protective cabin make it possible to use it as a pulsed laser deposition module. At the same time, the heat affected zone is minimal, which makes it possible to maintain the geometry and accuracy parameters of a fine instrument. Filler material is fed locally, the melting surface area reaches 90 %. Figure 9 shows the structure and quality of adhesion of deposited film with the plate made of VT6 alloy, measurement of microhardness of melted deposit at the substrate. The element-by-element composition of deposited metal on the VT6 (Ti-6Al-4V) plate is given in the table.

Element composition (%) in the areas of laser deposition at a depth of h Table

| h, μm | C | Al | Si | Ti | V | Cr | Fe | Ni |
|---|---|---|---|---|---|---|---|---|
| 400 | 13.14 | 1.28 | 2.65 | 9.27 | 0.18 | 7.17 | 3.07 | 63.25 |
| 525 | 7.29 | 4.18 | 0.91 | 54.33 | 2.41 | 2.72 | 1.40 | 26.74 |
| 600 | 5.056 | 5.84 | 0.08 | 85.30 | 3.96 | - | 0.11 | 0.13 |

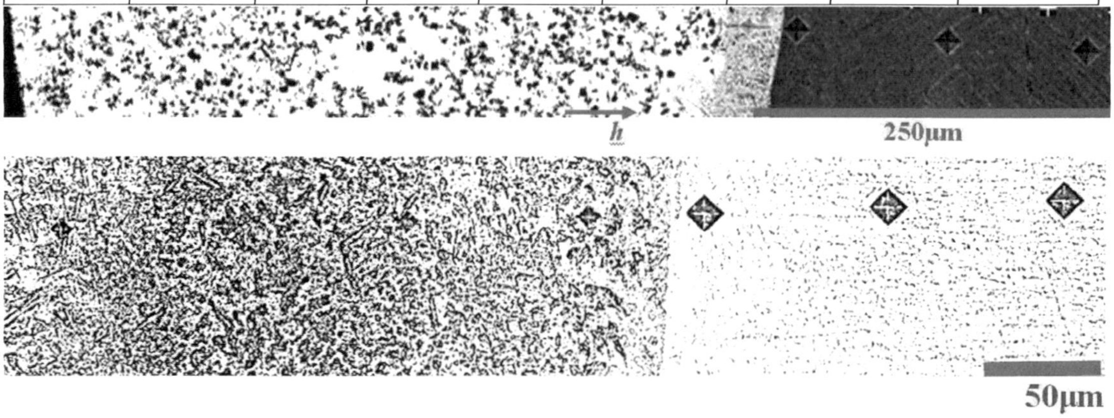

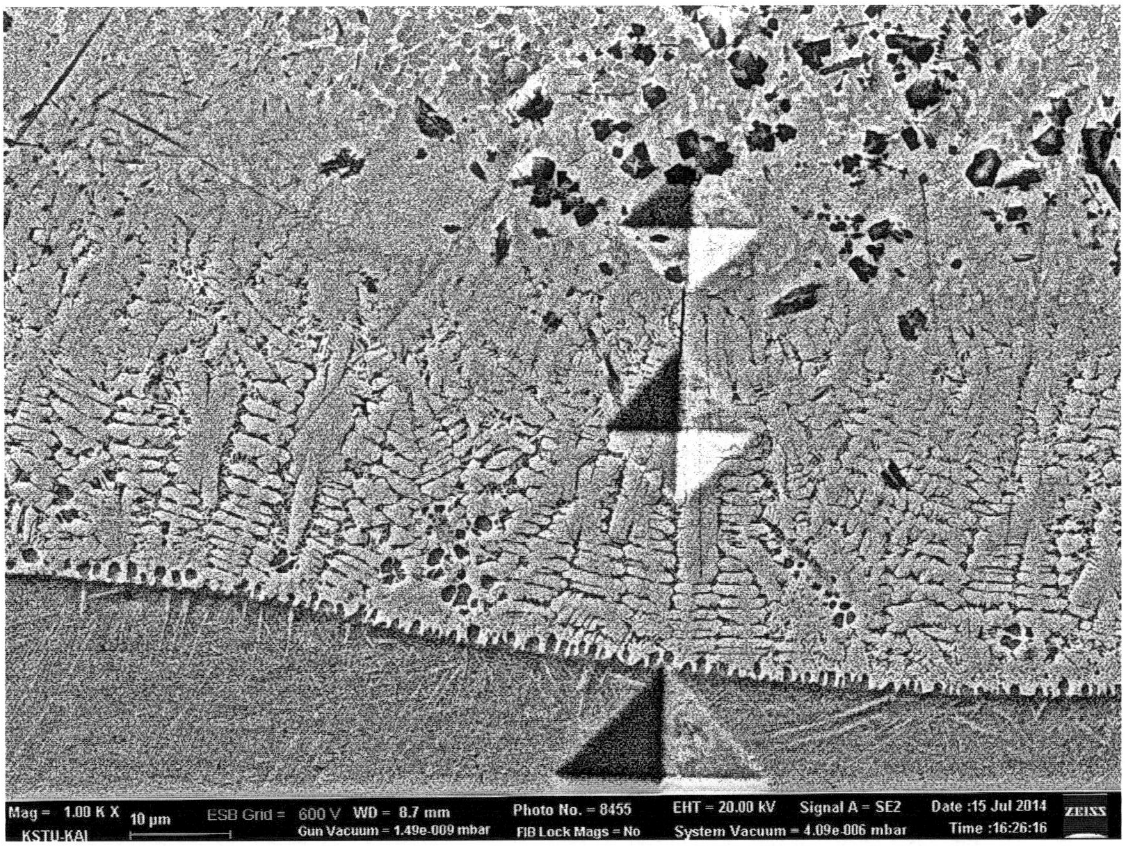


**Figure 9.** Measurement of microhardness and structure of deposited metal on the VT6 (Ti-6Al-4V) plate at the substrate

The deposited metal in the melted area has an austenitic-carbide microstructure. No melted metal of the substrate made of VT6 base alloy was found. Pulsed laser deposition on VT6 (Ti-6Al-4V) plate increases a metal hardness within the range from HRC 70 to 80 at a depth of 560 µm, which exceeds the characteristics of working surfaces of the Russian and foreign needle holders. For instance, a hardness of HRC 73…75 is provided in the plates made of hard alloy (Germany).

## Conclusions

1. The common deposited films on the working surfaces of needle holders with tungsten carbides on martensitic steels and VT6 alloy have a non-homogenous composition, thickness, and they are discontinuous.
2. Deposited "diamond coatings" of KMIZ needle holders are quite brittle, peel off, and their hardness is very low.
3. Plates welded to an instrument are not easy to manufacture, and they are impractical.
4. Therefore, the conventional methods of forming the working surfaces in needle holders are not always reliable, long lasting, and functional. A steel instrument has a higher weight and wears out users.
5. An instrument made of titanium alloy has a lower weight and higher corrosion resistance.
6. In terms of hardness and weight factor, proposed VT6 titanium alloy (Ti-6Al-4V) and variants of laser technologies (shock hardening and deposition) usually exceed the characteristics of working

surfaces of the needle holders made in Tatarstan, Russia and foreign countries, they can make microsurgery conditions more comfortable.

**References**

1. Murataev F.I., Zharzhanazi M.A. Provision of structural strength of titanium alloys according to criteria of ultimate plasticity and fatigue resistance // Vestnik of KSTU, 2013. No. 1. pp. 50-54. (in Russian).
2. Murataev F.I., Mukhamadeev I.M. Patterns of structural conditions and structural strength characteristics of pressed parts made of titanium alloy // Vestnik of KSTU, 2020. No. 2. pp. 50-58. (in Russian).
3. Murataev F.I., Klabukov M.A. Features of laser shock hardening for steels and titanium alloys // Vestnik of KSTU, 2012. No. 4. pp. 82-84. (in Russian).
4. Klabukov M.A., Danilov E.V. Rationale for composition of materials and technologies for working surfaces of microsurgical instrument // Vestnik of KSTU, 2019. No. 2. pp. 66-73. (in Russian).
5. F I Murataev et al Ranking materials technologies by limiting characteristics of heat-resistant alloys and their longevity in the problems of import substitution 2019 IOP Conf. Ser.: Mater. Sci. Eng. 570 012070.
6. Cao X et al (2009) Effect of welding speed on butt join quality of Ti-6Al-4V alloy welded using a high-power Nd: YAG laser. Optics and Lasers in Engineering. 47, pp. 1231-1241.
7. Murataev F.I., Khakimov S.Sh. Substantiation of microstructure and properties relationships of titanium alloys for prediction problems of fatigue resistance // Vestnik of KSTU, 2014. No. 3. pp. 110-113. (in Russian).
8. F I Murataev et al Substantiation of domestic material and welding technology for improving properties and competitiveness of pyrolysis furnace coils 2019 IOP Conf. Ser.: Materials Science and Engineering 570 012071.
9. Murataev F.I., Murataev A.F. Influence of composition and morphology of intermetallic phase of IN-738LC alloy on damageability of gas turbine blades // Vestnik of KSTU, 2015. No. 3. pp. 43-48. (in Russian).
10. Bratukhin A.G., Pogosyan M.A., Tarasenko L.V. Engineering and functional materials for today's aircraft industry. M.: MAI Publ., 2007. 301 p.
11. Shkanov, I.N., Braude, N.Z., Murataev, F.I. Model of optimization of basic mechanical properties with respect to static and fatigue strength criteria in complex stress state conditions. // Soviet Aeronautics (English translation of Izvestiya VUZ, Aviatsionnaya Tekhnika) Volume 25, Issue 1, 1982, pp. 79-84.
12. Gabdrakhmanov A.T., Shafigullin L.N., Galimov E.R., Ibragimov A.R. Surface thermohardening by the fast-moving electric arch. IOP Conf. Series: Journal of Physics: Conf. Series. Volume 789 (2017). Article number 012010.
13. Gabdrakhmanov A.T., Israphilov I.H., Galiakbarov A.T. The study the erosion of the electrodes under the influence moving electric arc // Journal of Physics: Conference Series, Volume 567, Issue 1, 2014, Article number 012013.
14. Gabdrakhmanov A.T., Galiakbarov A.T., Samigullin A.D., Galiakbarov R.T. The calculation of a thermal field in the surface of a processed partunder the influence of a low-temperature plasma // IOP Conf. Ser.: Mater. Sci. Eng., Volume 134, Issue 1, 2016, Article number 012040.
15. Gabdrakhmanov A.T., Israphilov I.H., Shafigullin L.N. The metal surface cleaning using a vapor-gas discharge // IOP Conf. Series: Journal of Physics: Conf. Series. Volume 1058 (2018). Article number 012009.
16. Galiev I.G., Khafizov K.A, Khusainov R.K. Increase of efficiency of tractors use in agricultural production // 17th International Scientific Conference Engineering for rural development Proceedings, Volume 17 May 23-25, 2018. pp. 373-377.
17. Khusainov R.N., Khafizov K.A, Nurmiev A.A., Galiev I.G. Optimization of main parameters of tractor and unit for seeding cereal crops with regards to their impact on crop productivity // 17th

International Scientific Conference Engineering for rural development Proceedings, Volume 17 May 23-25, 2018. pp. 168-175.
18. Alexandr Belinsky, Bulat Ziganshin, Ayrat Valiev, Damir Haliullin, Ilgiz Galiev. Theoretical investigation of increasing efficiency of combine harvester operation on slopes // 18th International Scientific Conference Engineering for rural development Proceedings, Volume 18 May 22-24, 2019. pp. 206-213.

IOP Conf. Series: Materials Science and Engineering **915** (2020) 012043 doi:10.1088/1757-899X/915/1/012043

# To the question of automation of the injection molding process

**D L Pankratov[1]and R V Gavariev[2]**

[1]Kazan Federal University, Naberezhnye Chelny Institute, 423812, Russia, Naberezhnye Chelny, Prospekt Syuyumbike 10A,
[2]Kazan national research technical university named after A.N. Tupolev-KAI,

Gavarievr@mail.ru

**Abstract**. The article discusses the features of the injection molding process (IMP) from the point of view of the possibility of process automation. The main factors of the casting process that provide a good opportunity for process automation are indicated. The range of tasks of the injection molding process is determined, for the solution of which it is advisable to use an expert system (ES). The ES model for injection molding is proposed, which provides high-quality castings.

**1. Introduction** Injection molding (IM) is one of the most productive and efficient methods for producing blanks with a material utilization rate of up to 95%. Another important advantage of IM is the possibility of producing castings of complex configuration with a minimum wall thickness of 0.6 mm. However, one of the limiting factors is the complexity of the processes that occur in the mold cavity of the mold when pouring the melt. The simultaneous influence of temperature and power factors leads to the likelihood of various casting defects, such as: porosity, cold laps, shells, various inclusions, heterogeneity, and others [1].

The elimination of these defects is achieved by optimizing the process parameters based on the experience of a technologist or operator of foundry equipment [2]. At the same time, there is subjectivity of the decisions taken to eliminate defects, which ultimately can negatively affect the quality of the products obtained.

## 2. Results

One of the possible solutions to this situation is the use of computer systems based on artificial intelligence (AI). There are various approaches to understanding the tasks of AI and creating intelligent information systems: genetic algorithm, neural networks, intelligent agent, artificial neural networks, fuzzy logic (expert system). At the same time, the ES is the most suitable for the injection molding process, since, unlike other artificial intelligence systems, they are designed to solve problems in a specific, narrow area [3].

A correctly selected competent expert can provide ES with unique and valuable knowledge [4]. Consequently, the value of the entire expert system in the form of a finished product is characterized by the quality of the created knowledge base.

Table 1 shows the relationship between the applications of ES and the casting process with pressure application, which confirms the advisability of using ES.

Table 1. The tasks of the injection molding process, for those solution it is advisable to use ES.

| Types of ES application | Explanation | ES objectives |
|---|---|---|
| Interpretation | Description of situations based on observed data. | Recognition, understanding of speech, image analysis, determination of the structure of chemical compounds. |
| Forecasting | Conclusion of probable consequences from given situations. | Prediction of the state of machines and mechanisms. Prediction of casting quality and probability of defects. |
| Diagnostics | Conclusion about violations in the system, based on observations. | Hardware troubleshooting. |
| Design | Constructing objects with restrictions. | The allocation of resources to support the IMP process. Design of the IMP process. |
| Monitoring | Flexible management of certain system functioning. | Monitoring the steps of the IMP process. Final result control. |
| Training | Transfer knowledge of any field to the user or its verification. | Testing of knowledge and advanced training of workers. Training in finding solutions to problematic situations. |

The information base includes permanent and variable information. Permanent information consists of an indicator of the state of the furnace heating element, data on the temperature of melt heating in the crucible, an indicator of the state of the mold lubrication process, verification of melt pressing in half-molds, crystallization and cooling control [5]. Variable information includes data on the number of defects, data on technical diagnostics, laws describing any change in technical condition, and more.
The expert system consists of a large number of algorithms that are combined into computational and logical models. These models help to find a solution to the problem and form the content of the model base. In the process of operation, ES databases can be corrected and updated with new information, previously unknown problem situations and models for their solution [6].

The decision-making function is aimed at choosing the optimal solution to problematic situations in case of malfunction in the normal operation of the control object. The implementation of the recommended measures is carried out by the regulatory function. The function of the planning unit is to calculate the resources necessary for the smooth operation of the system as a whole [7].
The control function compares the control object's planned and accounting indicators and analyzes the impact of decisions on the system. This block forms a control loop. In case of data disagreement, the transition to the control loop begins, then additional recommendations to the control object are made.

The implementation of ES is based on the constant interaction of three main parts: a person (a decision maker), a hardware-software complex, and a control object, which is main part in decision making process [8].

The interaction of the decision maker and the system occurs in a dialogue mode, in which, when describing the problem, the ES proceeds to the automatic search for a solution. ES decision making is concluded in a series of logically interconnected steps corresponding to the tasks that the system solves:

- evaluation of the process system;
- definition of goals and performance criteria;
- solution development;
- decision making;
- decision implementation;
- evaluation of results.

The expert system should apply a policy of direct withdrawal and splitting into subtasks. The essence of crushing into subtasks is to refine the fragments. Fragments are considered as the achievement of intermediate goals on the way to the final goal. An example would be the situation of deflection of the rod. The main factors, that can affect this defect, are formulated (uneven heating by thickness or width, mismatch of the axial load level, initial twist) [9], [10], this allows us to reduce the volume of technological methods of searching for regulatory parameters.

## 3. Conclusions

Thus, injection molding is a complex manufacturing process for producing castings, and there are many factors that affect the quality of their production. The correctness of the decisions taken to eliminate defects in each specific situation is directly related to the experience of the staff, which negatively affects the efficiency of the IMP process. Taking into consideration the good background of the IMP process for automation, the most promising is the use of computer systems utilizing artificial intelligence based on expert systems. This makes it possible to increase the efficiency of the injection molding process by improving the quality of the obtained castings, as well as saving various resources.

## References

[1] Gavarieva K N, Simonova L A, Pankratov D L and Gavariev R V 2017 *Development of expert systems for modeling of technological process of pressure casting on the basis of artificial intelligence* IOP Conf. Series: Materials Science and Engineering 240 012019 DOI: 10.1088/1757-899x/240/1/012019

[2] Yan Q S, Yu H, Lu G, Xiong B W, Lu B P and Zou X 2014 *Effect of crystallization pressure on secondary dendrite arm spacing of vacuum counter-pressure casting aluminum alloy* School of Aeronautic Manufacturing Engineering Nanchang Hangkong University (Nanchang) 330063 Vol 24 No 5 pp 1194-1199

[3] Gavariev R V, Savin I A and Leushin I O 2017 *Improvement of zinc castings surface quality by laminated protective coating* Tsvetnye Metally No5 pp 84-88 DOI 10.17580 / tsm.2017.05.13

[4] Hu X P, Zhao G Q and Wang W M 20*10 Solidifying pressure and microstructure of alsi10cu3 in die sleeve in high pressure die casting* International Journal of Cast Metals Research vol 23 № 5 pp 289-295

[5] Gavariev R V and Savin I A 2017 *Improvement of surface quality of casting produced by casting under pressure* Solid State Phenomena 265 pp 988-993 DOI: 10.4028/www.scientific.net/SSP.265.988

[6] Gavariev R V and Savin I A 2018 *Research of the mechanism of destruction of compression molds for casting under pressure of color alloys* Solid State Phenomena Vol 284 pp 326-331 https://doi.org/10.4028/www.scientific.net/SSP.284.326

[7] Pankratov D L and Gavariev R V 2019 *Improving the quality of castings made of non-ferrous metal alloys when casting in metal molds* IOP Conf. Series: Materials Science and Engineering 570 012072 doi:10.1088/1757-899X/570/1/012072

[8] Deev V B, Prusov E S and Kutsenko A I 2018 *Theoretical and Experimental Evaluation of the Effectiveness of Aluminum Melt Treatment by Physical Methods* La Metallurgia Italiana №2 pp 16-24

[9] Gavarieva K N, Simonova L A, Pankratov D L, Shibakov V G and Gavariev R V 2018 *Application of multi-agent system for control of parameters of precision stamping process of bevel gears* IOP Conf. Series: Materials Science and Engineering 412 012020 DOI: 10.1088/1757-899X/412/1/012020

[10] Gavariev R V, Savin I A and Leushin I O 2019 *To the question of casting alloys of non-ferrous metals in the metal mold* Materials Science Forum 946 pp 631–635. https://doi.org/10.4028/www.scientific.net/msf.946.631

IOP Conf. Series: Materials Science and Engineering **915** (2020) 012044 doi:10.1088/1757-899X/915/1/012044

# About computer calculation error of vermicular graphite share in microstructure of CGI

**A G Panov[1,2], I F Shaekhova[1], S A Ryzhkova[1], N G Degtyaryova[2] and V A Ivanova[3]**

[1]Naberezhnye Chelny Institute (branch), Kazan Federal University
Sjujumbike street 10a, Naberezhnye Chelny, 423800 Tatarstan, Russia
[2]LLC RC Modificator, Naberezhnye Chelny
[3]Yaroslav State Technical University, Yaroslav, Russia;

irrra1603@yandex.ru

**Abstract.** With each passing year, digital technologies are increasingly being used in industry, including for quality control of various materials. The article deals with the possibilities and problems of application of digital methods of microstructure control in CGI. The analysis of the main sources and grade of error of calculating the share of vermicular graphite in the microstructure of the CGI is made. The error remains unacceptably high for some material science tasks and requires further improvement.

1. **Introduction** It is well known that graphite inclusions (GI), which are usually characterized by dimensional-topological parameters (size, shape, distribution and quantity), have a significant influence on the properties of cast iron. It is well known also that specific properties of CGI are achieved due to the presence of the so-called vermicular graphite (VG) particles in its structure. The particles in the microsection plane are elongated and arranged in random order as in grey cast iron, but at the same time they are shorter and thicker and have rounded edges [1], which provides unique properties of CGI.

Determination of microstructure characteristics of CGI in Russia is regulated by GOST 3443-87, which defines only three parameters of vermicular graphite - its form, distribution and share - by the method of comparative analysis of real microstructures with standart microstructures [2].

2. **Body text** Domestic standard GOST 3443-78 was developed more than 30 years ago and obviously does not meet modern requirements of material science and mechanical engineering and requires updating. The analysis of world standards concerning requirements to cast iron with vermicular graphite, including those containing regulations for description of their microstructure, confirms it [3]. The analysis has shown that the first similar foreign standards was developed approximately at the same time as domestic standards - in the 1980s, but unlike domestic standards, they have been repeatedly updated to take into account constantly emerging new data on both properties and microstructure of CGI. Updating has been particularly active over the last 13 years. Initially, both in the USSR (Russia) and abroad, requirements to the properties and microstructure of a CGI were regulated by separate standards. However, in 2006 the first edition of the Specialized Complex International Standard on CGI "ISO 16112:2006 (E). Compacted (vermicular) graphite cast irons - Classification", which included both the requirements for the properties and microstructure of as well as a description of improved regulations for the control of the proportion of VG in the microstructure of a CGI. The standard was updated every 5 years, the last edition was released in

2017, the 2006 version was used as a basis for the development of standards GB/T 26655-2011 (China) and JIS G 5505:2013 (Japan). This situation with regulations on CGI is apparently a consequence of its uniqueness and has not yet been sufficiently studied, including in the area of vermicular graphite control in its structure.

Modern metallographic complexes include automatic image analyzers (AAI), which allow simplifying considerably of metallurgists work, processing at the same time a large array of data and generating reports according to established requirements, etc. [8]. By the AAI it is possible to describe structural components with a larger set of parameters than form, share and distribution. In addition, the AAI always reproduces mathematical criteria assigned to determine the parameters in the same way, which allows to correctly comparing products manufactured by different technologies. Modern AAI are able to calculate a large number of geometric characteristics of objects, both proposed before the appearance of computers [9] and new, including: the area of the object, the percentage of the area occupied by the object from the area of the image, the perimeter of the object, the maximum and minimum of the Feret and Martin diameters, the coefficient of the region filling, the elongation, the average diameter, the shape parameters, the orientation angle, eccentricity, etc. [4].

However, as shown by the authors industrial experience of AAI application and as repeatedly noted by other researchers [5], the combination and form of structural components on metallographic images can provide a very complex picture, for the objective interpretation of which a computer program currently requires the metallurgist's help in control of available parameters [6]. In order to eliminate the human factor that allows for an unpredictable error in the analysis result, actual metallographic task of of the CGI microstructure analysis is to develop a fully automatic mode without the need to control of AAI parameters by the metallurgist. To achieve this objective, it is necessary:

- to establish requirements to sharpness and resolution allowed on quality and quantity of defects of sample preparation, and also the area of micro surface (panorama size) of the analyzed image;

- to develop objective mathematical criteria for the most objective identification and division of graphite inclusions into types, providing the evaluation of morphology and geometric dimensions of each particle with the minimum error in the automatic without interference from the metallurgist.

3. **Conclusions** The tasks of developing methods for controlling structural components of materials based on modern geometric principles [7-12] in particular methods for estimating graphite phase parameters are still relevant and require urgent solutions. At the same time, the use of program complexes allows in some cases to significantly reduce the error of determining the VG share, which indicates the need for further development of these methods for assess the structure of CGI.

**Acknowledgements** The authors would like to thank Tatiana Alexandrovna Sivkova, SIAMS LLC specialist, for her valuable comments and recommendations on this work.

**References**

[1] Dawson Steve 2017 The special aspects of CGI structure and properties defining perspectives of its application in mechanical engineering *Mechanical engineering details from cast iron with vermicular graphite: properties, technology, control: abstracts of the ISTC* (Nab. Chelny, 17–18 oct. 2017 г.) edited by A G Panov, Kazan: Publishing house Kazan Federal University pp12-13.

[2] GOST 3443-87 Cast iron castings with graphite of different form. Methods of structure determination.

[3] Ivanova V, Pobegalova E, On the necessity of updating requirements for vermicular graphite iron 2018 *Science and technologies modifications of iron ISTC* (Nab. Chelny, 15–19 oct. 2018) edited by A G Panov, Kazan: Publishing house Kazan Federal University pp 33-34.

[4] Shaekhova I F, Panov A G, Nazipova A I, Chentaeva A E Microstructure image processing with automatic analyzer tools Image Expert Pro 3 *Astra Salvensis* vol. 2017 pp 471-480.

[5] Kazakov A, Kovalev P, Ryaboshuk S Metallurgical expertise as the base for determination of nature of defects in metal products, 2007 *CIS Iron and Steel Review* №1-2, pp 17-23.

[6] Saltykov S A Stereometric metallography 1970, *3rd ed. M.: Metallurgy* 376 p.

[7] Kuts Yu V, Povstyanoy O Yu Reviev of modern materials research software for computers in processing metallographic images 2014 *Interuniversity collection «Naukovi Notatki» Issue №45. Lutsk* pp 329-336.

[8] Sivkova T A, Gusev A O, Gubarev S V, Britsheva A V, Samoilova A Yu, Kadushnikov R M Peculiarities of the graphite microstructure control in cast iron by automatic methods (in Russian) 2018 *Metallurgia Mashinostroenia* №2 pp34-38

[9] ISO 16112:2006 (E). Compact (vermicular) graphite in cast iron. Classification.

[10] Panteleev V G, Slaev V A, Chunovkina A G Metrological provision for image analyzers 2008 *Measurement Techniques* vol. 51. № 1 pp 107-112.

[11] Makarenko K V Identification of graphite inclusions in cast iron 2009 *Liteynoe Proizvodstvo.* №4 pp 2-6.

[12] Makarenko K V Simulation of the process of crystallization of spheroidal graphite cast iron (in Russian) 2009*Metal Science and Heat Treatment* № 11 pp 16–20.

IOP Conf. Series: Materials Science and Engineering **915** (2020) 012045 doi:10.1088/1757-899X/915/1/012045

# Evaluation of the wear resistance and resource hardened by electrospark machining of the cutting surfaces of the fingers of the headers of combine harvesters

**S A Velichko[1], S N Sharifullin[2], I S Kuznetsov[3], A V Kolomeichenko[4], R Y Solovyev[4], I N Kravchenko[5], N R Adigamov[6]**

[1]National Research Mordovian State University named after N.P. Ogarev, Russia, e-mail:
[2]Kazan Federal University, Russia, e-mail:
[3]Orel State Agrarian University named after N.V. Parakhin, Russia, e-mail:
[4]Center for Agricultural Engineering "NAMI", Russia,
[5]Institute of Engineering A.A. Blagonravova RAS, Russia, e-mail:
[6]Kazan state agrarian University, Russia, e-mail:

velichko2005@yandex.ru, Saidchist@mail.ru, ivan-654@yandex.ru, a.kolomiychenko@nami.ru, rudsol@mail.ru, kravchenko-in71@yandex.ru, n-adigamov@rambler.ru

**Abstract:** The article presents the technology of hardening the fingers of the reapers of combine harvesters by electrospark processing. To strengthen the cutting surfaces of the fingers, it is recommended to use an amorphous alloy grade 84KXSR. The operational tests presented in the work showed that wear and dulling of both serial products and experimental parts with electric spark coatings is characterized by two stages: running-in and natural wear. The running-in occurs during the running time of the fingers in 1 ha. The results of operational tests showed that the application of a wear-resistant electrospark coating of amorphous alloy grade 84KXSR with a thickness of 30 μm and a track width of 2 mm on the cutting surfaces of the fingers of the headers of grain harvesters makes it possible to increase the resource from 1.4 to 2 times.

## 1. Introduction

One of the most often required elements of agricultural machinery are cutting elements of grain combine harvester headers [1]. To increase their lifetime is necessary to create hardened layers or apply coatings with high physical and mechanical properties [2]. One of the most advanced methods of hardening coatings application on the working surfaces of machine elements is electric spark machining.It is based on the phenomenon of electric erosion and transfer of the anode material (electrode) to the cathode (element) at electric discharges passing in gas chamber. Resulting from this electric spark coating is formed on the metal surface [3, 4].

One of the possible ways of increase productivity of electric spark machining and wear resistance of electric spark coating is the creation of amorphous structure in them [5 -6]. The obtaining of such electric spark coatings is possible at the usage of the corresponding electrode materials. Now, the larger number of amorphous alloys is manufactured in the band form with thickness of 50 μm, obtained with quenching out of liquid state with speed of $10^6$ Fps. Such technology of amorphous alloy manufacturing gives the opportunity to study the materials a set of unique properties, which in many factors exceed conventional alloys. Application of amorphous alloys as electrode material allows increasing of the process productivity, working surface wear resistance [7-10]. Herewith the increase of energy consumption of electric spark machining does not occur.

The goal of the research was the comparative estimation of wear resistance of the cutting surfaces of the experimental elements, hardened by electric spark machining, and stock produced items from headers of grain combine harvesters.

## 2. Materials and methods

Wear-resistant electrospark coating was applied to the surface facing the knife segment. For applying an electric spark coating, a modernized installation of the UR-121 brand manufactured by LLC PELM was used, which has a capacitance C = 56 μF and an electrode vibration frequency of f = 100 Hz. Coating was carried out in mode No. 5 (I = 17.5 A; U = 73 V; f = 100 Hz; t = 6 min / cm2; W = 0.16 J). The width of the trace left by the electrode was 2 mm, and the thickness was 30 μm. The electrode material was selected according to the results of wear resistance tests, according to which the amorphous alloy 84KXSR was the most suitable for hardening the fingers of grain harvesting machines. The fingers P230.21.000, DQ11499, H213405 and KG35 379720 participated in the tests. 40 serial products and 40 experimental parts with electric spark coatings were installed for operational tests. During testing, through each hectare of operating time of the studied objects, the considered working surfaces of the cutting parts were measured. The obtained values were processed using a personal computer using the PTC Mathcad 15 software package.

## 3. Results and discussion

To estimate the cutting surfaces wear resistance in the Orel region, in the conditions of common exploitation the operational tests of the experimental elements with electric spark coatings from amorphous alloy of trademark 84KXCP and stock produced items were carried out. After experimental data processing, the regression equations of wear from service hours and the numerical values of wear speed and wear resistance of the tested experimental elements and stock produced items were obtained (Table 1).

**Table 1.** Tribotechnical characteristics of cutting surfaces, investigated in the course of operational tests of objects.

| Finger model | Speed of wear, *(mm/ha)* | Wear resistance, *(ha/mm)* | Relative wear resistance | Equation of logarithmic regression |
|---|---|---|---|---|
| H213405 with electric spark coating | 0,063 | 15,87 | 1,91 | $I(T) = 0{,}178 \cdot ln(T + 0{,}554) + 0{,}107$ |
| H213405 | 0,12 | 8,33 | 1 | $I(T) = 0{,}43 \cdot ln(T + 0{,}933) + 0{,}037$ |
| KG 35 379720 with electric spark coating | 0,1 | 10 | 1,7 | $I(T) = 0{,}394 \cdot ln(T + 1{,}633) - 0{,}198$ |
| KG 35 379720 | 0,17 | 5,88 | 1 | $I(T) = 0{,}364 \cdot ln(T + 0{,}357) + 0{,}377$ |
| DQ11499 with electric spark coating | 0,09 | 11,1 | 2,05 | $I(T) = 0{,}388 \cdot ln(T + 2{,}13) - 0{,}281$ |
| DQ11499 | 0,185 | 5,41 | 1 | $I(T) = 0{,}453 \cdot ln(T + 0{,}485) + 0{,}322$ |
| P230.21.000 with electric spark coating | 0,094 | 10,64 | 2,24 | $I(T) = 0{,}456 \cdot ln(T + 2{,}299) - 0{,}38$ |
| P230.21.000 | 0,21 | 4,76 | | $I(T) = 0{,}56 \cdot ln(T + 0{,}658) + 0{,}248$ |

According to the investigation results, it was proved that wear, blunting of the stock produced items and tested experimental elements are characterized with two stages: aging and natural wear. Aging takes place, first, during cutting surfaces service hours in 1 ha. During this period, the most intensive wear of the examined surfaces is observed. On the cutting surfaces of the experimental elements, small concavity appears and on the analogical surfaces of the stock produced items, simultaneously the explicit convexity is observed. The edges of stock produced items after aging had spherical radius 0,07...0,1 mm. Herewith the similar rounding on the experimental elements was not observed.

The process of natural wear of the cutting surfaces was observed from service hours in 1 ha before the beginning of limit state and has monotonous character. It was found that the reasons of the investigation objects wear are: plant crop containing in its tissue silica crystals. The photo of the hardened and unhardened crossbars of finger P230.21.000 after service hours 6,5 ha is presented in figure 1 for comparison. It shows that the thickness of the hardened crossbars of finger exceeds the thickness of bar without hardening with electric spark machining. Crossbar without hardening has deflection and is unfit for further use. Herewith the hardened crossbar has regular shape and has remaining lifetime.

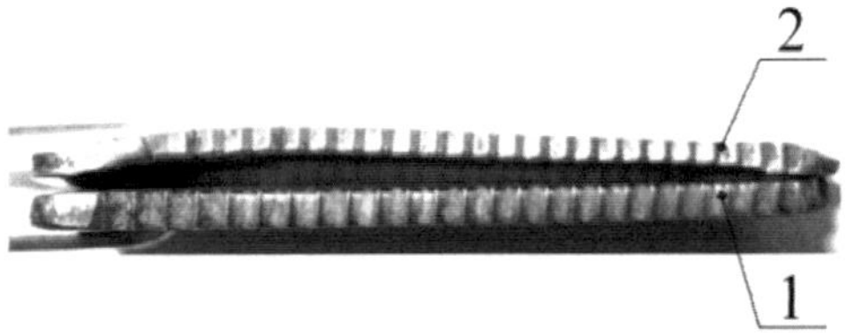


**Figure 1.** General view of cross bars after service hours 6,5 ha: 1 – hardened with electric spark coating, 2 – unhardened.

Performance tests were carried out until the research objects reach certain state. At the end of them, lifetime of the investigated objects was determined (Fig. 2). It was found, that at the cutting surfaces hardening by electric spark coatings of amorphous alloy of trademark 84KXCP, the fingers lifetime increases from 1.4 to 2 times, depending on the trademark.

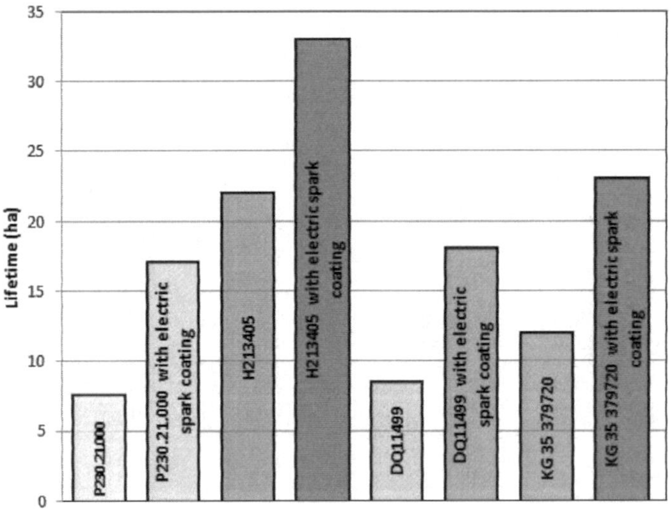


**Figure 2.** Lifetime of the investigated objects participating in operational tests, ha.

While investigating of wear of blades segments operating together with stock produced items or experimental elements a substantial difference was not found out. The loss of operational condition of the blades segment at the performance tests was due to stones getting into the cutting pair and natural wear connected with friction.

Experimental investigations provided manufacturing enterprises with recommendations on hardening of fingers of the header cutter bar of grain combine harvesters by means of electric spark machining. Wear resistant electric spark coating is applied on the finger surface, exposed to blade segment with electrode from alloy of trademark 84KXCP, in two passes (I=17,5 A; U=73 V; f=100 Hz; t= 6 min/sm$^2$; $W_и$=0,16 J). The trace width left with the electrode must be 2 mm, and electro spark coating thickness must be 30 μm.

**4. Conclusions:**

- wear and blunting of stock produced items and experimental elements with electric spark coatings are characterized with two stages: aging and natural wear. Aging takes place during service hours of the fingers cutting surfaces in 1 ha;
- in the process of exploitation on the elements cutting surfaces selected for the investigation convexity is observed. Herewith the application of electric spark coating on the cutting surfaces results in small concavity. It provides, the self-sharpening effect of the cutting edges;
- application of wear resistant electric spark coating from amorphous alloy of trademark 84KXCP on the finger cutting surfaces of grain combine harvesters allows increasing of elements lifetime from 1,4 to 2 times.

**References:**

[1]. Krupin, A.E. Calculation of reliability indicators for the DON-1500 B /A.E combine harvester. Krupin, E.A. Lisunov, A.A. Kalashov //Bulletin of the State Institute of Nuclear Power Engineering, 2018 .- No. 2 (81). - Pp. 42–53.

[2]. Kolomeychenko, A.V. Theory and practice of electrospark hardening of cutting parts of machines with amorphous and nanocrystalline alloys: textbook. monograph /A.V. Kolomeychenko, I.S. Kuznetsov. - Orel: Publishing House Orel GAU, 2015 .- 174 p.

[3]. Kuznetsov, I.S. Process of mass transfer of amorphous alloys under low-voltage electric spark treatment / I.S. Kuznetsov, A.V. Kolomeichenko, V.Z. Pavlov //Surface Engineering and Applied Electrochemistry, 2017 .- Vol. 53. - No. 4. - Pp. 333 - 338.

[4]. Velichko, S.A. Formation of thick layer electro-spark coatings for restoring worn-out parts of power hydraulic cylinders /S.A. Velichko, P.V. Senin, V.I. Ivanov, A.V. Martynov, P.V. Chumakov //Surface Engineering and Applied Electrochemistry, 2017 .- Vol. 53. - No. 2. - Pp. 116 - 123.

[5]. Kolomeichenko, A.V. Investigations of nanocrystalline alloy electrospark coating made of nanocrystalline alloy based on 5BDSR Ferrum /A.V. Kolomeichenko, I.S. Kuznetsov, A.Yu. Izmaylov, R.Yu. Solovyev, S.N. Sharifullin //IOP Conference Series: Materials Science and Engineering. - 2017 .- Vol. 240. - No. 1. - Art. No. 012047. DOI: 10.1088 / 1757-899X / 240/1/012047

[6]. Kolomeichenko, A.V. Investigation of Finemet nanocrystalline alloy coating obtained by the electric spark method /A.V. Kolomeichenko, I.S. Kuznetsov, A.Yu. Izmaylov, R.Yu. Solovyev, S.N. Sharifullin //Int. J. Nanotechnol., 2018 .- Vol. 15. - No. 4/5. - Pp. 380 - 387.

[7]. Kolomeichenko, A.V. Tribotechnical properties the electrospark coating of amorphous and nanocrystalline alloys based on iron /A.V. Kolomeichenko, I.S. Kuznetsov //Friction and wear, 2014 .-- Vol. 35. - No. 6. - Pp. 501 to 504.

[8]. Kolomeichenko, A.V. Investigation of the thickness and microhardness of electrospark coatings of amorphous and nanocrystalline alloys /A.V. Kolomeichenko, I.S. Kuznetsov, I.N. Kravchenko //Welding International, 2015. - Vol. 29. - No. 10. - Pp. 823 - 825.

[9]. Kudryashov, A.E., Structure and properties of Cr-Al-Si-B coatings produced by pulsed electrospark deposition on a nickel alloy / A.E. Kudryashov, A.Y. Potanin, D.N. Lebedev, I.V. Sukhorukova, D.V. Shtansky, E.A. Levashov //Surface and Coatings Technology, 2016 .- Vol. 285. - Pp. 278 - 288.

[10]. Penyashki, T. Investigation of properties and wear of WC, TiC and TiN based multilayer coatings applied onto steels C45, 210CR12 and HS6-5-2 deposited by non-contact electrospark process /T. Penyashki, G. Kostadinov, I. Mortev, E. Dimitrova //Journal of the Balkan Tribological Association, 2017 .- Vol. 23. - No. 2. - Pp. 325 - 342.

IOP Conf. Series: Materials Science and Engineering **915** (2020) 012046 doi:10.1088/1757-899X/915/1/012046

# Energy saving in mechanical engineering through energy management

**O A Plugatyreva[1], E M Khusnutdinova[1] and A N Khusnutdinov[2]**

[1]Kazan (Volga Region) Federal University 18 Kremlyovskaya Str., Kazan 420008 Russia
[2]Kazan State Power Engineering University 51 Krasnoselskaya str., Kazan 420061, Russia

rr-088@mail.ru

**Abstract:** The paper considers a range of issues of improving energy efficiency and saving with by means of a qualitatively new approach, which includes the widespread introduction of effective management techniques, namely energy management systems based on the ISO 50001 standard.

The enhancement of energy efficiency is the most promising and affordable way to achieve economic results without harming other systems, unlike, for example, the introduction of new industrial technologies that require large initial investments.

The industrial sector is one of the largest consumers of fuel and energy resources accounting for about 18% of the total structure of the Republic of Tatarstan. That is why it is important to manage the national agenda at the level of individual fuel and energy companies in making this agenda happen in the terms of energy conservation and improving energy efficiency.

It is worth noting that companies that operate in the industrial sector of the economy and have their energy management system (EMS), achieve a mean annual increase in energy efficiency of 2-3% against 1% when doing business without introducing an energy management system.

Energy management applied in a company provides for the performance of a number of functions that can provide unbiased information about the key consumers of resources and energy, energy efficiency of various processes and individual types of products, and opportunities for reducing energy consumption. All this constitute an energy management system (EMS) [1].

The energy management system is based on GOST R ISO 50001-2012 Energy Management Systems. Requirements and Application Guidelines [2], sourced by ISO 50001, which replaced the European EN 16001 standard in 2011. The application of this standard facilitates in solving the issue of improving energy efficiency, using the tools of an integrated management system (IMS) of an enterprise or organization.

The energy management system, introduced as required by GOST R ISO 50001-2012, is aimed at integrating the issues of energy efficiency in the general management concept of the company [3].

The standard related to the energy management makes the company management to constantly enhance the performance of the energy management system by defining the company's energy policy, energy objectives and how to achieve them; establishing boundaries and areas within which the EMS should be implemented, act and be applied by all company's units; determination of the necessary methods and criteria that guarantee stable efficiency and functioning of processes, and control of these processes; communicating to all interested parties inside and outside the company the importance of energy management; determining the necessary resources; and regular analysis of the system [4].

Designing an energy management system requires the company to take four steps [5]:

1. Pre-audit in the management system.
2. Introduction of the EMS in line with GOST R ISO 50001.
3. In-house audits of the company, performance analysis of the energy management system, and pre-certification audit.
4. Certification audit to be performed as per contract agreement with the certification body.

The organization of an energy management system requires the following: [6]

1. Document all energy-intensive equipment and energy-consuming systems of the company.
2. Ensure that the company completes the forms of energy efficiency indicators, which include:

a. Reporting for each item of energy-intensive equipment and energy-consuming system.

b. For each item of energy-intensive equipment and energy-consuming system: indicate the process procedure that determines the technical requirements for energy-consuming production processes; describe energy efficiency ratings.

c. In the form of energy efficiency indicators for each energy-consuming system and item of energy-intensive equipment, give the deviation (in percent) of the actual technical data having an effect on the resource efficiency other than data sheet.

d. In the form of energy efficiency indicators for each energy-consuming system and item of energy-intensive equipment, give the deviation (in percent) of the actual technical data having an effect on the resource efficiency other than the requirements of the technological regulation or any binding production document [7].

3. Update the form of energy efficiency indicators every 1-2 years.
4. Tighten the requirements for energy efficiency and data accuracy every 2–3 years.

It is not worth mistaking energy saving as a costly process. Improving the energy efficiency of even the entire country is not a financially costly project, since energy conservation is primarily setting up the existing system (production and energy consumption) [8]. Note that meeting the above requirements is not labor consuming. This work can be done by 1-2 people in large-scale production, or the in-house power engineer can do it as part of his/her regular job.

Despite the fairly high popularity of the energy management system in national companies, a number of organizations are faced with factors that impede or hamper the implementation of measures for the designing and introduction of EMS, or do not plan to introduce the system due to the following reasons [9]:

- insufficiently high awareness of the performance and efficiency of the energy management system among the management and personnel of companies;
- imperfection of the methodological basis for substantiating the economic efficiency of energy saving and energy efficiency measures, including the introduction of the EMS resulting in lack of funding for energy saving and energy efficiency measures;
- insufficient information on best practices and methodologies for the design and introduction of the energy management system;
- insufficient details of companies providing integrated quality services (energy analysis, recommendations on energy-efficient measures, accounting and automation) on the design and introduction of the energy management system;
- lack of motivation of the management and personnel of companies in the design and introduction of the energy management system.

To understand the relevance of this work for companies of the Republic of Tatarstan, let us analyze whether GOST R ISO 50001 is used in the integrated management systems of local companies.

Let us consider the results of the performance of KAMAZ PJSC, the country's largest automotive corporation. As sourced by the company's open data, it can be concluded that KAMAZ PJSC does not adhere to the standard completely, only some of its subsidiaries, for example, Chelnyvodokanal LLC, have introduced GOST R ISO 50001. Using the currently challenging approaches to the management of energy systems and resource conservation, the company has put an Energy Saving Program in place that allowed reducing electricity consumption by 30% for five years. However, reducing energy intensity and consumption is still among the priorities of this KAMAZ subsidiary.

Thus, it is worth concluding that for owners of companies interested in reducing energy costs and lowering production costs, it is important to determine by what criteria the energy efficiency indicators of individual departments and production lines will be tracked [10,11]. Any company, regardless of its type of activity, can provide statistics on energy consumption during the year or quarter with details that will allow senior management to compare the company with other industry undertakings [12].

Improving the energy efficiency of the economy, with minimal (even zero) costs and the absolute support of management, is able to increase revenue in the company's budget faster than the majority of the most breakthrough technological projects.

**References**

[1] Romanov G A 2010 *Organizational Basis for Improving Energy Efficiency. Energoaudit.* No. 2. p 14.

[2] Khafizov I I, Nurullin I G and Sadykov Z B 2018 *Problems of development of electrochemical production of Russia and possibility of their decision.* IOP Conference Series: Materials Science and Engineering. Vol.412, Is.1. - Art. № 012042.

[3] Kashapov N F, Khafizov I I, Nurullin I G and Sadykov Z B. *Influence of introduction of robotics on increase in efficiency of electrochemical production.* IOP Conf. Series: Materials Science and Engineering 412 (2018) 012034

[4] Andrizhievskii A A and Volodin V N 2005 *Energy Saving and Energy Management.* (Minsk: Higher School). p 294

[5] Aniskina N N 2013 *Towards Energy Efficiency Through Energy Management. Competence.* No. 1. pp 40-43.

[6] Aktsorayeva N G and Sboyev P E 2018 Quality management in the heat power industry. *Scientific and practical electronic journal Science Alley.* (Tomsk: Quantum Publ.). Vol. 5. No. 4(20). pp 559-602.

[7] Marimon F, Llach J and Bernardo M 2011 *Comparative analysis of diffusion of the ISO 14001 standard by sector of activity.* (J. Clean. Prod). No. 19. pp 1734–44

[8] Kosmodemyanskaya S S, Gilmanshin I R and Azimov Y I Features of the formation of interdisciplinary competencies in engineering education // IOP Conference Series: Materials Science and Engineering. - 2019. - Vol.570, Is.1. - Art. № 012063.

[9] Kudrin B I 1995 *Power Supply of Industrial Undertakings.* (Moscow: Energoatomizdat). p 510

[10] Lee D and Cheng C C 2016 *Energy savings by energy management systems: A review. Renew. Sustain. Energy.* No. 56. pp 760–777

[11] Gulbrandsen T Kh, Padalko L P and Chervensky V L 2010 *Energy Efficiency and Energy Management.* (Minsk: Belarusian State University of Agricultural Technology). p 240

[12] Kashapov L N, Kashapov N F, Kashapov R N, Denisov D G *Plasma electrolytic treatment of products after selective laser melting* Journal of Physics: Conference Series, 2016 Vol. 669, Is. 1, Art. № 012029

IOP Conf. Series: Materials Science and Engineering **915** (2020) 012047 doi:10.1088/1757-899X/915/1/012047

# Methods of Optimizing the Troubleshooting Parameters of Electric Power Facilities

**L S Sabitov [1], P P Pavlov[2], V P Fandeyev[2], V M Butakov[2], A N Khusnutdinov[2] and E S Siyetinskaya[2]**

[1]Kazan (Volga Region) Federal University 18 Kremlyovskaya Str., Kazan 420008 Russia
[2]Kazan State Power Engineering University 51 Krasnoselskaya str., Kazan 420061, Russia

rr-088@mail.ru

**Abstract.** The paper discusses the methods of optimizing the troubleshooting parameters, based on a combination of heuristic algorithms with the linear integer programming method. The methods are aimed at minimizing costs when creating troubleshooting support for healthy condition monitoring and troubleshooting for in-service and recoverable electronic devices and equipment of electric power systems.

Electric power system are engineering systems consisting of electric components and machinery. In-service safety and cost reduction is ensured by various measures, including the determination of the healthy condition of components, by troubleshooting and monitoring their key performance parameters. The solution to this problem lies in determining the reference points, choosing the optimal troubleshooting parameters and aids, developing algorithms and troubleshooting processes. Methods of combinatorial optimization (abridged enumeration) and heuristic (without enumeration) algorithms are used [1-4] to optimize the parameters.

Optimization of the parameters by the methods of abridged enumeration with a large power of set of many types of condition (hereinafter referred to as the types of condition) and component parameters may require unacceptable expenses of computing resources (time, memory). The use of heuristic optimization algorithms saves computational resources, but does not guarantee obtaining the extreme value of the target function.

The paper proposes a methodology for combining the linear integer programming method with binary variables and heuristic algorithms for optimizing the controlled parameters, aimed at minimizing costs of troubleshooting aids of condition monitoring and troubleshooting of the serviced and recovered electronic devices, and equipment of power systems.

Electronic devices and equipment of power systems are analog and digital features that require routine troubleshooting to facilitate their healthy operation mainly by means of acceptable methods of parameter monitoring, including comprehensive test methods (transitions and digits count and signature analysis) [1]. The object of troubleshooting is set by a state view set $E$ and a parameter set $U$ with known limit values and expenses for troubleshooting aids. The parameters for performance monitoring and troubleshooting are usually redundant.

The authors suggest that the solution to the problem of optimizing the excessive parameters lies in using the unified diagnostic model of continuous and digital electronic objects that defines a binary relation between state types and alternative results of parameter monitoring, for example, in the form of a bipartite digraph:

$$G=(E,U_2,\varphi), \qquad (1)$$

where $G$ is the digraph designation; $E$ is the set of vertices associated with the types of state; $U_2$ is the set of vertices associated with valid and invalid parameter values; is the set of arcs connecting the vertices from the sets $E$ and $U_2$.

Arcs of a digraph set the binary relation

$$\varphi: E \rightarrow U_2, \quad (2)$$

according to which alternative types of parameters are compared to each type of state. If the form of the state $E_i$ is manifested by a valid value $u_{j1}$ or an invalid value $u_{j0}$ of the parameter, then an arc connects the corresponding vertices of the digraph (1).

The digraph (1) satisfies the following constraint:

$$\varphi^{-1}(u_{j0}) \cup \varphi^{-1}(u_{j1})=E, j=1, 2, m, \quad (3)$$

$$\varphi^{-1}(u_{j0}) \cup \varphi^{-1}(u_{j1})=\varnothing, j=1, 2, m, \quad (4)$$

$$\varphi (E_i) \neq \varphi (E_k), i\neq k, i, k=1, 2, n, \quad (5)$$

where $\varphi^{-1}(u_0)$, $\varphi^{-1}(u_1)$ are complete inverse images, and $\varphi(E_i)$, $\varphi(E_k)$ are the images of the corresponding vertices; $\varnothing$ stands for an empty set; $m$ is the number of parameters; $n$ is the number of failures.

Conditions (3) - (5) mean that the value of each parameter is known for any kind of state; no kind of state is manifested simultaneously with an invalid and valid value of the parameter; parameters are sufficient to distinguish between types of state in pairs.

The digraph (1) is formed in expert methods and based on circuit simulation of the object. The dimension of the model is determined by the power of many kinds of state and parameters.

Condition (5) in relation to the detection of failure is written as follows:

$$\varphi (E_0)\neq\varphi(E_i), i=1, 2,\ldots, n, \quad (6)$$

where $\varphi(E_0)$, $\varphi(E_i)$ is the image of the vertex associated in digraph (1) with the healthy condition and the failure, respectively.

The sets $\varphi(E_i)$ that make up the failure images that differ with the healthy condition are not empty:

$$|\varphi(E_i)|>0. \quad (7)$$

The choice of the parameters for detecting failures with minimal cost for the troubleshooting aids of monitoring the healthy condition of the object simulated by a digraph (1) is reduced to a linear integer-programming problem with binary variables:

$$\min \sum_{j=1}^{m} c_j x_j , \quad (8)$$

$$\sum_{j \in S_i} x_j >0, \quad (9)$$

where $x_j$ is a variable taking the value 1 if the parameter $u_j$ is selected to detect failure and the value 0 otherwise; $c_j$ is the cost of troubleshooting aids for monitoring the parameter $u_j$; $S_i$ is the set of parameter numbers, the invalid values of which manifest a failure $E_i$.

Constraints (9) are formed based on conditions (7). The number of constraints is equal to the number of failures.

The problem of linear integer programming with binary variables is solved by the additive algorithm with the Balas filter (*Balas*) [5, 6, 7]. As a filter, the authors take the costs of the troubleshooting aids for monitoring parameters selected by the heuristic exclusion algorithm.

The set of parameters sufficient to detect failure is generated using the heuristic algorithm by excluding from the digraph (1) the vertices associated with the parameters with the highest troubleshooting costs, the exclusion of which does not breach conditions (7).

Parameters are ranked in descending order of troubleshooting costs. The parameter with the highest troubleshooting costs is selected. The satisfaction of condition (7) is checked for all failures without taking into account the values of the selected parameter. If conditions (7) are satisfied, then the vertex associated with this parameter is excluded from digraph (1). In the event of default of conditions (7) or after excluding the vertices from the digraph, the actions are repeated for the next parameter.

The upper bound on the number of computational operations for excluding vertices from the digraph and checking conditions (7) is estimated by the formula

$$(n+1)m. \tag{10}$$

With an increase in the number of failures and parameters, the upper limit of the number of calculations increases slowly and almost linearly [8, 9, 10, 11].

The proposed method is applicable for optimizing the controlled parameters by combining heuristic exclusion algorithms with both the Balas method and other methods of mathematical programming, for example, with the branch and bound method [12, 13, 14, 15]

The obvious idea of combining the heuristic algorithm with the method of combinatorial optimization, which has quantitative substantiation in the paper, seems to be one of the possibilities of a compromise between the requirements of reducing complexity and achieving optimal decision-making. A small increase in computing resources in terms of the implementation of the heuristic decision-making algorithm is offset by a significant decrease in computing resources in terms of achieving the optimal solution using mathematical programming. As a result, the cost of computing resources to achieve the optimal solution is significantly reduced. At the same time, combining the heuristic algorithm with the method of combinatorial optimization only partially solves the problem of saving computing resources while optimizing the composition of parameters. Optimization of the object's parameters with a large, albeit limited by heuristic algorithm, power of the set of failures and parameters is achieved at the cost of high computational resources.

**References**

[1] Fandeyev V P 2001 *Technologies and means of failure analysis of recoverable electronic instruments*: monograph. (Penza: Penza State University Press) p 240

[2] Khusnutdinov A N, Khusnutdinova E M and Butakov B M 2019 Operation of brushes with increased service life on traction machines of electric vehicles *IOP Conf. Series: Materials Science and Engineering* Vol. 570, Iss. 1 № 012053

[3] Fandeyev V P 2016 T*roubleshooting, repair and maintenance of control systems for household machines and devices*: textbook / Zh. Romanovich A, Scriabin V A, Fandeyev V P and Tsypin B V - M.: Dashkov and Co. Publishing Corp. p 316.

[4] Fandeyev V P, Tutushkin V I A methodological approach to the selection of troubleshooting parameters for predicting the condition of electronic devices and equipment // *News of higher educational institutions. Volga region. Engineering Sciences*. - **No. 3**. - 2011. - P. 124 - 131.

[5] Kashapov N F, Sabitov L S, Litvinenko R S, Auhadeev A E and Pavlov P P 2019 New methodological approach to the development of the theory of electric traction of urban electric transport *IOP Conference Series: Materials Science and Engineering*. Vol. 570, Iss. 1, № 012041

IOP Conf. Series: Materials Science and Engineering **915** (2020) 012047 doi:10.1088/1757-899X/915/1/012047

[6] Hizbullin Rob and Hizbullin Rad 2019 Ways to improve safety in the power industry: an automated hardware complex for preshift inspection of personnel of power enterprises *E3S Web of Conferences.* vol. 124. – 05037
[7] Filina O A and Tsvetkov A N 2019 Evaluation of the operational life of direct current motors IOP *Conference Series: Materials science and engineering* Vol. 489 № 012016
[8] Khusnutdinov A N, Khusnutdinova E M, Khamidullina G R, Pavlov P P, Aukhadeyev A E, and Litvinenko R S 2018 Training transport specialists based on a contemporary view of self-organization of sophisticated developing systems *IOP Conf. Series: Materials Science and Engineering* Vol. 412 № 012047
[9] Safarov D T, Kondrashov A G and Khafizov I I 2018 The method of diagnosing machine systems by measuring the accuracy of manufactured parts. *IOP Conf. Series: Materials Science and Engineering* Vol. 412 № 012065
[10] Auhadeev A E, Idiyatullin R G, Pavlov P P, Butakov V M, Kisneeva L N and Tukhbatullina D I 2019 Improving the theory for calculating the rational modes of traction electrical equipment. *E3S Web of Conferences.* vol. 124. – 05077
[11] Sabitov L S, Gatiyatov I Z, Kashapov N F, Gilmanshin I R and Kiyamov I K 2018 Increase of reliability of contact networks of electric transport, due to increase of strength of the joint unit of pipes of different diameters *IOP Conference Series: Materials Science and Engineering.* Vol. 240, № 012058
[12] Filina O A, Tsvetkov A N, Pavlov P P, Radu D. and Butakov V M 2019 Vibration model as a system of coupled oscillators in a direct current electric motor *E3S Web Conf.* Vol. 124, № 02002
[13] Aukhadeev A E, Idiyatullin R G, Kisneeva L N, Zalyalov R.R. Pavlov P P and Litvinenko A R 2019 Development of the theory and methods for calculating the rational operation modes of electric traction equipment of urban electric transport *IOP Conference Series: Materials Science and Engineering.* vol. 643(1). - 012027.
[14] Kashapov L N, Kashapov N F., Kashapov R N, Semushin V D 2016 Changes of thermal balance in a plasma-electrolyte system according to the shape of the applied voltage *Journal of Physics: Conference Series.* vol. 669(1). - 012028.
[15] Kashapov L N, Kashapov N F, Kashapov R N, Denisov D G 2019 Plasma electrolytic treatment of products after selective laser voltage *Journal of Physics: Conference Series.* vol. 669(1). - 012029.

IOP Conf. Series: Materials Science and Engineering **915** (2020) 012048 doi:10.1088/1757-899X/915/1/012048

# Reliability study for traction electrical equipment of urban electric transport

**L S Sabitov [1], A E Aukhadeyev[2], R G Idiyatullin[2], R S Litvinenko[2], L N Kisneeva[2], A N Khusnutdinov[2] and O V Ilyushin[1,2]**

[1]Kazan (Volga Region) Federal University 18 Kremlyovskaya Str., Kazan 420008, Russia
[2]Kazan State Power Engineering University51 Krasnoselskaya str., Kazan 420061, Russia

rr-088@mail.ru

**Abstract.** The paper discusses the matter of reliability of the windings of the asynchronous traction motors being part of the urban electrically powered vehicles, contains the characteristic curve of the stator failure-free operation probability, and refers to experimental data on the failure rate. The results can be used both at the design stage of asynchronous traction motors and while in operation to optimize the preventive maintenance.

Generally, the companies of urban electric transport (UET) operate rolling stock (RS), providing efficient transport services to the urban population, namely: maximum reduction of time spent on movement and reduction of transport fatigue; maximum release of RS for utilization, high regularity of movement and transport comfort of passenger traffic [1-4].

The reliability of traction electric equipment of RS, providing the healthy operation in a given period, when there is an impact of operational factors, directly or indirectly influences the high quality performance of UET [5-7].

Taken together, the influence of design quality and deviations from the specifications for operation leads to failures and malfunctions of the RS traction components [5, 6].

Nowadays, traction electric equipment based on squirrel-cage asynchronous traction motors (ATM) is actively used on the existing RS. This motor type demonstrates increased reliability as determined by its simple design. In this case, the most typical cause of failure (up to 80 % of all cases) for ATM is the destruction of the stator insulation due to its overheating. The main types of insulation failures are turn-to-turn short circuit, coil insulation failure, coil short circuit to frame. With a continuous load on the ATM and deviations in the process of operating the RS, the load current increases, resulting in overheating of the insulation, and, as a result, its accelerated aging [8, 9].

Currently, the existing RS is required to increase the operating speed, causing the traction electric equipment to run longer in traction mode, and, consequently, it increases the heating time of the ATM insulation. The allowable amount of heating of the traction motor components is calculated at the design stage. The permissible heating of the traction motor components is set so that the insulation in contact with them or located in the immediate vicinity of them can perform during the estimated lifetime. With this in mind, the ultimate level of heating of the ATM components depends on the heat resistance class of the insulation [3, 6].

The probability of failure-free operation of the stator components as a function of the RS run was determined to analyze their reliability based on the statistical data of failures of the traction electric equipment. In this case, $\lambda_k$ = 0.08, where the failure rate of the coil insulation and the housing insulation of the traction motor stator winding slot is summed. The failure rate $\lambda_e$ = 0.24 for the previous value of the overheating of the winding, equal to 10 $^0$C [10, 11].

Design and technological solutions that determine the value of $\lambda_k$ also influence the operation of stator insulation. As the number of in-service defects increases, the failure rate climbs as well, i.e. becomes more than the value specified in the design specifications. If the load conditions of the ATM in operation exceed the specifications limits, then this will cause an increase in the value of $\lambda_k$. The failure rate λ depends on the quality of design, manufacturing technology and the influence of external operational factors $\lambda_e$ , i.e. $\lambda = \lambda_k + \lambda_e$. If we accept the condition that the failure flow obeys the Poisson stationary flow, then the equation for the reliability of the ATM stator slot will be as follows [12, 13]:

$$P(t) = e^{-(\lambda_k+\lambda_e)l}$$

where $\lambda_e$ is the flow of failures caused by design and technological factors; is the flow of failures caused by operational factors.

Figure 1 shows the various factors vs. the RS run characteristic curves. The highest failure-free operation probability is determined by factors contingent on the quality of the design. Lower reliability is caused by in-service factors – in this event, failure to comply with the specifications for operation leads to overloading of the ATM, overheating of the insulation, its thermal breakdown, and, as a result, to failure [13, 14].

The constructed characteristic curves demonstrate the probability of failure-free operation as a function of time. As is seen from the above curves, the influence of operating conditions on the reliability of the traction electric equipment is stronger compared to failures caused by the quality of design and workmanship [10, 14, 15]. In view of the above, the following important features can be noted:

First, if there is $\lambda_k < \lambda_e$, then the influence of design and technological defects on the reliability of the ATM will be lower [16,17,18]. This is a case where the high quality of design and workmanship is ensured, but at the same time the specifications is not met in operation. Based on the results of the data obtained, it can be noted that the operational reliability of the ATM insulation, taking into account the effect of factors at the design and fabrication stages$P_k(t) = 0{,}45$, taking into account the effect of operational factors $P_e(t) = 0{,}09$. Thus, it can be said that the reliability of the ATM insulation is lower when exposed to operational factors and higher when exposed to design and production factors.

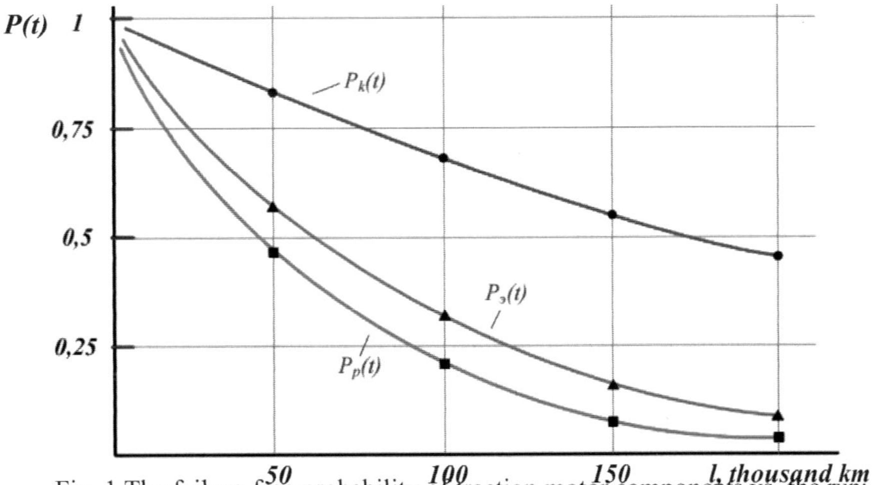


Fig. 1 The failure-free probability of traction motor components vs. the run:

$P_k(l)$ - design factors; $P_e(l)$ is the operational factors;

$P_r(l)$ - resulting reliability

The second option is the release of defective ATM to operation if $\lambda_k > \lambda_e$ (i.e. due to insufficient design quality and workmanship). Moreover, let us assume that the specifications are strictly observed in operation.

And the last, third option, if $\lambda_\delta = \lambda_k = \lambda_e$, i.e. there are practically no cases of releasing the defective ATM to operation providing that the specifications for operation are met. This option is the most favorable, the reliability characteristics of the ATM correspond to the design $\lambda_\delta$ that were specified in the design specifications.

Studies show that in practice the most likely option is the first, although the second and third options cannot be excluded. This can be explained by the fact that specialists with a high level of education and experience are engaged in the stages of designing the specifications, sketch design, detailed drawings, fabrication and finalization of the design of ATM.

Although the failure rate is regarded constant during the period of steady wear of ATM components, insulation aging occurs due to operational factors. Therefore, for $\lambda_i$, it is necessary to take an average value, which is the arithmetic mean sum of the initial and final values.

The characteristic curves show how the failure rate can bias the ATM failure-free operation probability: with increasing $\lambda_i$, the failure-free operation probability of motor components reduces. Then, at the design stage, for a specific value of $P_i(t)$ (Fig. 1), the measure $\lambda_i(t)$ (Fig. 2) can be determined. The in-service failure rate of insulation may depend on both external factors, such as temperature, humidity, vibration, etc., and internal factors, such as load, number of starts and stops, etc.

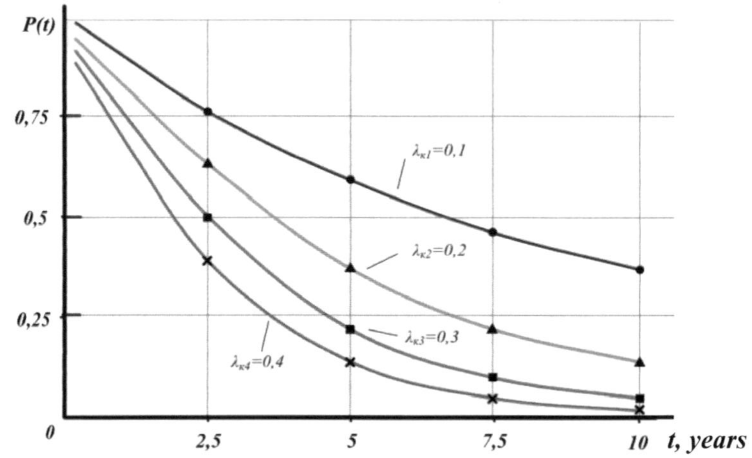


Fig. 2 Reliability characteristics of the ATM stator slot insulation for various failure rates

Of course, these factors increase the failure rate if the operating modes of the ATM can vary over wide ranges.

Calculations of the reliability parameters show that in the design and operation of ATM it is necessary to ensure that the failure rate asymptotically tends to zero. In this event, with the Poisson flow of failures, it can be assumed that the failure-free operation probability will be close to 0.98.

**References**

[1] Kashapov N F, Sabitov L S, Auhadeev A E, Litvienko R S and Gatiyatov I Z 2019 Description of a complex technical system of urban electric transport from the standpoint of synergistic methodology *IOP Conference Series: Materials Science and Engineering.* vol. 570(1). – 012040

[2] Kashapov N F, Sabitov L S, Litvinenko R S, Auhadeev A E and Gatiyatov I Z 2019 The approach to the study of the reliability of the electric transport system of the city as a complex technical system *IOP Conference Series: Materials Science and Engineering.* vol. 570(1). – 012043

[3] Sabitov L S, Gatiyatov I Z, Kashapov N F, Gilmanshin I R and Kiyamov I K 2018 Increase of reliability of contact networks of electric transport, due to increase of strength of the joint unit of pipes of different diameters *IOP Conference Series: Materials Science and Engineering.* Vol. 240, № 012058

[4] Hizbullin Rob and Hizbullin Rad 2019 Ways to improve safety in the power industry: an automated hardware complex for pre-shift inspection of personnel of power enterprises *E3S Web of Conferences.* vol. 124. – 05037

[5] Auhadeev A E, Litvinenko R S, Fandeev V P, Pavlov P P, Butakov V M and Litvinenko A R 2019 Research of reliability of urban electric transport system *E3S Web of Conferences.* vol. 124. – 05078

[6] Auhadeev A E, Idiyatullin R G, Pavlov P P, Butakov V M, Kisneeva L N and Tukhbatullina D I 2019 Improving the theory for calculating the rational modes of traction electrical equipment. *E3S Web of Conferences.* vol. 124. – 05077

[7] Nechiporenko V I 1977 *Structural analysis of systems* (Moscow. Sovetskoye radio) p 214

[8] Khafizov I I, Nurullin I G and Sadykov Z B 2018 Problems of development of electrochemical production of Russia and possibility of their decision *IOP Conference Series: Materials Science and Engineering* Vol. 412, Iss. 1, № 012042

[9] Biryukov V V and Porsev E G 2018 *Traction electric drive* (Moscow. Yurait) p. 315

[10] Idiyatullin R G 1987 *Reliability of traction electric machines* (Tashkent. MekhNAT) p. 152

IOP Conf. Series: Materials Science and Engineering **915** (2020) 012048 doi:10.1088/1757-899X/915/1/012048

[11] Kotelenets N F and Kuznetsov N L 1988 *Tests and reliability of electric machines* (Moscow. Vysshaya shkola) p. 232
[12] Barlow R and Proschan F 1984 *Statistical Theory of Reliability and Reliability Tests* (Moscow. Nauka) p. 326
[14] Popov A V, Idiyatullin R G and Aukhadeev A E 2016 *Calculation method of reliability of winding of high-voltage asynchronous electric motors*. Izvestiya vuzov. Problemy energetiki. No. 5-6. pp. 94-98
[13] Popov A V, Idiyatullin R G and Aukhadeev A E 2016 *Experimental study of load modes of asynchronous electric motors under operating conditions*. Energetika Tatarstana. No. 2 (42). pp. 76-79
[15] Pavlov P P, Fandeev V P, Aukhadeev A E, Litvinenko R S, Butakov V M 2019 Technique for optimization of diagnostic parameters composition for power systems objects *IOP Conference Series: Materials Science and Engineering*. vol. 643(1). - 012013.
[16] Filina O A and Tsvetkov A N 2019 Evaluation of the operational life of direct current motors IOP *Conference Series: Materials science and engineering* Vol. 489 № 012016
[17] Filina O A and Salnikova O.V. 2020 Construction of verification and diagnostic tests for the functional diagram of the object of diagnosis *IOP Conference Series: Materials Science and Engineering*. Vol. 747, № 012111
[18] Kashapov R N, Kashapov L N, Kashapov N F, 2017 Formation of cracks in the selective laser melting of objects from powdered stainless steel 17-4 PH *IOP Conference Series: Materials Science and Engineering*. Vol. 240, Is.1 № 012074

IOP Conf. Series: Materials Science and Engineering **915** (2020) 012049 doi:10.1088/1757-899X/915/1/012049

# Medical and Biological Problems in the Control of Therapeutic Exposure Parameters in Sports Medicine and Ways of Their Solution in New Medical Instruments and Systems

**L S Sabitov[1], Rob N Khizbullin[2], Rad N Khizbullin[2], P P Pavlov[2], A N Khusnutdinov[2] and A R Abdullina[2]**

[1]Kazan (Volga Region) Federal University 18 Kremlyovskaya Str., Kazan 420008 Russia
[2]Kazan State Power Engineering University 51 Krasnoselskaya str., Kazan 420061, Russia

rr-088@mail.ru

**Abstract.** High-quality treatment and recovery of athletes-patients depends on reliable informational diagnosis of injuries and diseases, but also on adequate therapeutic effect. In most cases, the question remains of further predicting the condition of the patient-athlete as a result of the treatment. The article discusses the main biomedical problems in monitoring the therapeutic parameters of exposure and ways to solve them in new devices and systems based on thermometric and optoelectronic methods.

1. Existing biomedical problems in healthcare and in sports medicine. The problem in light-laser therapy: registration of the function "exposure - response of the body, i.e. response ", is becoming particularly relevant. Despite objective and subjective difficulties, one can find ways to solve this problem, based on the isolation of the signal of the photobiological effect. The tasks of implementing these paths are to develop adequate optoelectronic information and metrological techniques as applied to medical technologies.

In general, the problem of determining the set of correlated parameters of the body's response to low-intensity laser exposure is an integral part of the problem, the determination of the functional relationship between the totality of the exposure parameters and the corresponding parameters of the body's response and its use for exposing therapeutic doses of the primary exposure parameters.

In this regard, the proposed: laser plethysmography, laser pulse oximetry and digital precision thermometry are universal methods on which the exposure control of the therapeutic dose will be based on feedback informative links. The development of new biocontrolled methods for monitoring the therapeutic parameters of exposure in electro-light-vacuum-laser therapy is an urgent problem. Carrying out laser-laser therapeutic procedures is always closely related to the concept of compulsory perception of the impact of their bio-object, for example, when conducting electro-laser stimulation, you can use the temperature of the target organ, which is targeted at, allowing you to estimate the amount of electro-laser exposure perceived by a particular organism.

The study of the effects on the human body is the first urgent research problem.

Changes: organ temperature, body blood flow, conductivity, and other factors in the zone of physiotherapeutic exposure can be measured and used to control exposure parameters. However, for these purposes, it is necessary, within 1 ° C to an accuracy of 0.1 ° C, to measure the temperature of the internal organ in hard-to-reach places (urology, andrology, gynecology, etc.). Existing non-invasive radiothermometric and thermal imaging methods do not provide the accuracy required for measuring temperature.

The first research problem was implemented using the contact method of temperature measurement using thermistors with negative TCR, providing in the range + 5 ° C ... + 50 ° C an error of ± 0.1 ° C. Therefore, thermometers (ТМЦП-2) [1] based on thermistors can be classified as precision. Based on the thermometer, it is possible to implement temperature control for biocontrolled thermometry.

The second biomedical problem is the study of the relationships between the therapeutic parameters of exposure to PPG [2], with the help of which it is possible to isolate many therapeutic parameters that can be calculated from it to quantify the reaction of the body [3]. The use in existing single-channel laser plethysmographs (PPG) of a light emitting diode (LED) (920 ... 960 nm) inevitably leads to "physiological interference" when recording a PPG signal by 30%. It is impossible to obtain an acceptable accuracy of less than 5%; the repeatability and adequacy of the PPG signal is impossible with LED PPG. It is necessary to develop a new single-wave laser ($\lambda = 805 \pm 0.75$ nm, P = 0.1 ... 0.2 mW) of two-channel FPG1-2KL [4,5]. The wavelength of the laser diode (LD) 804.25 ... 805.75 nm is the "isobestic point" for oxy- and deoxyhemoglobin. It provides an error of no more than 5% and is not only an information optical-electronic device, but also a device for automated feedback “impact-response”.

The third problem is the biomedical problem associated with the treatment of oxygen in the blood. Using a three-wave two-channel laser pulse oximeter (PSO3-2KL) [6], it is possible to optimize the dose parameters of electro-laser and other quantitative assessment of the body's response, which allows us to consider the positive fulfillment of the third problem.

2. Ways to solve biomedical problems and practical implementation in new medical devices and systems. To solve the posed biomedical problems, new principles, techniques, methods and methods have been developed for constructing optoelectronic devices for healthcare and sports on their basis.

1. The principle of constructing FPG1-2KL [4, 5, 7] taking into account LD in the laser mode with an operating wavelength of $\lambda = 805 \pm 0.75$ nm, which provides:

1) the implementation of a new parameter for registering the PPG signal at the “isobestic point” of 805 nm, in order to eliminate “physiological interference” by 30% compared to the single-channel prototype “Akutest” on an LED (920 ... 960 nm);

2) the use of the second LD-FD channel as a reference and rationing of the optical signal from the first channel relative to the reference error of up to 5% of the registration of the PPG curve;

3) instrumental implementation of FPG1-2KL [4,5,7] accuracy class 2.5, recording not only the parameters of blood flow in hemodynamics, but also allowing the use of PPG as a biological feedback as a corrector of therapeutic parameters in order to eliminate overdose of exposure.

2. The developed method for controlling temperature changes as a biological parameter of the feedback between the action and the temperature reaction of the “target organ” to it allows you to:

1) use the temperature response of organs to physiotherapeutic effects as secondary effects, and adjust the therapeutic parameters of exposure as primary effects;

2) instrumentally implement thermometers precision TMCP-2 [1,7], providing measurement with an accuracy of 0.1 ° C; changes in the temperature of the investigated organ within + 5 ... 50 ° C;

3) adjustment of exposure parameters in accordance with the applicable procedure, including registration of the skin temperature of the patient in the area of the surface sensors of the pulse oximeter.

3. The developed construction principle and instrumentally implemented on its basis PSO3-2KL [6], which records PPG signals with an error of 5% (PSO “Spectrotest” has 10%) from two LD-PDs ($\lambda = 640$ nm, $\lambda = 940$ nm) , instead of one LED-PD optocoupler, and one calibration LD ($\lambda = 805$ nm) in the laser sensor, using the effect of "backscattering" of light from blood and biological tissue, allows you to [8-11]:

1) to identify in the diagnostic local pathology of blood vessels, superficial and deep, simultaneously in different parts of the cardiovascular system with the possibility of early and high-precision assessment of the state of arterial vessels and regional blood flow;

2) the introduction of a third calibration LD with a radiation wavelength $\lambda = 805$ nm, which is the “isobestic point” for the absorption spectra of hemoglobin, oxyhemoglobin and other hemoglobin compounds, into the design of the PSO3-2KL laser sensor [6,7], to ensure the measurement reliability of the patient’s content blood oxygen by 30 percent relative to PIC with sensors on the LED;

3) constructive construction of surface sensors using selected three-axis accelerometers in order to record the magnitude of mechanical acceleration along three mutually perpendicular axes to eliminate the influence of patient movement artifacts and simplify the software PSO3 -2KL [12,13];

4) by introducing into the design of surface sensors a temperature sensor, record the temperature of the patient’s skin surface in the area of the sensors and provide control of the final functional dependence of the output on the thermistor of the power sensor from changes in skin temperature to adjust the treatment process [14,15];

5) confirm the results obtained experimentally and made by theoretical calculation of the optimal location of the LD and PD surface sensors 10 ... 20 mm.

The above characteristics and the principles of constructing new medical devices described above, and on their basis medical automated systems, allow solving current medical and biological problems facing sports medicine.

**References**

[1] Laryushin A I, Galkin M A, Khizbullin R N and Novikov V A 2010 Digital precision medical thermometer *World of Measurements* N. 9. pp 9-14

[2] Khizbullin R N and Laryushin A I 2014 Automated medical hardware complex for pre-shift inspection of personnel of energy enterprises *News of higher educational institutions "Problems of Energy"* №1-2 pp 125-133

[3] Zmievskoy G N, Galkin M A and Kostyukov D V 2008 Biodosimetry in low-intensity phototherapy *Laser-inform* Iss. 17 (392) pp 6-9

[4] Laryushin A I, Galkin M A, Mishanin E A, Kuzmich A P, Novikov V A and Khizbullin R N 2010 The use of a two-channel laser photoplethysmograph in urology *World of Measurements* Iss. 9. pp 28-33.

[5] Laryushin A I, Galkin M A, Khizbullin R N and Novikov V A 2010 Two-channel laser photoplethysmograph *World of measurements* Iss. 7. pp 22-28

[6] Khizbullin R N 2017 Optical two-channel pulse-oximeter based on laser sensors for solving urgent problems in medical practice *Photonics* Iss. 1. pp 144-157

[7] Hizbullin Rob and Hizbullin Rad 2019 Ways to improve safety in the power industry: an automated hardware complex for preshift inspection of personnel of power enterprises *E3S Web of Conferences.* vol. 124. – 05037

[8] Kashapov N F, Sabitov L S, Auhadeev A E, Litvienko R S and Gatiyatov I Z 2019 Description of a complex technical system of urban electric transport from the standpoint of synergistic methodology *IOP Conference Series: Materials Science and Engineering.* vol. 570(1). – 012040

[9] Kashapov N F, Sabitov L S, Litvinenko R S, Auhadeev A E and Gatiyatov I Z 2019 The approach to the study of the reliability of the electric transport system of the city as a complex technical system *IOP Conference Series: Materials Science and Engineering.* vol. 570(1). – 012043

[10] Kashapov N F, Khafizov I I, Nurullin I G and Sadykov Z B 2018 Influence of introduction of robotics on increase in efficiency of electrochemical production *IOP Conf. Series: Materials Science and Engineering* Vol. 412 № 012034

[11] Gilmanshina S I, Bendjukevich K G, Gilmanshin I R, Glushkova V A, Gadirova E.M. 2019 Methods of identifying talent in the field of engineering and natural science *IOP Conference Series: Materials Science and Engineering* Vol.570, Is.1. № 012028

[12] Khusnutdinova, E M, Khamidullina, G R, Pavlov, P P, Aukhadeyev, A E, Litvinenko, R S, Khusnutdinov, A N 2018 Training transport specialists based on a contemporary view of self-organization of sophisticated developing systems *IOP Conference Series: Materials Science and Engineering* 412(1), 012047

[13] Filina O A and Salnikova O V 2020 Construction of verification and diagnostic tests for the functional diagram of the object of diagnosis *IOP Conference Series: Materials Science and Engineering.* Vol. 747, № 012111

[14] Kashapov L N, Kashapov N F, Kashapov R N, Pashaev B Y 2016 The application of additive technologies in creation a medical simulator-trainer of the human head operating field *IOP Conference Series: Materials Science and Engineering.* 134(1), № 012011

[15] Kashapov L N, Kashapov N F, Kashapov R N 2017 Investigations of the growth of the vapor-air shell of a gas discharge with a liquid electrolytic cathode of sodium hydroxide solution *Journal of Physics: Conference Series.* 927(1), № 012085

IOP Conf. Series: Materials Science and Engineering **915** (2020) 012050 doi:10.1088/1757-899X/915/1/012050

# Digitalization of education: transformation of a natural science experiment

**R N Sagitova[1], S I Gilmanshina[1], I R Gilmanshin[1,2], A I Galeeva[3], G M Abyzbekova[4], K Saduakaskyzy[4]**

[1]Kazan Federal University, Kazan, Russia, 420008
[2] Kazan National Research Technical University named after A N Tupolev, Kazan, Russia
[3]Kazan State Power Engineering University, Kazan, 420066, Russian Federation
[4]Korkyt Ata Kyzylorda University, Aiteke bi, 29A; 120014, Kyzylorda, Republic of Kazakhstan

is-er@yandex.ru

**Abstract.** Under the conditions of digitalization of education, a natural science experiment is analyzed on the example of the development and transformation of a chemical experiment. The influence of the development of information and communication technologies and their implementation in education on the transformation of a natural science experiment and the effectiveness of using various forms of education, such as real and virtual experiments, in digital laboratories is examined.

The development of information and communication technologies and their implementation in the educational process has led to the emergence of a new concept - digitalization of education. The problem of digitalization of teacher education is considered in ([1], [2]), engineering education is touched upon in works ([3], [4], [5]), scientific engineering research is addressed in ([6], [7], [8], [9]).

In this paper, we consider the effect of digitalization of education on the conduct of a natural science experiment using the development and transformation of a chemical experiment as an example.

A chemical experiment is an essential part of the study of natural phenomena, at the same time it is a specific method of understanding nature and teaching chemistry. The theory and practice of chemical experiment continues to develop in the works of leading scientists in the field of chemical education, starting from the beginning of the 20th century. These are the classic works of V.N. Verkhovsky. For example, "Technique and methodology of a chemical experiment at school: a manual for teachers and students."

Analysis of the content of V.N. Verkhovsky's first half of the 20th century shows that most laboratory chemical experiments used in the education of that time are not a combination of quantitative, but qualitative chemical experiments. Quantitative chemical experiments were, but they are relatively smaller. This is a laboratory work to establish the chemical formulas of some compounds, determining the quantitative composition of metal oxides. The technique of quantitative determination in those years was mainly based on measuring the mass and volume of a substance. In subsequent years, there was an improvement in the technique and methodology of a chemical experiment in education. More attention has been paid to the use of health-saving and resource-saving technologies.

The situation with a chemical experiment in education has fundamentally changed since the beginning of the 21st century. Dramatic changes have occurred in the technical equipment of universities and schools, which is favorable for the formation of competencies of the 21st century and environmental literacy (for more details, see [10], [11]).

Currently, in accordance with the order of the Ministry of Education of the Russian Federation No. 465 of September 3, 2019, the ideology of the natural science experiment in general education institutions has changed dramatically. In addition to the traditional set of dishes and equipment, it is necessary to have digital laboratories, a digital microscope, an oxygen gas analyzer with digital indicators, and a tablet computer with educational content. In addition, the availability of electronic teaching aids, such as interactive manuals, a set of videos, is mandatory. Presentation of the results of demonstration experiments should be provided by an interactive whiteboard complete with a projector (or an interactive panel), as well as a document camera.

The introduction of new digital equipment, methods for presenting and processing the results of a natural science experiment led to its transformation and the emergence of its new forms. At present, a natural science experiment can be classified as real, virtual, computer simulation of natural science phenomena, videos of a real experiment.

The most widespread in recent years have received videos demonstrating the laws of science, properties and production of chemical compounds. According to surveys, almost all teachers use video of real experiments as an auxiliary material in the lessons of natural sciences.

The development of computer technology has led to the use of virtual reality in education. Modeling of natural science phenomena is carried out using virtual laboratories and augmented reality laboratories. A natural science experiment is considered virtual, in which the simulation of the experiment and its result are obtained through a computer program.

In Russia, in general science education, relatively simple virtual laboratories are available for teachers and students to simulate a number of science experiments.

For example, the ENKA virtual laboratories, the digital educational resource "Chemistry. Virtual Laboratory "([12], [13]).

In general, virtual science experiment by teachers in educational institutions is rarely used, mainly to increase interest in the subject. Perhaps this is due to a simplified visualization of the course of the laboratory experiment and equipment, as well as the fact that the theoretical and methodological foundations of the virtual experiment are currently only being formed.

However, in connection with the pandemic (March 11, 2020), many schools and universities were forced to switch to distance learning technologies. One of the most important problems was the organization and conduct of laboratory work. The basis for solving this problem may be the use of video clips, virtual laboratories and augmented reality laboratories for the formation of science literacy.

Video fragments of natural science experiments have very high visibility, but suggest some passivity of students. You can increase the cognitive activity of students through a clearly thought-out lesson plan, in which students watch video clips combined with the need to answer questions prepared by the teacher.

Virtual laboratories in the presence of a schematic visual series suggest the active participation of students in the implementation and planning of the experiment. The combination of two forms of experiment, virtual and demonstration of video clips, allows to optimize the educational process, both school and university.

It should be noted that such aspects of using virtual laboratories and video fragments of natural science experiments as reagent saving and the safety of a virtual experiment compared to real ones.

With the introduction of digital laboratories, the possibilities of a real (natural) experiment have expanded. Digital laboratories are a set of sensors with signals for changes in the physical characteristics of solutions (electrical conductivity, temperature, redox potential, optical density) complete with processing and displaying these values units (computer, tablet or liquid crystal display).

The set of sensors is quite extensive. This is a temperature sensor, a pH sensor, a conductivity sensor, an optical density sensor of a solution, a drop counting sensor, gas sensors and others. The presented set of sensors expands the capabilities of digital laboratories in conducting natural science experiments. There is an opportunity to work on a qualitative and quantitative level.

At the same time, the active use of digital laboratories requires the teacher to have good natural science knowledge to accurately understand the capabilities and limitations of the sensors used.

For example, a potentiometric titration requires a pH sensor and ion-selective sensors. To conduct conductometric titration - conductivity sensor. When voltammetry - a sensor for determining dissolved oxygen. To determine the carbon dioxide content in the air - a carbon dioxide sensor. Determination of the concentration of a substance in colored solutions is carried out by means of an optical density sensor. Detection of suspended particles in dispersed systems - by nephelometric and turbidimetric sensors.

The most effective in organizing a chemical experiment, in our opinion, is a mixed approach, which provides the optimal combination of a real experiment, virtual laboratories and video clips of chemical experiments and phenomena. When performing a real chemical experiment, students develop practical experimental skills. Viewing video clips of a real experiment before performing a virtual experiment consolidates theoretical knowledge and methodological skills.

Thus, the digitalization of education leads to the transformation of the natural science experiment, and the methodology of its implementation in the context of digitalization involves the use of a set of digital equipment, which requires the teacher to master new digital competencies.

**References**

[1] Gilmanshin I.R., Gilmanshina S.I. (2018). Technology of e-learning in the university education. IOP Conference Series: Materials Science and Engineering, 412, (2018) 012022. doi:10.1088/1757-899X/412/1/012022. https://iopscience.iop.org/article/10.1088/1757-899X/412/1/012022.

[2] Gilmanshina, S.I., Gilmanshin, I.R., Bendjukevich, K.G. (2019). Digital learning technologies in the University education system. IOP Conference Series: Materials Science and Engineering, Volume 570, Number 1. https://iopscience.iop.org/article/10.1088/1757-899X/570/1/012027.

[3] Kosmodemyanskaya S.S, Gilmanshin I.R, Azimov Y.I. (2019). Features of the formation of interdisciplinary competencies in engineering education. IOP Conference Series: Materials Science and Engineering. 2019. Vol. 570, Is.1. Art. № 012063.

[4] Gilmanshin, I.R., Gilmanshina, S.I., Galeeva, A.I., Gadirova, E.M., Galeev, I.A. (2019). Methods of training specialists in the field of sustainable resource management and waste management. IOP Conference Series: Materials Science and Engineering, Volume 570, Number 1. https://iopscience.iop.org/article/10.1088/1757-899X/570/1/012025

[5] Gilmanshin I.R., Gilmanshina S.I. (2017). Competence formation of engineering directions students in the field of energy saving as a way to create new generation technologies. IOP Conference Series: Materials Science and Engineering, Volume 240, Number 1, 2015, pp. 012022. doi:10.1088/1757-899X/240/1/012022.

[6] Kashapov N.F., Gil'manshin I.R., Konahina I.A. et al. (2014). System analysis of the energy complex of engineering enterprise as a basic tool of effective energy management. IOP Conference Series: Materials Science and Engineering, Vol. 69, 2013. DOI: 10.1088/1757-899X/69/1/012024.

[7] Sabirzyanov, R.G., Gilmanshin, I.R., Kashapov, N.F., (...), Galeev, I.A., Krainova, D.R. (2018). Material and technical base of the laboratory for the thermal destruction of production and consumption waste for the training of specialists in the field of waste management. IOP Conference Series: Materials Science and Engineering, 412, (2018) 012061.

doi:10.1088/1757-899X/412/1/012061. http://iopscience.iop.org/article/10.1088/1757-899X/412/1/012061.

[8] Duarte, A.J., Malheiro, B., Arno E., Perat, I., Silva, M.F., Fuentes-Dura, P., et al. (2020). Engineering Education for Sustainable Development: The European Project Semester Approach. IEEE Transactions on Education. Vol. 63(2), P. 108-117.

[9] Azimov Yu.I., Gilmanshin I.R., Gilmanshina S.I. (2016). Modern technologies of waste utilization from industrial tire production. IOP Conference Series: Materials Science and Engineering, Volume 134, Number 1, 2015, pp.12003. DOI: 10.1088/1757-899X/134/1/012003.

[10] Gilmanshina, S.I., Gilmanshin, I.R., Dyatlova, E.V. (2019). Formation of competencies of the 21st century by means of project activities. IOP Conference Series: Materials Science and Engineering, Volume 570, Number 1. https://iopscience.iop.org/article/10.1088/1757-899X/570/1/012026.

[11] Gilmanshina S.I., Gayfullina A.Z., Fedotova N.R., Gilmanshin I.R. (2018). Building the Environmental Literacy Through School Natural Science Education. Astra Salvensis, VI (2018), Special Issue, p. 245-253. https://astrasalva.files.wordpress.com/2018/07/astra-salvensis-vi-2018-special-issue.pdf.

[12] Inorganic chemistry. Electrochemistry. Virtual laboratories of ENKA. - URL: http://www.int-edu.ru/content/neorganicheskaya-himiya-elektrohimiya-virtualnye-laboratorii-enka (accessed: 09.03.2020).

[13] Chemistry. Virtual lab. - URL: http://www.mmlab.ru/products/chemlab/chemlab.shtml (accessed March 9, 2020).

IOP Conf. Series: Materials Science and Engineering **915** (2020) 012051 doi:10.1088/1757-899X/915/1/012051

# Recycling components made of polyurethane foam

**L N Shafigullin[1], N V Romanova[1], A T Gabrakhmanov [1], O P Korchagin[1], I R Shamsutdinov and D A Agashkin[1]**

[1]Naberezhnye Chelny Institute (branch), Kazan Federal University
Sjujumbike street 10a, Naberezhnye Chelny, 423800 Tatarstan, Russia

misharin_82@mail.ru

Abstract. The paper studies the influence of recyclable materials and waste (that are sandwich structures of polyurethane foam products) on physical mechanical properties of polyurethane foam products. It was established that indentation load deflection changed by 11 percent, thermal conductivity coefficient - by 12.5 percent, and the strength of bond between face layer and flexible PU foam remains the same (0.3 N/mm) when compared with PU foam products.

**1. Introduction** Today, polyurethane (PU) foam is widely used in construction and machine building industry. PU foam has a good usability: high sound absorption and heat insulation properties, possibility to manufacture parts of a complex shape, high adhesive capability, low manufacturing cost of end products. Advanced PU foam has a wide range of uses in production of interior parts for automotive industry: floor mats, sun visors, upholstery components of cabins, doors, roof, and seats, bunk mattresses, overhead consoles, heat and sound insulation, heat insulators [1-8].

Sandwich structures are used due to high hygroscopic properties of PU foam, for better product esthetics and soft feel, higher wearing properties during operation, UV protection. One of the significant challenges for flexible foam manufacturers is to recycle foam products and waste. Block cutting and fabrication waste can be up to 15 percent of total output [9]. Therefore, the paper proposes the concept of recycling flexible foam waste and investigation of products based on this technology in order to reduce environmental contamination with foam waste.

**2. Body text** In order to investigate recyclability of PU foam waste, we studied the compounds based on PU foam sandwich structures of floor mat [8], including manufacturing waste such as PVC film and rigid PU foam; synthetic leather (vinyl synthetic leather, fancy, TR type) and rigid PU foam; synthetic leather (industrial PVC material for vehicles, VID NT, type 02, TU 8729-156-05790484-2001), flexible PU foam and corona-treated film; semi-rigid PU foam and non-woven fabric; metallized polyethylene terephthalate film; flexible PU foam.

It was established that synthetic leather and film were joined together by foam waste. Its size varies from 100x100 mm$^2$ to 300x700 mm$^2$, and its edges are irregular.

Waste is baled under hydraulic pressure, stored in a plant storage yard, and moved to a municipal solid waste landfill by vehicles. A landfill receives approximately 34 tons of above mentioned waste every month. So, manufacturers and industry as a whole are facing an acute problem of industrial waste utilization.

It has been established that milled PU foam waste can be refoamed using heating, pressure, and binder. Refoaming is widely used to manufacture anti-vibration sound absorbing carpets, flooring, gym mats, stuffing materials, and carpet supports.

Comminuted polyurethanes are used as a filler for production of PU foam and elastomers. When comminuted powder is used as a filler, it is added first to polyol component during production. Molded PU products such as car headrests can contain up to 20 percent of re-milled material without loss in quality and performance.

It has been established that milling should be very fine as air pockets formed in the porous structure would shrink at a higher pressure and tend to restore its earlier volume under normal conditions. Using coarse recyclable particles (larger than 20 mm) for production of PU foam without sifting showed that resulting products had a non-uniform hardness in terms of area. It was observed that there was layer separation and poor adhesion between PU foam layer and outer layers.

In order to study the influence of recyclable materials on physical mechanical properties of PU foam products, the key properties of the studied samples made from waste and initial products were subjected to comparative analysis according to the specification TU [10]. The results of studies are shown in Table 1.

Table 1. Results of comparative assessment of physical mechanical properties

| No | Property name | Test results for initial products | Test results for products with added recyclable material | Reference for parts [11] | Test methods |
|---|---|---|---|---|---|
| 1 | Strength of bond between face layer and flexible PU foam, kN/m | 0.30 | 0.30 | 0.25 | GOST 17317 [11], paragraph 4.4. [10] |
| 2 | Heat resistance | No layer separation and blistering, no changes in front coating and changes in geometry less than 5 % | No layer separation and blistering, no changes in front coating and changes in geometry less than 5 % | No layer separation and blistering, no changes in front coating and changes in geometry more than 5 % are permissible | paragraph 4.6 [10] |
| 3 | Indentation load deflection, mm, minimum:- absolute indentation | 1.64 | 1.46 | 1.20 | GOST 11529 [12], paragraph 4.9 [10] |

| | | | | | |
|---|---|---|---|---|---|
| 4 | Sound absorption coefficient, conv. units at frequencies, HZ, minimum | from 0.015 at a frequency of 315 Hz to 0.24 at a frequency of 1600 Hz | from 0.005 at a frequency of 315 Hz to 0.42 at a frequency of 1600 Hz | from 0.07 at a frequency of 315 Hz to 0.5 at a frequency of 1600 Hz | procedure ISO 10534-2 transfer function method [13], paragraph 4.3 [10] |
| 5 | Thermal conductivity coefficient, W/(m·K) | 0.081 | 0.072 | No data | GOST 7076-99 [14] |
| 6 | Odor intensity, points, maximum | 2.0 | 2.0 | 3.0 | paragraph 4.13 [10] |

The studied samples have the same appearance as the samples of the initial material. However, using recyclable materials in sandwich PU foam structures changes physical mechanical properties such as indentation load deflection (by 11 percent), thermal conductivity coefficient (by 12.5 percent), and the strength of bond between face layer and flexible PU foam remains the same (0.3 N/mm). It should be noted that the most significant properties (sound absorption (up to 1000 Hz) and thermal conductivity) changed only slightly. Therefore, PU foam products with added recyclable materials have satisfactory physical mechanical properties, and they can be used for production of vehicle components, resulting in better environmental aspect of production.

**3. Conclusions** Therefore, the paper showed that the coarse particles of flexible foam waste could be used in vehicle parts such as supports for floor mats or inserts for molded foam. It is necessary to perform further exploratory studies to define optimum size of waste particles, their quantity in a product, spatial distribution of shape-forming elements, optimization of process conditions using recyclable materials and PU foam production waste.

**References**

[1] Gottlieb E M, Galimov E R, Zenitova L A and others 2016 *Polyurethanes: synthesis, properties and application in mechanical engineering* (Kazan: Kazan publishing house UN-TA) p 149

[2] Shafigullin L N, Yurasov S Yu, Shafigullina G R, Shafigullina A N and Zharin D E. 2017 Sound-Absorbing Polyurethane Foam for the Auto Industry *Russian Engineering Research* vol 37 Issue 4 pp 38-40

[3] Shafigullin L N, Gumerov I F, Gumerov A F, Lakhno A V, Yurasov S Yu and Shafigullina G.R. 2017 Glass-Filled Polyurethane Materials Produced by Fiber Composite Spraying *Russian Engineering Research* vol 37 Issue 11 pp 925-28

[4] Shafigullin L N, Astashchenko V I, Romanova N V, Ibragimov A R, Shafigullina G R and Shafigullina A N 2017 Investigation of physical-mechanical and performance properties of glass fiber reinforced polyurethane materials used in the machine building industry *IOP Conference Series: Materials Science and Engineering* vol 240 Issue 1 012063

[5] Kashapov N F, Nafikov M M, Gazetdinov M X, Nafikova M M and Nigmatzyanov A R 2016 Innovative production technology ethanol from sweet sorghum *IOP Conference Series: Materials Science and Engineering* vol 134 issue 1 8 number 012012

[6] Fedyaev V L, Galimov E R, Galimova N Ya, Gimranov I R and Siraev A R 2017 Dynamics of coalescence and spreading of liquid polymeric particles during coating formation *IOP Conf. Series: Journal of Physics: Conf. Series* vol 789 number 012006
[7] Fedyaev V L, Galimov E R, Galimova N Ya, Takhaviev M S and Siraev A R 2017 Mathematical modeling of processes occurring during deposition of sprayed particles of polymeric powder *IOP Conf. Series: Journal of Physics: Conf. Series* vol 789 number 012007
[8] Official website of Avtotekhnik, OOO [Electronic resource]. - Access mode: http://автотехник.net/ru/produkts/ (accessed 20.09.2019)
[9] Klempner D 2009 Polymeric Foams and foam technology (Saint Petersburg: Professiya Publ.) p 600
[10] TU 2292-015-14682925-2017 2017 Molded parts for vehicle sound insulation (Nizhny Novgorod: Zavod avtomobilnykh komponentov Avtokomponent, OOO)
[11] GOST 17317-88 1988 Artificial leather. Method for determination of interlayer bonding strength (Moscow, Izdatelstvo Standartov Publ.)
[12] GOST 11529-86 1986 PVC materials for floors. Methods of control (Moscow, Izdatelstvo Standartov Publ.)
[13] GOST 16297-80 1980 Sound insulation and sound absorption materials. Methods of testing (Moscow, Izdatelstvo Standartov Publ.)
[14] GOST 7076-99 2000 BuIlding materials and products. Method of determination of steady-state thermal conductivity and thermal resistance (Moscow, Gosstroy Rossii Publ., GUP TsPP)

IOP Conf. Series: Materials Science and Engineering **915** (2020) 012052 doi:10.1088/1757-899X/915/1/012052

# Recommendations for making sound absorbing polyurethane materials

**L N Shafigullin [1], N V Romanova [1], S V Zonina [1], I E Valinova[1] and I Yu Karasev[1]**

[1]Naberezhnye Chelny Institute (branch), Kazan Federal University
Sjujumbike street 10a, Naberezhnye Chelny, 423800 Tatarstan, Russia

misharin_82@mail.ru

**Abstract.** We performed the comprehensive studies of physical mechanical properties of KAMAZ vehicle mats from three manufacturers. It was found that structure of PU foam cells was the key structural parameter that influenced heat- and sound-absorbing properties of products. The paper provides manufacturers with the recommendations for managing structural parameters of a PU foam layer in a mat.

**1. Introduction** Today, human environment is surrounded by multiple noisemakers such as vehicles, plumbing systems [1, 2]. Noise is harmful to a man: it increases the concentration of stress hormones; disrupts the function of a nervous system; leads to cardiovascular diseases; reduces hearing sensitivity.

There are two effective methods of noise reduction: sound insulation and sound absorption. The easiest one is to use sound absorbing materials. Two types of sound absorbing materials are typically used in the automotive industry: fibrous materials and polyurethane foam. The first ones have high sound absorbing properties due to a large quantity of interfaces on the noise path. The latter ones have a porous structure that consists of interconnected pores where sound waves impinge against walls of a material; wave energy transforms to oscillation of walls, heating of a material, and its heat exchange with the environment [3-9].

## 2. Body text

There is a strong trend in the automotive industry to use products that perform several related functions. For example, a floor mat performs the following functions in KAMAZ vehicles: heat- and sound insulation (sound absorption), protection, and decor. A floor mat consists of three layers: PVC film, flexible polyurethane (PU) foam, polyethylene film. Outer layers protect an inner layer against dirt, water, and UV radiation.

The main functional part of these products is flexible polyurethane foam which is formed as a result of reactions between liquid polyisocyanates and liquid polyols over a number of other ingredients necessary for product formation. Foams are three-dimensional agglomerations of gas bubbles that are separated from each other by thin polymer walls/strands. In the resulting flexible polyurethane material, empty spaces (cells) are remnants of gas bubbles which have been added to the reaction mixture at the beginning [10].

The comparative studies were performed for the samples of KAMAZ vehicle floor mats from three manufacturers: sample 1 is manufactured by DAIMLER (thickness of PU foam layer is 15 mm); sample 2 - by NPP Ikar (thickness of PU foam layer is 22 mm) [8]; sample 3 - by AVTOTEKHNIK (thickness of PU foam layer is 18 mm) [9].

The following properties of the samples were analyzed: normal sound absorption coefficient according to GOST 16297-80 [11]; thermal conductivity coefficient according to GOST 7076-99 [12]; apparent density of PU foam layer; thermal properties of PU foam layer according to GOST R 55134-2012 (ISO 11357-1:2009) and GOST R 55135-2012 (ISO 11357-2:1999) [13, 14]; macrostructure properties of PU foam layer.
The results of studies are shown in Tables 1 and 2.

Table 1. Normal sound absorption coefficient from 100 to 6300 Hz

| Frequency | Sample 1* | Sample 2* | Sample 3* |
|---|---|---|---|
| 100 | 0.021 | 0.061 | 0.027 |
| 200 | 0.013 | 0.064 | 0.019 |
| 250 | 0.038 | 0.115 | 0.050 |
| 315 | 0.056 | 0.162 | 0.066 |
| 400 | 0.082 | 0.188 | 0.080 |
| 500 | 0.114 | 0.189 | 0.105 |
| 630 | 0.157 | 0.204 | 0.157 |
| 800 | 0.168 | 0.058 | 0.195 |
| 1000 | 0.274 | 0.583 | 0.434 |
| 1250 | 0.337 | 0.558 | 0.534 |
| 1600 | 0.443 | 0.513 | 0.655 |
| 2000 | 0.614 | 0.488 | 0.855 |
| 2500 | 0.816 | 0.262 | 0.939 |
| 3150 | 0.977 | 0.167 | 0.979 |
| 4000 | 0.964 | 0.173 | 0.902 |
| 5000 | 0.830 | 0.220 | 0.742 |
| 6300 | 0.694 | 0.209 | 0.582 |

* arithmetic mean values of tests of three samples

Table 2. Results of physical, thermal, and texture characteristics of flexible PU foam layer

| Property | Sample 1 | Sample 2 | Sample 3 |
|---|---|---|---|
| Thermal conductivity coefficient, $\lambda$ (W/(m*K)) | 0.085 | 0.090 | 0.095 |
| Apparent density, g/l | 58 | 55 | 51 |
| $T_{max}$ of decomposition, °C | 397.5 | 391.5 | 389.2 |
| Remaining mass at 550°C, % wt. | 15.6 | 19.4 | 15.5 |
| Cell size distribution range, mm | 0.04-0.33 | 0.13-0.76 | 0.07-0.50 |
| Cell shape coefficient (degree of cell elongation) | 1.0-1.8 | 1.1-2.6 | 1.0-1.8 |

The studies showed that samples 1 and 3 had similar values of physical acoustic properties (tables 1, 2), and sample 3 had a higher sound absorption coefficient in the frequency range from 630 to 2000 Hz with a lower thickness (by 3-4 mm) than sample 2. All the samples have a comparable thermal conductivity coefficient in the range from 0.085 to 0.095 W/(m*K). The apparent density is comparable for all the samples, and it varies within 51-58 g/l. The thermal studies showed that temperature of maximum decomposition rate ($T_{max}$) was 389-398 °C, which is typical for polyurethane, and sample 3 had a lower value (389 °C) when compared with other samples, and its additional decomposition temperature was 327.7 °C, which could be typical for formation of bimodal polyurethane structures.

It is well known that end performance properties of flexible PU foam is the result of a complex combination of factors, which are associated with macroscopic geometry of cells, i.e. with foam texture [10].

The main characteristic of macrostructure of expanded foams is the structure of their cells, which depends on the parameters of gas-structural elements [15]. The study of parameters of PU foam macrostructure (determination of linear cell dimensions) showed that sample 1 and sample 3 had the smallest cell size distribution range (0.04-0.33 mm and 0.07-0.50 mm respectively) with a cell shape coefficient of from 1.0 to 1.8, sample 2 with an elongation degree of 1.1-2.6 had more elongated cells with a bigger cell size distribution range (0.13-0.76 mm) when compared with samples 1 and 3. The elongation of cells affects anisotropy for not only mechanical but also physical properties of PU foam, and sizes of cells and strands define their acoustic properties. Therefore, sample 1 features a fine-cellular structure of PU foam. It may result from substances in a compound, which make it easier to release solution gases as a separate phase as a change in the nature and concentration of such additives leads to the formation of cellular materials with various structures.

## 3. Conclusion

The studies showed that structure of PU foam cells was the key structural parameter that influenced heat- and sound-absorbing properties of products.

Therefore, in order to ensure the required physical mechanical and performance properties for floor mats, it is necessary to be able to manage and modify dimensions and nature of cellular structure of flexible PU foam, on the other hand, the structure of cells depends on a number of factors active at the stage of PU foam formation.

## References

[1] Gottlieb E M, Galimov E R, Zenitova L A and others 2016 *Polyurethanes: synthesis, properties and application in mechanical engineering* (Kazan: Kazan publishing house UN-TA) p 149

[2] Shafigullin L N, Yurasov S Yu, Shafigullina G R, Shafigullina A N and Zharin D E. 2017 Sound-Absorbing Polyurethane Foam for the Auto Industry *Russian Engineering Research* vol 37 Issue 4 pp 38-40

[3] Shafigullin L N, Gumerov I F, Gumerov A F, Lakhno A V, Yurasov S Yu and Shafigullina G.R. 2017 Glass-Filled Polyurethane Materials Produced by Fiber Composite Spraying *Russian Engineering Research* vol 37 Issue 11 pp 925-28

[4] D.E. Zharin, S.Y. Yurasov, M.I. Gumerov, L.N.Shafigullin, Vibration-and noise-absorbing polymer composites used in manufacturing *Russian Engineering Research* vol 30 Issue 2, pp 194-96

[5] Kashapov N F, Nafikov M M, Gazetdinov M X, Nafikova M M and Nigmatzyanov A R 2016 Innovative production technology ethanol from sweet sorghum *IOP Conference Series: Materials Science and Engineering* vol 134 issue 1 8 number 012012

[6] Fedyaev V L, Galimov E R, Galimova N Ya, Gimranov I R and Siraev A R 2017 Dynamics of coalescence and spreading of liquid polymeric particles during coating formation *IOP Conf. Series: Journal of Physics: Conf. Series* vol 789 number 012006

[7] Fedyaev V L, Galimov E R, Galimova N Ya, Takhaviev M S and Siraev A R 2017 Mathematical modeling of processes occurring during deposition of sprayed particles of polymeric powder *IOP Conf. Series: Journal of Physics: Conf. Series* vol 789 number 012007

[8] Official website of NPP IKAR, OOO [Electronic resource]. - Access mode: http://ikar-ufa.ru/about (accessed 20.09.2019)

[9] Official website of Avtotekhnik, OOO [Electronic resource]. - Access mode: http://автотехник.net/ru/produkts/ (accessed 20.09.2019)

[10] Klempner D 2009 *Polymeric Foams and foam technology* (Saint Petersburg: Professiya Publ.) p 600

[11] GOST 16297-80 1981 Sound insulation and sound absorption materials. Methods of testing (Moscow, Izdatelstvo Standartov Publ.)

[12] GOST 7076-99 2000 Bullding materials and products. Method of determination of steady-state thermal conductivity and thermal resistance (Moscow, Gosstroy Rossii Publ., GUP TsPP)

[13] GOST R 55134-2012 (ISO 11357-1:2009) 2012 Plastics. Differential scanning calorimetry (DSC). Part 1. General principles (Moscow, Standartinform Publ.)

IOP Conf. Series: Materials Science and Engineering **915** (2020) 012052 doi:10.1088/1757-899X/915/1/012052

[14] GOST R 55135-2012 (ISO 11357-2:1999) 2014 Plastics. Differential scanning calorimetry (DSC). Part 2. Determination of glass transition temperature (Moscow, Standartinform Publ.)
[15] Dementyev A G 1997 *Structure and properties of expanded polymers* Dissertation of doctor of Technical Sciences (Moscow: VNII Sinteticheskikh Smol Publ.) p.409

IOP Conf. Series: Materials Science and Engineering **915** (2020) 012053 doi:10.1088/1757-899X/915/1/012053

# Rationality and possibility of automation of precision calibration of gears

**V G Shibakov, D L Pankratov, R S Nizamov and R A Hairullin**

Kazan Federal University, Naberezhnye Chelny Institute, 423812, Russia, Naberezhnye Chelny, Syuyumbike Ave., 10A

vladshib50@gmail.com, pankratovdl@gmail.com, nizamovrs@yandex.ru, guanava@mail.ru

**Abstract**. The main factors that determine the accuracy and quality of the ring gear during precision calibration are described. The corresponding ACS (automated control system) schemes are proposed.

Currently in the process of precision calibration of gear teeth, there are no automation elements. The operation is performed by the employee in the press, and quality control is carried out periodically, by measuring the finished product. Such a process leaves a high probability of manufacturing non-conforming products and does not allow the calibration process to be controlled.

To identify the main factors affecting the accuracy and quality of precision calibration, a matrix of significance of the system parameters {M, W, E, T, I, S, E} for cold calibration has been developed [1].

**Table 1.** Significance Matrix

| System parameters M, W, E, T, I, S, E | |
|---|---|
| (M) Material | |
| The chemical composition of the material | +▲3 |
| Mechanical properties of the material (hardness, strength) | +▲○4 |
| Shrinkage during cooling | - |
| (W) Workpiece | |
| The accuracy of the size of the initial semi-finished product | +○5 |
| (E) Equipment | |
| The rigidity of the press | +○3 |
| Press power | +▲3 |
| (T) Technological process | |
| Changing the workpiece temperature during the stamping process | ○ 1 |
| Lubrication (type) | ▲ 3 |
| (I) Instrument | |
| Elastic deformation of the stamp | ▲4 |
| Expansion of dies as a result of heating during operation | ○3 |
| Surface roughness | ▲5 |
| (S) Staff | |
| Putting a semi-finished product in a stamp | ▲Δ2 |
| Lubrication of stamps (application of grease) | + Δ3 |
| (E) Environment | |

| | | |
|---|---|---|
| Environment temperature | | +○1 |
| Total | ▲ | 31 |
| | Δ | 5 |
| | ○ | 10 |
| | ● | - |

Based on the ability to manage parameters, all factors are divided into: ▲ - managed proactively (at the design stage of the technology); Δ - managed by setting; ● - managed adaptively; ○ - not managed. For ranking, indicators of the target importance of process factors are used: 1-not important; 2-less important; 3-not important enough; 4-important; 5-very important [1].

According to the assessment carried out by experts in table 1[2,3], the factors that have the greatest impact on the accuracy and quality of products obtained by precision calibration are the following:

1. Precision of die tooling manufacturing;
2. Elastic deformation of the stamp;
3. Mechanical properties (strength, hardness) of semi-finished products;
4. Accuracy of the height dimensions of stamped semi-finished products before the calibration operation.

Consider the variability and control capabilities of each factor.

1. The accuracy of the die tooling is set by the designer at the design stage of the die tooling. Production of die tooling for precision calibration must be performed at high-precision machining centers with an accuracy of 0.01 mm or higher. According to the results of 3D scanning of the die tooling (figure 1), the actual profile deviation in the zone of teeth in the die tooling was less than 0.03 mm on the dies for the gear and less than 0.01 mm for the stamp of smaller dimensions for calibration of the satellite.

The considered factor with high significance is managed proactively and does not require adaptive control during the stamping process.

2. Elastic deformations according to the Q-form simulation program are up to 0.05 mm in the contact spot and are critical in terms of the accuracy of the calibrated pinion contact spot. However, the ability to determine the values of elastic deformations using simulation allows you to take this factor into account at the design stage of the tooling by adjusting the geometry of the tooth profile. Therefore, this factor does not require adaptive management.

3. The mechanical characteristics of the calibrated semi-finished products depend on the chemical composition of the metal, which varies within GOST4543 [4] and heat treatment modes.

During the production of a pilot batch of gears and satellites with a calibrated tooth, hardness measurements were performed after heat treatment:

The range of hardness for forgings made:
- 162-187 HB for steel satellites 18ХГР;
- 229 – 241 HB for steel satellites 25Х5М.

Taking into account the fact that the pilot stamping was made from rolled metal of one metallurgical melting, fluctuations in the chemical composition within the stamped batch can be considered insignificant. Heat treatment of forgings within the same group of forgings was carried out in one heat unit, according to a single mode. However, despite the low variability of the above factors, the spread of hardness reaches 15%. Taking into account the direct dependence on the strength, it is possible to predict a 15% variation in the deformation force during calibration.

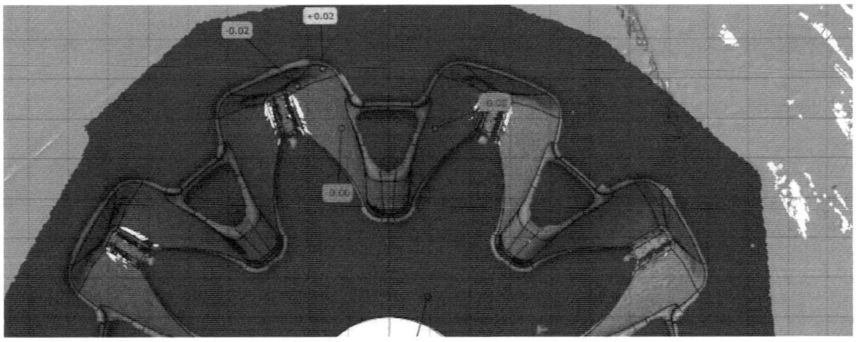


Figure. 1. Results of 3D scanning of a stamp for satellite calibration

The scatter of the values of the deformation force causes various elastic deformations of the stamp and the calibrated forging surfaces, which negatively affects the accuracy of the calibrated contact spot of the gears.

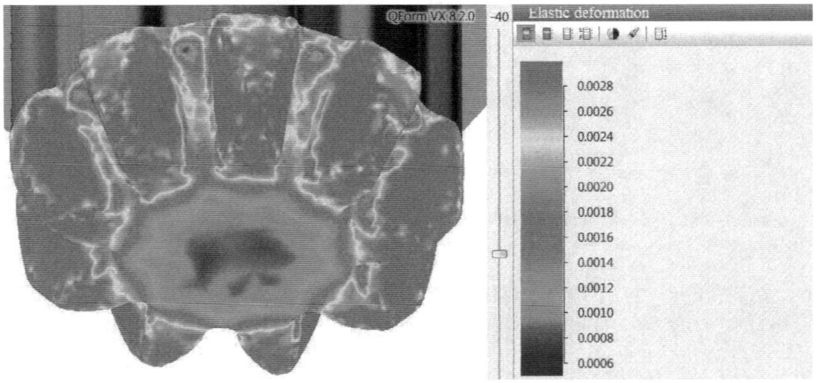


Figure 2. The result of modeling elastic deformations after calibration of a satellite with a hardness of 204 n V

To determine the values of elastic deformations of the forging and tool depending on the strength characteristics of the hot-stamped semi-finished product, a simulation was performed in the Qform program. The modeling results are shown in Figure 2. Elastic deformation of the forging at the minimum permissible, according to the Engineering Documentation, the hardness value of 156NV is 0.132 mm; at the maximum permissible value of 207NV is 0.154 mm. Thus the difference in elastic deformations of the forging in the entire permissible range of hardness is 0.022 mm. The difference in elastic deformations of the tool is less than 0.002 mm, so they can be ignored. The dependence of the elastic deformation of the "Press-stamp" system on the deformation force is presented in the paper [5].

According to the modeling results, the total variability of elastic deformations of the forging is more than 0.02 mm. Taking into account that the tolerance to the surface of the contact spot is 0.1 mm, changing the mechanical properties of the semi-finished product makes an error of more than 20% of the tolerance field.

4. During the hot volumetric stamping of gear forgings, the accuracy of the obtained products is influenced by many factors, such as the accuracy of the sizes of the initial workpiece, the variability of the heating temperature of the workpieces, the accuracy of the manufacture of die tooling, etc. [6,7,8]. As a result, there is a significant variation in the height dimensions of the semi-finished products before calibration.

Figure 3 shows a histogram of the scatter in the altitude size of the batch of forgings "Satellite" obtained by experimental stamping in various conditions.

The difference between the highest and lowest values is 0.9 mm. During carrying out precision calibration, the indicated spread has a significant effect on the quality of the calibrated surface. Figure 3 shows photographs of calibrated surfaces of the forgings "Satellite" with a nominal height size of 33.5 mm and a size at the lower tolerance limit of 33.0 mm.

During calibration, equipment and tools were adjusted to the nominal size of the semi-finished product. Figure 4 (a) clearly shows the trace of the calibration tool on the surface of the teeth. In figure 4 (b), the calibrated spot was not formed over the entire required surface, and the tool footprint is uneven.

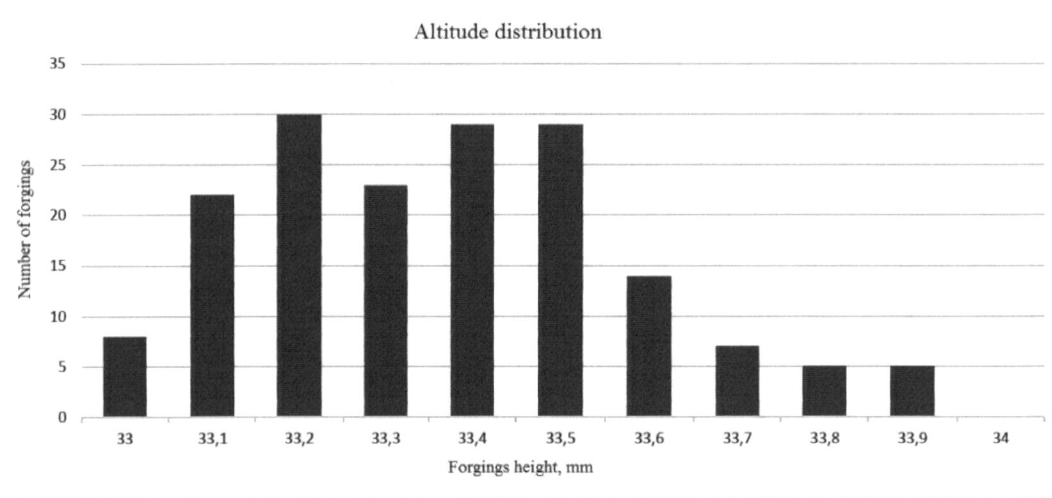


Figure 3. A histogram of altitude distribution size of the batch of forgings "Satellite" obtained during pilot stamping.

The analysis shows that the main factors affecting the accuracy and quality of calibration are the accuracy of the height dimensions and the mechanical characteristics of the hot stamped semi-finished product.

Automation of the sub-adjustment (adjustment) of the calibration press [9, 10] during operation, depending on the height dimensions and mechanical characteristics, will make it possible to solve the problem of ensuring the stability of the quality of the calibration performed.

To provide control of the calibration process, it is proposed to use the SCADA system (SCADA - Supervisory Control And Data Acquisition), which is the main and currently remains the most promising method for the automated control of complex dynamic systems (processes).

It is proposed to measure the hardness and height size of a hot stamped semi-finished product in an automatic installation consisting of the following modules:
- automated milling cutter - to prepare the surface for measuring hardness;
- automatic hardness tester (for example, model Q3000E from QNESS, Austria) – to determine the value of hardness;
- optical height meter – to determine the height value of a hot stamped semi-finished product.

The structural diagram of an automated control system for the precision calibration of gears depends on the type of press used for calibration. Figure 5 shows the structural diagram for a press with rigid kinematic connections (double-crank press). Figure 6 shows the structural diagram for a press without hard kinematic connections (hydraulic press).

IOP Conf. Series: Materials Science and Engineering **915** (2020) 012053 doi:10.1088/1757-899X/915/1/012053

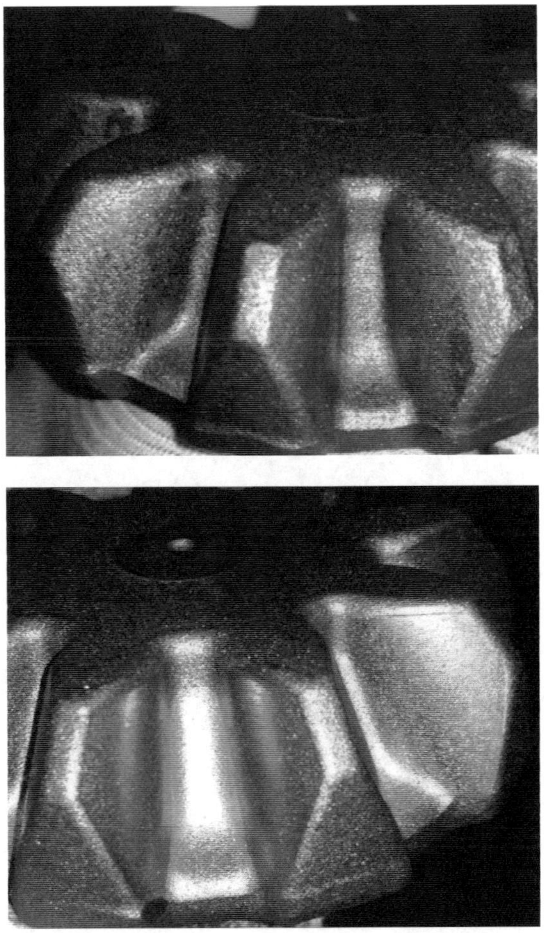

a)

b)

Figure 4. The calibration results of the teeth of the forgings "Satellite"
a) height of hot-stamped semi-finished product 33,5 мм.
б) height of hot-stamped semi-finished product 33,0 мм.

The proposed automated control system of technological process is based on a three-level principle and contains lower, middle and upper levels. The upper level is an intelligent add-on in the form of a decision-making module that corrects the control of the calibration process depending on the data received from the automatic device for measuring the hardness and height of the semi-finished product. At an average level, the controller, based on the state of the monitored parameters, provides control signals to the press actuators. At the lower level, the signals of the sensors are matched with the inputs of the control device, as well as the generated commands with actuators.

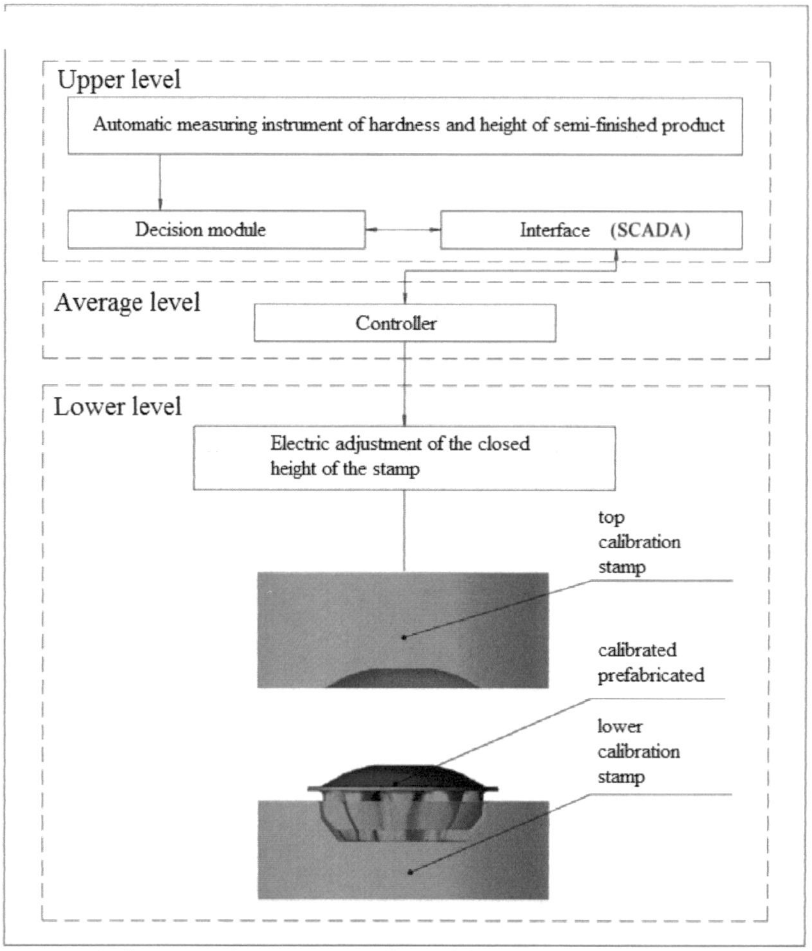


Figure 5. Structural diagram of an automated precision calibration process control system for a press with rigid kinematic connections.

Conclusions.

The main factors that determine the accuracy and quality of the gear ring during precision calibration are described: the accuracy of manufacturing die tooling, elastic deformation of the stamp, mechanical properties of semi-finished products, the accuracy of the height dimensions of stamped semi-finished products before the calibration operation.

Structural diagrams of automated control systems for precision gear calibration are proposed, depending on the type of press used.

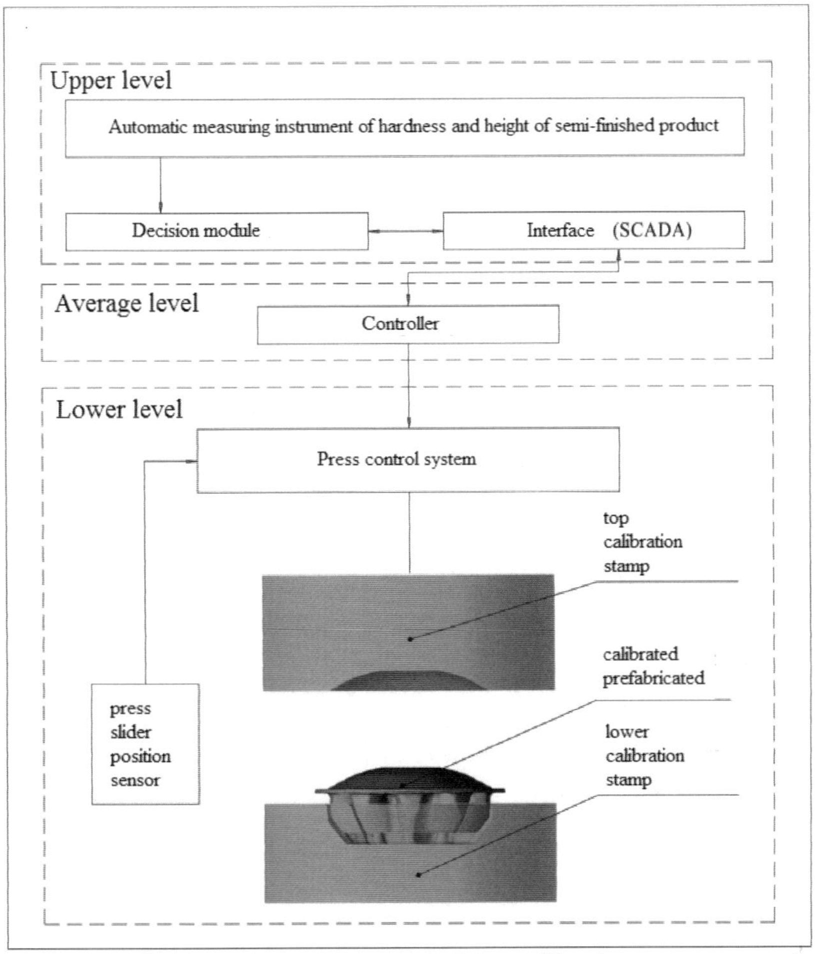


Figure 6. Structural diagram of an automated precision calibration process control system for a press without rigid kinematic connections (hydropress).

## References

[1] V G Shibakov, D L Pankratov, R A Hairullin. Systems Analysis of Process Control Capabilities for Forging Accuracy. Solid State Phenomena, Vol. 265, pp. 1110-1115, 2017

[2] Antoniuk F.I. Analysis and provision of high precision cold forging on presses with a scientifically based choice of the rigidity of the elements of the technological system: Abstract for the degree of Doctor of Technical Sciences. - Moscow, 2004. - 36 p.

[3] Kruk A.T. Development of scientifically-based technical solutions to improve the accuracy of forgings, the creation on their basis and the industrial introduction of heavy crank hot forging presses: Abstract for the degree of Doctor of Technical Sciences. - Voronezh, 2006. - 36 p.

[4] GOST 4543-2016. Structural alloy steel products. Specifications

[5] Dibner Yu.A. Option for subsetting of hot-pressing presses with a non-linear system “press - stamp - forging” / Dibner Yu.A., Prutskov R.N. // Forging and Stamping Production. Material Working by Pressure. - 2003 - №10 - C 29-33.

[6] V G Shibakov, D L Pankratov, R S Nizamov, R A Hairullin The choice of a rational transition system for precision punching of the gear ring gear. Proceedings of the VIII International Scientific and Technical Conference "Innovative machine-building technologies, equipment and materials - 2017" Kazan, 2017. - 388 p.

[7] V G Shibakov, D L Pankratov, R S Nizamov and R A Hairullin. Surface quality assurance of ring gear during precision stamping of gearwheels. IOP Conf. Series: Materials Science and Engineering 412 (2018) 012073 doi:10.1088/1757-899X/412/1/012073

[8] V G Shibakov, D L Pankratov, R S Nizamov, R A L M Nizamova, Hairullin Justification of

IOP Conf. Series: Materials Science and Engineering **915** (2020) 012053 doi:10.1088/1757-899X/915/1/012053

requirements for quality and accuracy of blanks for precision stamping. Proceedings of the IX International Scientific and Technical Conference "Innovative machine-building technologies, equipment and materials - 2018" Kazan, 2018. - 408 p.
[9] Valiev A.M., Shibakov V.G., Pankratov D.L. Automated Control System for the Assembly of a Composite Punching Tool // Russian Engineering Research, - 2016. - 36 (2). - P. 142-145.
[10] Pishchuhin A.M. Automation of technological processes based on flexible production systems: Tutorial. – Orenburg, 2004.- 111 p.

IOP Conf. Series: Materials Science and Engineering **915** (2020) 012054 doi:10.1088/1757-899X/915/1/012054

# Frame model of knowledge representation in the expert system of technological route selection

**L A Simonova**[1], **G M Kapitonova**[1]

[1] Kazan Federal University, Naberezhnye Chelny Institute, Prospect Mira 13A, Naberezhnye Chelny, 423800, Russian Federation

lasimonova@mail.ru

**Annotation**. In the process of technological preparation of production, a large amount of industrial data is used, such as the availability of production materials and components, the availability of production resources, the use of various tools and information systems for production management based on ERP and MES. In order to ensure operational management of production and help a technologist in making a decision while choosing a technological route for manufacturing a product, there is a need to use an expert system. Taking into account the above-mentioned requirements, a frame representation of knowledge becomes the most appropriate method of presenting knowledge in an expert system. The paper proposes a frame model for representing knowledge about the technological route of manufacturing an expert system product.

**1. Introduction** The process of technological preparation of production involves the use of a large amount of information, working with large amounts of data, such as the availability of production materials and components, the availability of production resources, the use of various tools and information systems for production management based on ERP and MES.

An important task in the process of technological preparation of production (TPP) is one of selecting equipment and forming product manufacturing routes. The task of selecting equipment in the machine-building industry to fulfill an order only by the capabilities of a specialist is characterized by the use of a subjective solution that depends on the experience of this specialist and requires a lot of time. For such tasks, it is advisable to use multi-agent technologies used for dynamic management of network resources, which have advantages when used in a distributed environment. The use of multi-agent technology in relation to the problem of automatic selection of equipment and formation of the product manufacturing route is due to the developed model of the external world due to the presence of a multi-agent system knowledge base, mechanisms for solving and analyzing actions. Since the production process involves a lot of people from different departments working in different directions, an important task of automating this process is to resolve contradictions in the construction of the optimal technological route for manufacturing the product at the stage of the TPP. With a multi-agent approach, this problem is solved by using methods for organizing interaction and coordinating the actions of intelligent agents. To achieve the maximum effect, the production management system must be fully automated [1].

## 2. Methodology for presenting knowledge in an expert system

The methodology of presenting knowledge in an information system should help to extract elements of knowledge in the right situation using a relatively simple and natural mechanism. Also provide translation of input data into a form suitable for storing in machine-readable form and have a knowledge base with advanced indexing and contextual addressing tools. Then the information system will be able to manage the sequence of application of certain knowledge, even without having accurate information about how they are stored.

The use of multi-agent technology requires a systematic method of describing at the machine level what the technologist who specializes in the field of TPP knows, and a certain organization of knowledge.

The main issue in the representation of knowledge is to determine the composition of knowledge and an adequate representation of the simulated system.

The information system consists of a database that stores data about the manufacturing process of the product; a knowledge base that accumulates the rules of the expert system; and a precedent base that generates new knowledge for the knowledge base.

The knowledge base includes the subject area of the TPP: a set of entities that describe the area of expertise, i.e. a set of objects that affect the choice of the optimal route for manufacturing the product, the values of their characteristics and the relationships that link them. The knowledge base includes entities-data structures and tasks solved on them, represented as executable statements in the form of rules, procedures, and formulas [2].

From the point of view of the architecture of knowledge in the knowledge base, it is advisable to divide it into interpreted and non-interpreted. The first type is the knowledge that the technologist is able to interpret. All other knowledge is of the second type.

One of the indicators of system intelligence in terms of knowledge representation is the ability of the system to use relevant knowledge at the right time. Knowledge connectivity is the main way to speed up the search for relevant knowledge. Knowledge should be organized around the most important objects (entities) of the subject area. All knowledge that characterizes an entity is linked and represented as a separate object. The most appropriate method of presenting knowledge in an information system, taking into account the abovementioned requirements, is the frame-representation of knowledge.

Firstly, a frame is a data structure designed to represent some standard situation. Unlike formal models, heuristic models have a diverse set of tools that convey the specific features of the manufacturing process. This is why heuristic models are superior to logical models both in terms of the ability to adequately represent the problem environment and in terms of the effectiveness of the inference rules used.

Secondly, frame representations of knowledge in expert systems differ from formal production systems in that they use more complex rule structures, often expressed as semantic structures, since this occurs when describing a technological process. The difference between frames and procedure descriptions is that frames can be called not by name, but by matching the current situation to the situation that this frame describes. Associated with each frame slot are descriptions of conditions that must be met. In the simplest cases, these conditions can be reduced to specifying semantic categories that the slot value must satisfy. In more complex cases, the conditions may relate to the relationship between values selected for multiple slots [3].

Thirdly, the most important property of frame theory is the property inheritance borrowed from the theory of semantic networks. Inheritance of properties can be partial or full borrowing of properties, which is convenient when forming typical technological processes for structurally similar products [2].

Fourthly, the main advantage of frames as a model of knowledge representation is that it reflects the conceptual basis of the organization of human memory, as well as its flexibility and visibility.

Figure 1 shows a general diagram of the frame representation of the subject knowledge base of the information system for selecting the technological route for manufacturing a production order using the relationships between the individual blocks of the process of technological preparation of production.

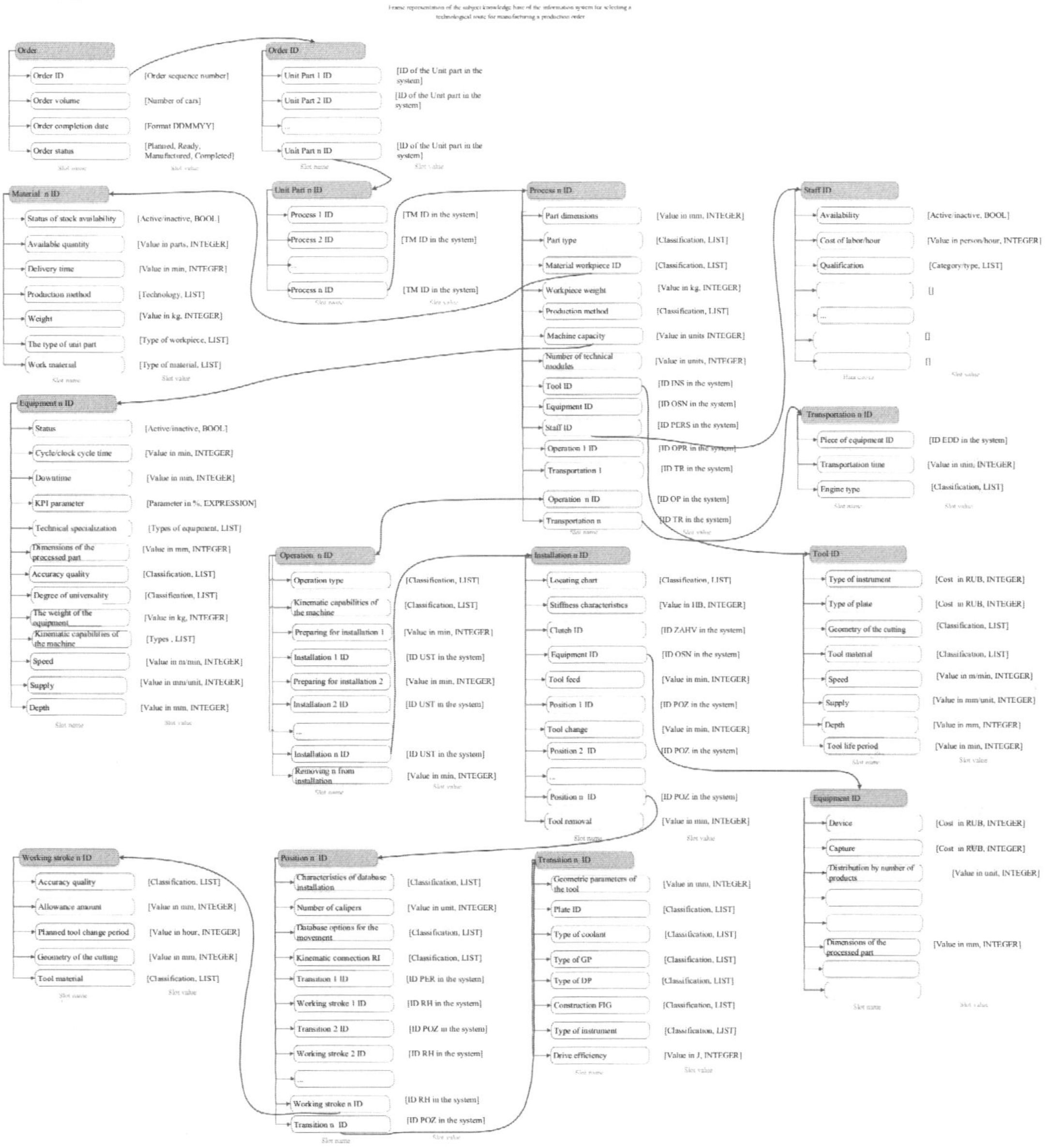


Figure 1 - Diagram of the frame representation of the subject knowledge base of the information system for selecting a technological route for manufacturing a production order

In the diagram the main identifier that determines the uniqueness of the route is the order frame to which the unit part frames are linked. Each unit part frame has a link to the process frame. A process frame contains a description of the dependencies between the material, equipment, operation, personnel, transportation, tooling, and tool frames that are included in it. The frame operation is further decomposed into the frames of the setting, position, transition, and working stroke.

Each of these frames has its own set of slots that allow you to describe the essence of the frame and ensure the relationship of frames to each other, forming a unified system. By using this representation of dependencies explicitly, it allows to predict the transition from one state A (expressed by frame Ar) to another state B (expressed by frame Br) that depends on it, and to perform this transition effectively, i.e., without re-calculating the values of all parameters that characterize state B, but listing only the changed or new parameters [2].

The choice of alternative variants of the technological route is made by comparing the corresponding source data of the sample frame with restrictions on the processed part of the frame instance.

**Reference**

[1] Simonova L A, Abramova V V and Kapitonova G M 2019 *Information Flow Management System in Alternative Technological Routes at the Pre-Production Stage Using a Multi-Agent System* (Scientific and Technical Bulletin of the Volga Region) No. 3 126-128

[2] Morozova V A, Pautov V I 2017 *Presentation of Knowledge in Expert Systems* (Yekaterinburg: Textbook of Ural Publishing House University) 120

[3] Popov E V, Fominyih I B, Kisel E B and Shapot M D 1996 *Static and Dynamic Expert Systems* (Studies. Handbook for universities, M.: Finance and Statistics) 320

IOP Conf. Series: Materials Science and Engineering **915** (2020) 012055 doi:10.1088/1757-899X/915/1/012055

# Physical basis of thermal control of the operating mode of the refrigeration unit

**I F Timerbaev[1], I R Gilmanshin[1,2]**

[1]Kazan Federal University, Kazan, Russia, 420008
[2] Kazan National Research Technical University named after A N Tupolev, Kazan, Russia

is-er@yandex.ru

**Abstract.** Diagnostics and output quality control of finished products in the context of digitalization of industrial enterprises can be considered as the most valuable source of objective data for compiling a digital passport of products and consumer goods. In modern conditions, confining ourselves exclusively to a performance check on the principle of "conforms / does not correspond" means breaking the causal relationship between products and the production process. In this article, we consider the engineering foundations of thermal imaging control of finished products on the example of a domestic refrigerator.

Studies of the current state of the level of development and the demand for non-destructive testing methods have revealed the possibility of monitoring and controlling the operating mode and technical condition of complex technical systems and objects in various industries and engineering to analyze their temperature fields. [1-6] This conclusion is based on the close relationship of individual temperatures elements of engineering systems and the efficiency of the energy conversion process in it. According to GOST R 56542-2015 "Non-destructive testing. Classification of types and methods": thermal non-destructive testing: type of non-destructive testing, based on the analysis of the parameters of the thermal fields of controlled objects caused by defects. The temperature field, the analysis of which is used in thermal imaging control methods, carries valuable data about the features of the technological process of energy conversion, individualized with respect to this object.

It is important to note that the thermogram is an exclusively graphical interpretation of the temperature field of the object of analysis. I'm studying the thermogram, it is possible to localize areas with a particular temperature and it is impossible to directly assess the quality of the process taking place in the system. At the initial examination, the result of the examination becomes a protocol with an array of data on the differential temperature difference of different parts of the surface of the system elements, and not a conclusion about the characteristic features, operation mode or detected defects. At the same time, thermal control has a number of significant advantages distinguishing it from other non-destructive and / or contact control methods [7]:

- continuous operation of the facility;
- lack of direct contact with the object, especially in cases of increased danger;
- informativeness of diagnosis;
- high speed of primary information processing;
- high performance tests;
- the ability to control a wide range of materials;

IOP Conf. Series: Materials Science and Engineering **915** (2020) 012055 doi:10.1088/1757-899X/915/1/012055

- the possibility of automated data processing;
- the possibility of organizing conveyor control.

The scope of thermal control is quite wide [1.8]:

- production, control of technological processes (production): printed circuit boards, welded joints, steel, cement, glass, paper industry;
- maintenance, technical diagnostics: electrical equipment, boilers and steam systems, mechanical equipment, buildings and structures, gas distribution systems, tanks with liquid products;
- medicine: circulatory diseases, oncology;
- traffic monitoring;
- forest fire fighting;
- biology;
- astronomy;
- military equipment.

Thus, the use of thermal control in solving problems of monitoring the quality of finished products in the production of refrigeration equipment has a high potential for implementation. A systematic approach to the design of thermal control devices was considered by scientists of Tomsk Polytechnic University D.A. Nesterukov and V.P. Vavilov. The algorithm adapted to the solution of the problem of monitoring the quality of finished products has the form presented in Fig. 1:

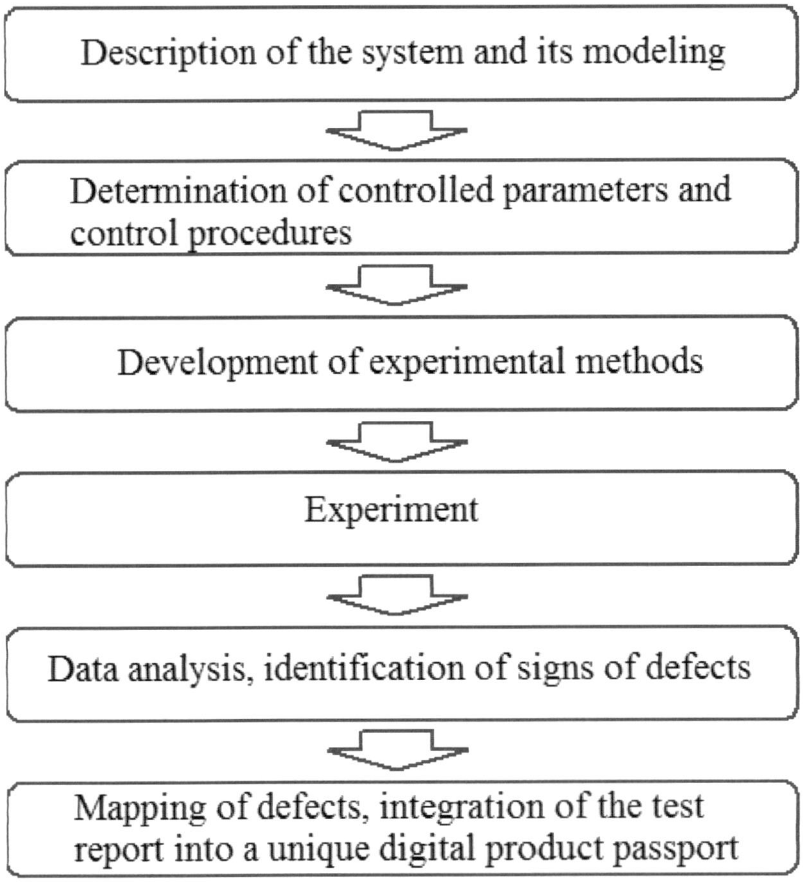


Fig. 1 Algorithm for solving the problem of quality monitoring by the finished method of thermal control.

The key point of the algorithm is a comprehensive analysis of the control object. Physical and technical characteristics of the system determine the amplitude of the temperature signal in the controlled area, its dynamics and the optimal detection time. [9-10] The limiting monitoring parameters are determined by the noise level of the signal, thus determining the hardware requirements (resolution and recording frequency of thermograms, duration of measurements ) In numerical modeling, a digital double of the object is formed using well-known theoretical models. [11-12] Further, analyzing the dynamics of temperature of the control points, they develop the experimental technique and optimize it taking into account possible practical limitations. The simulation result is a certain temperature function T (i, j, τ) defined for each point of the control object. The analysis of experimental thermograms is carried out using specialized or universal software, taking modeling data confirmed as a reference by a series of field measurements. At the end of the control procedure, a thermal control protocol is formed, which is integrated with a unique digital passport of the product.

The proposed algorithm was tested on a series of experiments on thermal control of refrigeration equipment. A Vestel MDD 317 VW two-chamber household refrigerator equipped with an ETY 95L motor compressor was chosen as an object. block diagram is presented in Fig. 2.

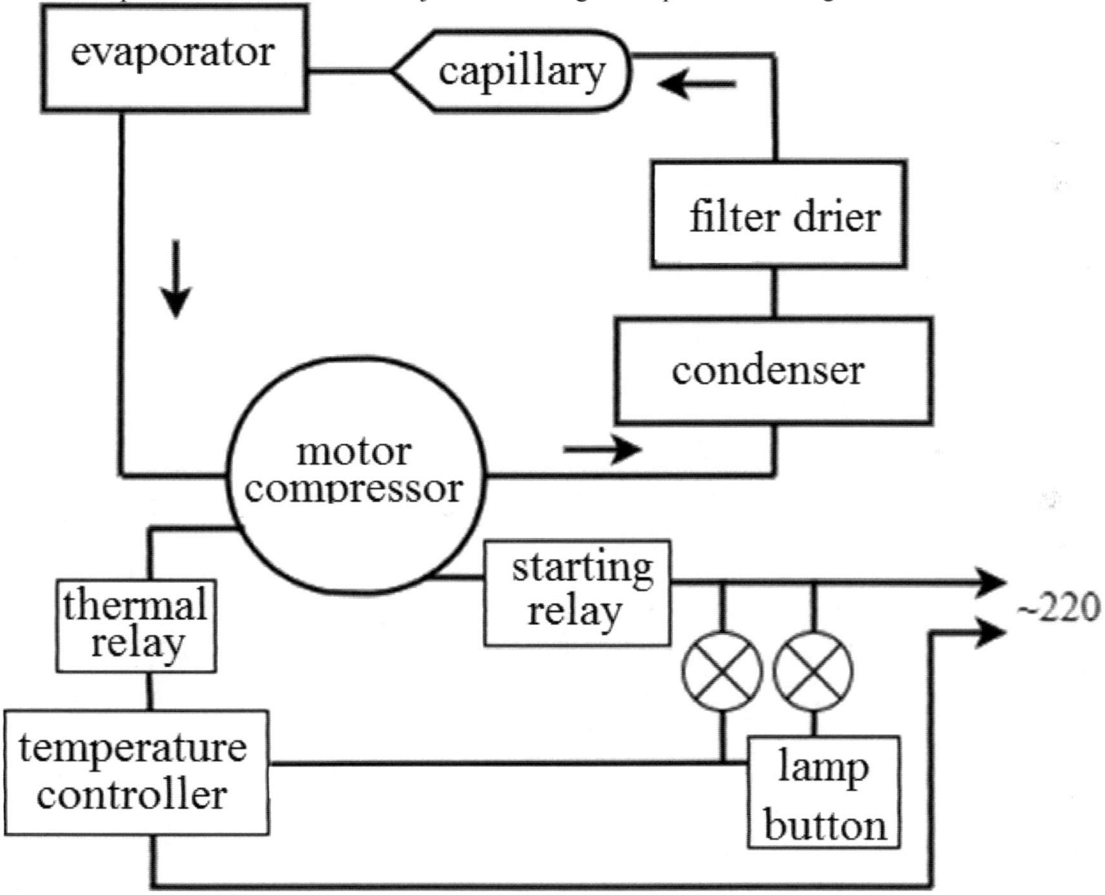


Fig. 2. The structural diagram of a two-chamber domestic refrigerator Vestel MDD 317 VW.

The experiment was carried out using a DALI, LT7 thermal imager. Technical characteristics are presented in table 1.

Ttable 1.

IOP Conf. Series: Materials Science and Engineering **915** (2020) 012055 doi:10.1088/1757-899X/915/1/012055

| Specifications | DALI LT7-P |
|---|---|
| Matrix size | 384 x 288 |
| Thermal sensitivity | ≤0,05 °C at 30 °C |
| Temperature measuring range | -20 °C ~ +650 °C |

During the experiment, thermograms of evaporators and condensers of the refrigerator were recorded in the cold start mode with time recording. The exit to the steady state is detected by the fact of establishing a stable temperature on the coil pipe throughout its length in the working area. The results of the experiment are presented in the form of thermograms and graphs of temperature gain.

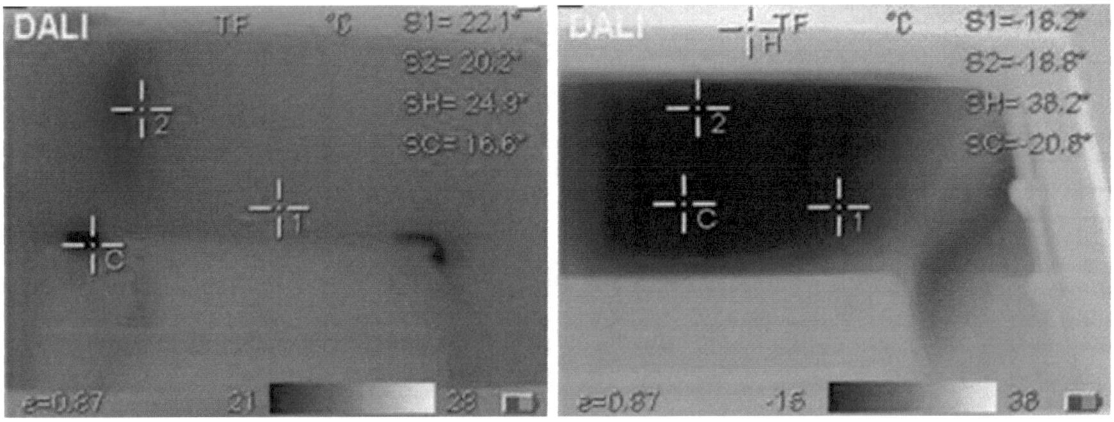


Fig. 3. Thermograms of the upper evaporator at the start of the experiment and for 150 sec.

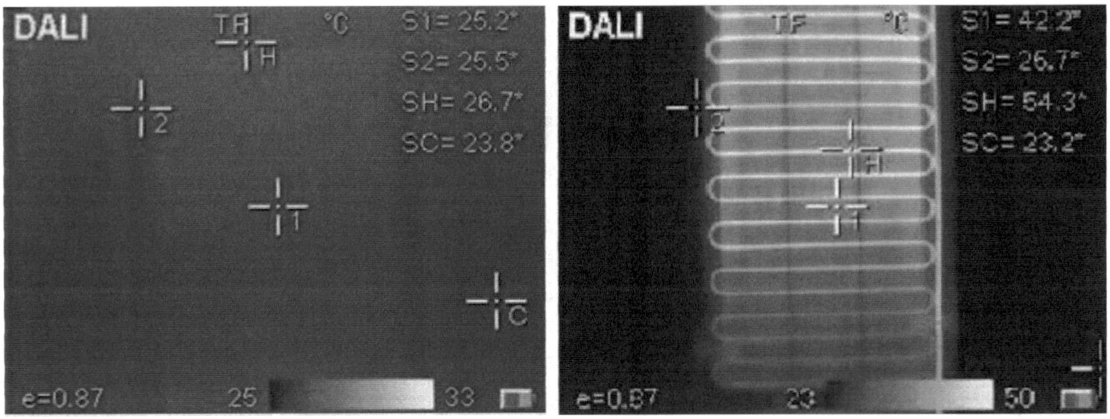


Fig. 4. Thermograms of the condenser at the start of the experiment and for 725 sec.

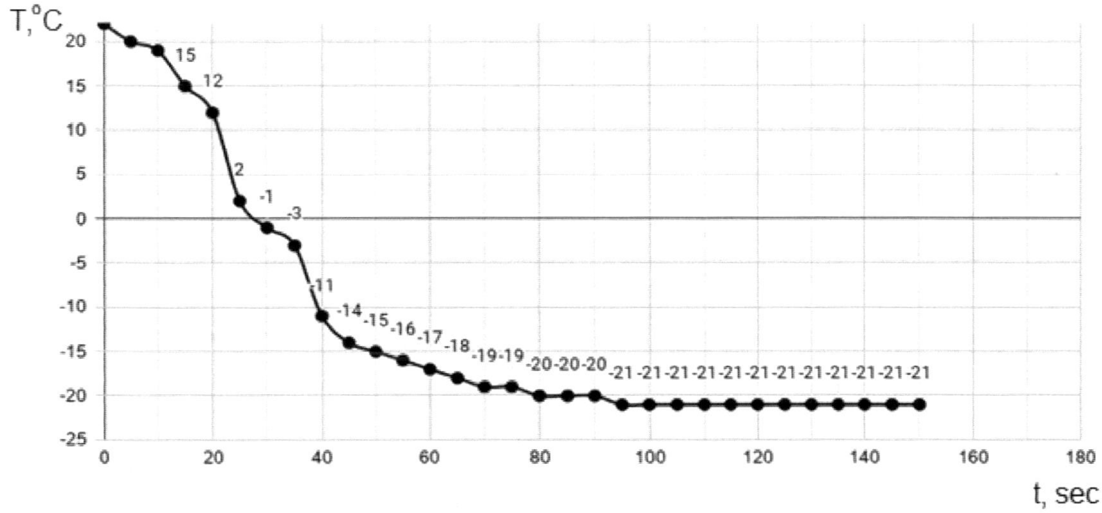


Fig. 5 Schedule of temperature set for the upper evaporator.

According to the results of the work, the high information capacity of the thermograms of the elements of the refrigeration machine was experimentally confirmed. The possibility of conducting output control of the finished product's health using the proposed method of thermal control is proved. The absence of dependence of the efficiency of thermal control on the design (closed / open) of the evaporator and condenser is established. It was established that the time for the refrigerant circuit to return to nominal mode is no more than 12 minutes, versus 180 minutes for typical control. Thus, the work experimentally confirmed the effectiveness of thermal control of the finished product.

**References**

[1] Vollmer, M., Möllmann, K.-P. Infrared thermal imaging: Fundamentals, research and applications (2017) Infrared thermal imaging: Fundamentals, research and applications, pp. 1-769.

[2] D.A. Nesteruk, V.P. Vavilov Thermal control and diagnostics. Textbook for the training of specialists of I, II, III level. - Tomsk :, 2007 - 104 p.

[3] Jin, X., Jiang, Q., Yao, S., Zhou, D., Nie, R., Hai, J., He, K. A survey of infrared and visual image fusion methods (2017) Infrared Physics and Technology, 85, pp. 478-501.

[4] Bates, D., Smith, G., Lu, D., Hewitt, J. Rapid thermal non-destructive testing of aircraft components (2000) Composites Part B: Engineering, 31 (3), pp. 175-185.

[5] Ciampa, F., Mahmoodi, P., Pinto, F., Meo, M. Recent advances in active infrared thermography for non-destructive testing of aerospace components (2018) Sensors (Switzerland), 18 (2), art. no. 609.

[6] García Márquez, F.P., Tobias, A.M., Pinar Pérez, J.M., Papaelias, M. Condition monitoring of wind turbines: Techniques and methods (2012) Renewable Energy, 46, pp. 169-178.

[7] Efficiency and methods of thermal control in production // RostBusinessConsult URL: https://rostbk.com/o-kompanii/stati/teplov-kontrol/ (accessed: 12/11/2019).

[8] Vavilov, V.P. Thermal imagers and their applications / Vavilov V.P., Klimov A.G. - M.: Intel universal, 2002. - 87 p.

[9] Gilmanshin, I., Shaimukhametov, R., Strekalov, V. Digital Recursive Filters for Building Thermal Modelling. Proceedings of 2018 IEEE East-West Design and Test Symposium, EWDTS 2018. 5 November 2018, #8524702

[10] Yakimov, I.M., Kirpichnikov, A.P., Mokshin, V.V., Yahina, Z.T. The Modelling and Optimization of Machine Management System with Computer Numerical Control. Lecture Notes in Electrical Engineering. Volume 641 LNEE, 2020, Pages 724-731

[11] Zijatdinova, D.R., Kashapov, N.F., Fatkulina, R.R., Fathutdinov, R.H. Optimisation of a degree of quality of isolating textile materials. Izvestiya Vysshikh Uchebnykh Zavedenii, Seriya Teknologiya Tekstil'noi Promyshlennosti. Issue 1, 2011, Pages 16-19

[12] Konahina, I.A., Kashapov, N.F., Gil'Manshin, I.R., Ganiev, R.R. Optimization of industrial steam supply and steam-and-condensate farming of machine building enterprise. IOP Conference Series: Materials Science and Engineering. Volume 69, Issue 1, 2014, #012025

IOP Conf. Series: Materials Science and Engineering **915** (2020) 012056 doi:10.1088/1757-899X/915/1/012056

# Industrial control system software patterns in manufacturing

**R A Valiev and Sh Sh Khuzyatov**

Kazan Federal University Naberezhnye Chelny Institute, 68/19 Mira Street, Naberezhnye Chelny, 423810, Russian Federation

rustvali@mail.ru

**Abstract**. The technique of pattern-designing software for an industrial control system for standard engineering processes has been implemented, where the same type of equipment is used and the same workpiece processing sequence is implemented. Patterns for the user controller program and supervisory control and monitoring software have been developed. Pattern elements are represented by the same type of sensors and actuators, which have the same algorithms for data processing and calculating of control actions. For them data types, data storage, functionality and organization of the exchange between programmable logic controllers and human-machine interface devices are defined. The necessary correspondence was established between template projects of various levels of the control system, which made it possible to organize a network exchange by data blocks.

## 1. Introduction

The diverse relationship of automation objects in mechanical engineering complicates the development of industrial control system (ICS) software [1-6].

The use of modern integrated software development systems for controllers and SCADA-systems simplifies the process of creating a control system greatly [7-9]. But with a large number of sensors and actuators and their complex relationship, the development of industrial control system software requires a lot of time. It should also be borne in mind that the organization of data exchange over the network between the user program of the controller and the software of the human-machine interface is carried out in real-time [10-13].

From the point of view of developing industrial control systems, typical technological processes are of particular interest, where the same type of equipment and the same workpiece processing sequence are used. When automating these typical technological processes, the developed system implements the same control and data processing algorithms, while the processes differ from each other only in the number of sensors and actuators [14-16]. In such cases, the use of template projects for both middle-level software and top-level software development becomes appropriate for the development of industrial control system software [17-18].

## 2. Methods

The creation of template projects for the development of ICS software for typical technological processes in mechanical engineering is proposed to be carried out on the basis of the pattern-design technique [19]. In accordance with this technique, the same type of sensors and actuators are adopted as pattern elements, which have the same data processing algorithms and control actions calculation. For them, it is necessary to determine the types of data, data storage, functionality and organization of the data exchange between programmable logic controllers and devices of the human-machine interface [20-23].

Template projects for the automation of typical technological processes of mechanical engineering were developed on the basis of Siemens hardware and software [24-26]. The programmable logic controller software was developed in Step 7 (PLC template project) and is intended for automatic control system. The human-machine interface software developed in the WinCC SCADA system (HMI template project) is intended for monitoring and supervisory control of the technological process.

## 3. Results and discussion

For pattern elements, data types, data storage, functionality of executive devices, and the organization of the data exchange between lower and upper level software are defined.

For each type of sensor or actuator, user-defined data types are created in the template project of the programmable logic controller, and structural types are created in the template project of the human-machine interface. In a PLC project, type elements are intended for storing object parameters, as well as output and input data. In the structural type of the HMI project, the number of elements is equal to the number of elements in the corresponding user type of the PLC project. Moreover, the types of these elements are similar to the types of data storage in memory in accordance with IEEE standards.

Storage of data related to the same sensors and actuators in a template PLC project is implemented as a data block. It is a one-dimensional array with user-defined data type elements. In a template HMI project for storing data of each type of sensor and actuator, the same type of structural tags are created, which are grouped into the corresponding tag groups.

Each array element in the data block in the user program of the controller must correspond to a structural tag in the application of the human-machine interface. For this, the offset of the array element in the data block is selected as the address of the structural tag.

Calculation of control actions for actuators and data processing of the same sensors in a template PLC project were performed using the corresponding arrays created in the project. The data processing algorithm for each group of objects of the same type is implemented as a function or function block.

In the HMI template project, the functionality of the same type of objects is implemented for supervisory control and monitoring of the process, for setting parameters during commissioning, as well as for alarm logging and tag logging. The graphic part of the template HMI project is a hierarchical system consisting of objects screen and status screens.

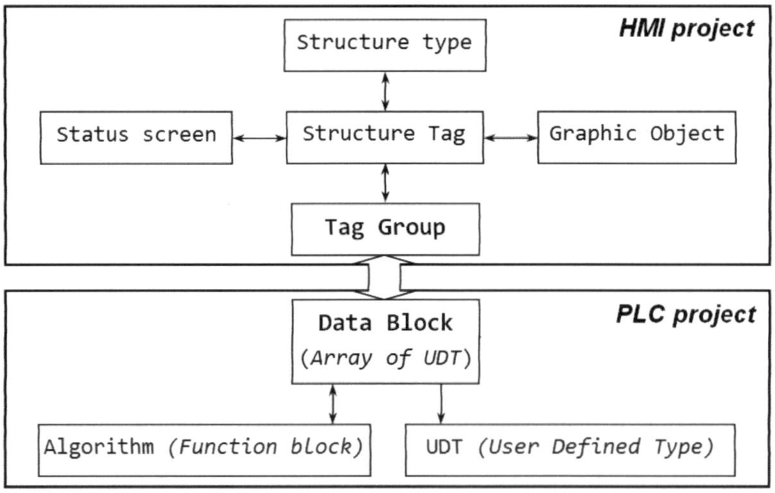

**Figure 1.** Pattern elements for one type of sensor or actuator.

On the objects screen graphical objects of typical actuators and sensors are presented. The current state of these objects is displayed as various visual effects. This is achieved by dynamizing objects using a dynamic dialog or by cyclic execution of VB- and C-scripts.

For each type of actuator or sensor, subordinate process screens are created. These status screens display the values of the object's output and input signals in real-time mode, alarm loggings and signal trends. During commissioning, the status screen allows you to set the parameters of the selected object. Also, the status screen makes it possible to operate and diagnose the process in real-time mode.

Due to the fact that Siemens software for different levels can implement the exchange of data blocks, the simplification of the organization of the exchange between the human-machine interface device and programmable logic controllers is carried out through the use of a data block for storing data [27].

When creating structural data types, the correspondence of elements in structural types in template PLC and HMI projects was already ensured, which allows the exchange of data blocks (Figure 1).

Correspondence of the structural tag of the HMI project to the array element in the data block of the user program of the controller was implemented at the stage of organizing data storage, which also ensures the correct network data exchange.

## 4. Practical Use

The proposed structure of template projects facilitates their adaptation to the specific requirements of the corresponding control system greatly.

When developing a PLC project for a specific automation object, depending on the number of sensors and actuators, the number of array elements in the data block for each type is determined. In the initialization section of the data block and the initial values are assigned to some elements of the structural type.

When developing an HMI project for a specific automation object, firstly a mimic diagram of the technological process is created by copying graphical objects of typical sensors and actuators from the screen of the template project objects. Next, the structural tags are created in the required quantity. For each structural tag, the data block number and the offset within this block are determined.

## 5. Conclusion

The developed technique greatly simplifies the software design process for industrial control system with a large number of automation elements of the same type.

In the template projects created on the basis of this technique, the structure and functionality of process control software are predetermined. The use of patterns becomes effective in the development of control system software for typical engineering processes. At the same time, the universality of approaches makes this technique universally valid for the automation of various industrial facilities.

The use of template design significantly reduces the cost of the project by reducing the time of development and implementation, as well as the use of less skilled labor of software developers.

## References

[1] Rigatos G G 2011 *Modelling and Control for Intelligent Industrial Systems: Adaptive Algorithms in Robotics and Industrial Engineering* (ISRL vol 7) (Springer-Verlag Berlin Heidelberg) p 379
[2] Sisbot S 2011 *Systems Engineering* **14** 193-207
[3] Valiev R A, Khairullin A Kh and Shibakov V G 2015 *Russian Engineering Research* **35** 662-5
[4] Sullivan L L and Oliver C R 1996 *ASEE Annual Conf. Proc.* 1943-6
[5] Chistyakova T B, Novozhilova I V and Kozlov V V 2019 *Proc. 2019 1st Int. Conf. on Control Systems, Mathematical Modelling, Automation and Energy Efficiency, SUMMA 2019* 523-6

[6] Khuzyatov Sh Sh and Galiullin L A 2019 *Journal of Advanced Research in Dynamical and Control Systems* **11** (S08) 1758-61

[7] Figueroa-Lorenzo S, Añorga J and Arrizabalaga S 2019 *Sensors* **19(20)** 4455

[8] Tcaciuc S-A 2017 *International journal of advanced computer science and applications* **8(11)** 456-60

[9] Mund J, Bougouffa S, Badr I and Vogel-Heuser B 2018 *At-Automatisierungstechnik* **66(10)** 784-94

[10] Quan Z, Yuan X, Zhu Y and Wang, Z 2019 Real-time flow control system based on Siemens PLC *Proc. of 2019 IEEE Int. Conf. on Mechatronics and Automation, ICMA* 1703-8

[11] Tcaciuc S-A (Gherasim) 2018 *Int. J. of Advanced Computer Science and Applications* **9(3)** 319-23

[12] Khuzyatov S S and Valiev R A 2017 Organization of data exchange through the Modbus network between the SIMATIC S7 PLC and field devices *Proc. Int. Conf. on Industrial Engineering, Applications and Manufacturing, ICIEAM 2017* 8076369

[13] Dremin R S, Bets Yu S, Tynchenko V S, Bukhtoyarov V V, Milov A V, Egorova D V and Kukartsev V V 2020 *IOP Conf. Ser.: Mat. Sci.Eng.* **734** 012134

[14] Khaziev E L and Galiullin L A 2019 *Journal of Advanced Research in Dynamical and Control Systems* **11(S08)** 1754-7.

[15] Zubkov E V and Khaziev M L 2020 Method for controlling tracking actuator *Advances in Automation. RusAutoCon 2019 (Lecture Notes in Electrical Engineering* vol 641) ed A Radionov, A Karandaev (Springer, Cham) pp 26-34

[16] Khaziev E L and Khaziev M L 2019 Intelligent diagnostic system for hydraulic actuator *2019 International Conference on Industrial Engineering, Applications and Manufacturing, ICIEAM 2019* 8742779.

[17] Bougouffa S, Vogel-Heuser B, Fischer J, Schaefer I and Li H 2019 Visualization of variability analysis of control software from industrial automation systems *Conf. Proc. - IEEE Int. Conf. on Systems, Man and Cybernetics, 2019-October* pp 3357-64

[18] Valiev R A, Khuzyatov Sh Sh and Galiullin L A 2019 *Journal of Advanced Research in Dynamical and Control Systems* **11(S08)** 1762-6

[19] Khalil M 2013 Pattern-based methods for model-based safety-critical software architecture design *Lecture Notes in Informatics (LNI), Proc.Ser. of the Gesellschaft fur Informatik (GI)* 493-9

[20] Valiev R A and Khuzyatov S S 2016 Pattern-design software of automated control systems *2016 2nd Int. Conf. on Industrial Engineering, Applications and Manufacturing, ICIEAM 2016 Proc.* 7910942

[21] Sitdikov I M, Farkhetdinov R S, Jakupov F A 2011 *Sovremennye Tehnologii Avtomatizatsii* **1** 54-9

[22] Iliukhin A N, Khuzyatov S S and Valiev R A 2018 *HELIX* **8(1)** 2455-9

[23] Khuzyatov S and Valiev R 2019 The method of automated configuration objects of the WinCC project for the oil and gas industry *Advances in Automation. RusAutoCon 2019 (Lecture Notes in Electrical Engineering* vol 641) ed A Radionov, A Karandaev (Springer, Cham) pp 986-93

[24] Sheng Z, Ji C and Hua S 2018 Application of Siemens PLC and WinCC in the monitoring-control system of bulk grain silo *Proc. 30th Chinese Control and Decision Conf., CCDC 2018* pp 4689-93

[25] Zhang W and Sun S 2018 Design of automatic feeding control system in tank area based on Siemens PLC *Proc. 30th Chinese Control and Decision Conf., CCDC 2018* pp 3622-7

[26] Guerra-Zubiaga D A, Bondar A, Escobedo G and Schumacher A 2019 Digital twin in a manufacturing integrated system: Siemens TIA and PLM case study *ASME Int. Mechanical Engineering Con. and Exp., Proc.*, vol 2B-2019 (ASME)

[27] Berger H 2013 *Automating with Simatic: Controllers, Software, Programming, Data*

*Communication, Operator Control and Process Monitoring* (Erlangen: Publicis Publishing) p 289

IOP Conf. Series: Materials Science and Engineering **915** (2020) 012057 doi:10.1088/1757-899X/915/1/012057

# Quantum chemical simulation for calculation of processes of sulfur composite materials

**A A Yusupova[1], A Y Akhmetova[1], L R Baraeva[1], R T Akhmetova[1], A I Khatsrinov[1] and A A Bobyshev[2]**
[1] Kazan National Research Technological University,
Karl Marx street 68, Kazan, 420015 Tatarstan, Russia
[2] Naberezhnye Chelny Institute (branch), Kazan Federal University
Sjujumbike street 10a, Naberezhnye Chelny, 423800 Tatarstan, Russia

misharin_82@mail.ru

**Abstract**. Sulfide composite materials are becoming more widespread as they have a number of advantages such as strength, density, water resistance, and resistance to aggressive environments due to a chemical interaction between components. To ensure this, amorphous silica and silicon tetrachloride activator are used (in our opinion) to encourage electrophilic opening of sulfur molecules and formation of reactive sulfur radicals in order to increase the probability of chemical interaction between components in a system. The paper provides the quantum chemical calculations to demonstrate the interaction mechanism for components in a system.

**1. Introduction** The known sulfur materials feature only mechanical mixing of sulfur and mineral components. In our opinion, chemical bonding of components with the formation of sulfides will have a positive influence on the strength properties of materials, and environmental and health industrial safety. In that respect, it is reasonable to use amorphous silica as a mineral component, which has extended specific surface area and a large number of active sites on the surface. However, additional activation is required to ensure a chemical interaction between components due to a high energy barrier. A Lewis acid can be used as an activator. It attaches to the surface of amorphous silica (atomic layer deposition), thus increasing the activity of a mineral component. Today, there is a lot of information about atomic layer deposition in academic literature. The principles of atomic layer deposition technique based on Aleskovsky matrix hypothesis were described by V.B. Aleskovsky and S.I. Koltzov [1, 2]. The fundamental idea behind atomic layer deposition technique lies in a gradual growth of monolayers of structural units with a given chemical composition and structure on the surface of solid phase matrix due to chemical reactions between functional groups of a solid body and added reagents as far as possible from equilibrium [1-3].

Atomic layer deposition technique makes it possible to synthetize nanostructures with various chemical compositions on the surface of solid phase matrix, and to perform atom-by-atom chemical assembly of surface nano-, micro-, and macrostructures by repeatable alternation of chemical reaction following a set program [4, 5]. It should be noted that atomic layer deposition technique ensures the formation of nanolayers on the surface of solid substrate within the accuracy of one monomolecular layer. And the thickness of nanolayer is defined not by reagent omission time but by a number of atomic layer deposition cycles that includes a set of certain chemical reactions. Therefore, there are elements of self-organization that are observed during formation of the first and next monomolecular layers in the process of atomic layer deposition.

The theoretical approaches and, in particular, quantum chemical methods become more significant when addressing critical challenges in the field of synthesis of new composites. The results of theoretical studies on inorganic compound structure and reactivity often complement data of experimental studies, simplify their interpretation, and make it possible to get important information.

For this paper, quantum chemistry methods were used to study the structures of reagents, geometry of systems, reaction mechanisms. A reaction mechanism is a sequence of unit steps that add to conversion of initial reagents to products. A structure and energy of reagents and reaction products, as well as transient state that separates them should be known for each unit step. Today's quantum chemical study methods for electronic structure and geometry of molecules, which are implemented in software such as Gaussian and Priroda, make it possible to conduct a simulation experiment and obtain data about the structure and energy of transient states, as well as the structure and geometry of reagents and reaction products. These software products can calculate a reaction path - lowest energy line that connects reagents and products and goes through a transient state [3].

The calculations were performed using Priroda application software (basis1.bas). The unrestricted Hartree-Fock method was often used to estimate radical decomposition reactions and study the structure of biradical transit state. It involved mixing of the highest occupied and lowest unoccupied molecular orbitals at the start of calculations. So, the Priroda can be used for qualitative evaluation of molecule geometry and energies. For studies of polyatomic systems, the Priroda software is more preferable as the calculations do not take a lot of computing time.

## 2. Body text

We know from the earlier papers that a direct interaction between sulfur and silica is impossible due to a very high activation energy for this reaction. Aluminum chloride can be used as an activator [6-8]. Our paper proposed that an activator of sulfur and amorphous silica surface should be used, and it would be silicon tetrachloride SiCl4.

We assumed that fragments of an electrophilic activator of silicon tetrachloride, which are attached to the surface of opal-cristobalite rock, would act as a bridge that connects opal-cristobalite rock and sulfur, and, therefore, ensure a chemical interaction between binder and filler, and high physical and mechanical properties.

The possible mechanism of polysulfide formation is shown in Figure 1.

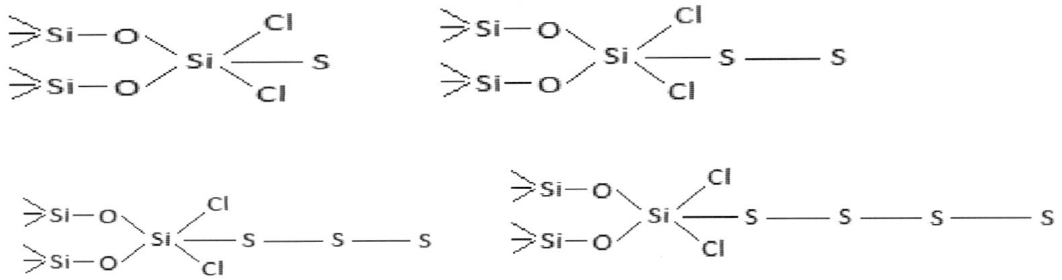


Figure 1. Mechanism of polysulfide formation with silicon tetrachloride activator

The quantum chemical calculations showed that the activation energy for addition of sulfur to amorphous silica decreased by 103.8 kJ/mol with silicon tetrachloride activator.

As a result, sulfide composite materials are formed, and they have high strength, density, water resistance, and resistance to aggressive environments.

## 3. Conclusions

IOP Conf. Series: Materials Science and Engineering **915** (2020) 012057 doi:10.1088/1757-899X/915/1/012057

Therefore, the theoretical quantum chemical methods confirmed that silicon tetrachloride activator decreased the consumption of energy for formation of sulfide composite materials with high physical mechanical and performance properties. These materials can be used for flagstones, curbstones, and poured composite floors resistant to aggressive environments for chemical plants.

**References**


[1] Foresman J B and Frish A 1996 Exploring Chemisnry with Electronic Stryctyre Methods (Pittsburgh PA: Gaussian Inc.) p 304

[2] Minkin V I, Simkin B Ya and Minyaev G M 1986 Quantum chemistry of organic compondounds. Reaction mechanisms (Moscow: Khimiya Publ.) p 248

[3] Stepanov N F 2001 Quantum mechanics and quantum chemistry (Moscow: Mir Publ.) p 390

[4] Sevryukov N N and Emelyanova V I 1981 About chemistry of hydrolysis of iron (III) sulfate in aqua solutions Journal of inorganic chemistry vol 26 Issue 5 pp 1281.

[5] Agaeva Z R, Annagiev M Kh and Zeynalov S D 2006 Influence of sulfate cathode part on their corrosion inhibition performance Condensed environments and interfaces vol 8 vol 1 pp 5-6

[6] Yusupova A A, Khatsrinov A I and Akhmetova R T 2018 Activating effect of aluminum chloride in the preparation of sulfur concrete from sulfur and silica Inorganic Materials vol 54 Issue 8 pp 809–14

[7] Kashapov N F, Nafikov M M, Gazetdinov M X, Nafikova M M and Nigmatzyanov A R 2016 Innovative production technology ethanol from sweet sorghum IOP Conference Series: Materials Science and Engineering vol 134 issue 1 8 number 012012

[8] Fedyaev V L, Galimov E R, Galimova N Ya, Gimranov I R and Siraev A R 2017 Dynamics of coalescence and spreading of liquid polymeric particles during coating formation IOP Conf. Series: Journal of Physics: Conf. Series vol 789 number 012006

IOP Conf. Series: Materials Science and Engineering **915** (2020) 012058 doi:10.1088/1757-899X/915/1/012058

# High-technology energy-efficient composite material made from waste of thermal power plants

**A A Yusupova[1], G A Medvedeva[2], R T Akhmetova[1] and A A Bobrishev[3]**

[1] Kazan National Research Technological University, Karl Marx street 68, Kazan, 420015 Tatarstan, Russia
[2]Kazan State University of Architecture and Engineering, ul. Zelyonaya 1, Kazan, 420043 Tatarstan, Russia
[3] Naberezhnye Chelny Institute (branch), Kazan Federal University, ul. Sjujumbike street 10a, Naberezhnye Chelny, 423800 Tatarstan, Russia

misharin_82@mail.ru

**Abstract**. Polysulfide composite materials made from high tonnage waste of oil and gas and thermal power sectors were developed and studied. The composition of these materials was optimized. It was shown that the components interacted with each other, and a compact homogeneous structure was formed when pyrite was added.

**1. Introduction** Today's machine building industry is a complex industry that involves production of various machines, tools, instruments, as well as consumer items and defense products. A wide range of aggressive environments is used in advanced processes in machine building plants. All these factors lead to accelerated mechanical, physical chemical failure, including corrosion, which results in a lower load capacity of individual structures (foundations for acid pumps, some floor sectors, etc.) and building as a whole. However, there is a high quantity of composite sulfur materials that have a number of advantages: high chemical inertness in aggressive environments with operating parameters, high mechanical performance, low cost, and accessibility to materials [1-3].

**2. Body text** The paper studies polysulfide composite materials made from high tonnage waste of oil and gas and thermal power sectors. Volumes of ash and slag waste are among the highest. They are generated with combustion of solid fuels (various types of coal, oil shale, peat) in thermal power stations. Enormous amounts of ash and slag waste are accumulated in dumping grounds, they take up valuable lands and are expensive to maintain. Meanwhile, ash and slag from thermal power plants can be effectively used to manufacture various building materials, which has been confirmed by many scientific studies and practical experience [4].

Sulfur from gas and oil processing sector is among high tonnage waste. In Russia, a significant amount of by-product sulfur is accumulated in the dumping grounds of Astrakhan gas processing plant. In Tatarstan, Minnibaevsky gas processing plant generates more than 300 tons of sulfur waste annually, and Nizhnekamsk gas processing plant - more than 30 k tons of sulfur. Developing sulfur composite material technologies is relevant today as these materials have a number of assets: strength, abrasion resistance, water resistance, acid resistance, etc. [5-8]. A proportion of low-cost by-product sulfur is increasing so production of materials with it is economically feasible.

IOP Conf. Series: Materials Science and Engineering **915** (2020) 012058 doi:10.1088/1757-899X/915/1/012058

For the purpose of the paper, the following materials were used: sulfur waste from the Nizhnekamsk oil refinery, with 99.9% of sulfur; pyrite $FeS_2$ (iron sulfide); masonry sand (GOST 8736-93); ash and slag waste from KAZAN thermal power plant (TETs-2) with the following composition (% wt.): $SiO_2$ – 47.7-52.2; $Al_2O_3$+$TiO_2$ – 21.24-25.28; CaO+ MgO –- 4.3; $Fe_2O_3$ – 5.2-5.9; $R_2O$ – 1.84-19.03; $SO_3$ – 0.2.

For more detailed study, the ash and slag waste was separated into ash and slag constituents by screening. For the purpose of the paper, ash constituents with a particle size of less than 1 mm were used. The samples of sulfur composite materials were made by hot mixing of initial components with different holding time (from 10 minutes to 4 hours). Then these mixtures were formed to make the samples. Physical mechanical properties of the resulting materials were tested, and the materials were studied using the methods of physical-chemical analysis. In terms of the material strength properties, the sulfur-ash and slag waste ratio of 1:1.5 is most suitable.

It is commonly known that crystalline sulfur materials have insufficient strength, impact resistance, etc. due to its shrinking deformation tendency. It can be assumed that modification of a sulfur component with metal containing compounds will encourage the formation of a compact homogeneous structure in composite material and, therefore, improvement of physical mechanical characteristics as a result of formation of sulfur (polysulfide) substances, elimination of irregularities of component densities, improvement of adhesion characteristics.

Sulfur (polysulfide) substances with metal-containing compounds (such as iron sulfide (pyrite)) are interesting from a point of view of scientific studies and practical application. Each atom of iron in a pyrite group is located at the center of octahedron made of six atoms of sulfur, and sulfur coordination is tetrahedral. The length of S-S bond is 217 pm. Pyrite is an electron donor, i.e. using it as a modifier makes it possible to get polysulfur and, therefore, to increase mechanical and impact product strength due to the formation of new bonds.

It is known that normally inert sulfur becomes reactive with nucleophilic and electrophilic reagents. As pyrite, iron polysulfide, is a compound with disulfide bonds, it can act as a nucleophilic reagent. Using pyrite as a nucleophilic additive makes it possible to increase reactivity of sulfur, and it can interact with filler components and additive, i.e. it will encourage the increase in strength of the samples.

In order to investigate the mechanism of interaction between components in a system and assess the influence of iron sulfide on activation of sulfur ring opening, quantum chemical calculations were performed for sulfur-iron sulfide system (with density functional method using Priroda software, with 3z.bas basis, including p- и d-orbitals in atoms). As there is hexatomic sulfur in the melt at a temperature of 180 °C the calculations were performed for ring hexatomic sulfur, its geometry is shown in Figure 1.

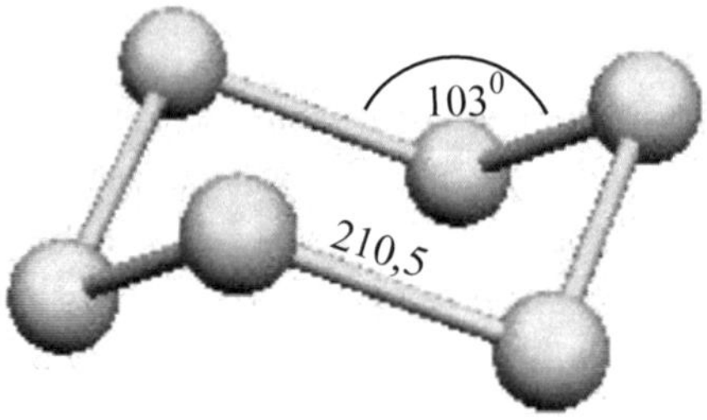


Figure 1. Geometry of $S_6$ ring

IOP Conf. Series: Materials Science and Engineering **915** (2020) 012058 doi:10.1088/1757-899X/915/1/012058

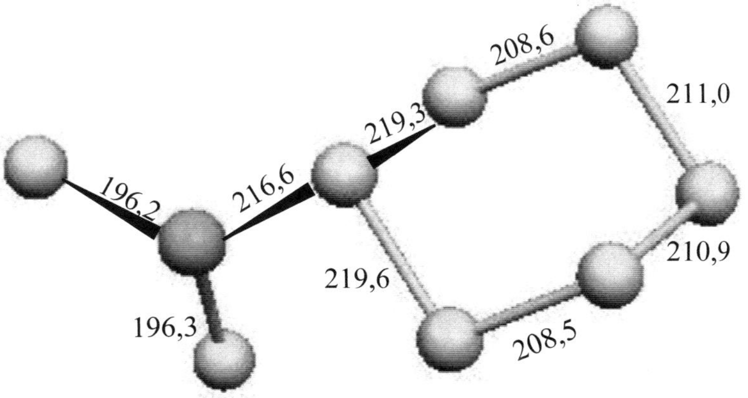


Figure 2. Scheme of $FeS_2$ addition to $S_6$ ring (iron sulfide $FeS_8$)

It was established that addition of pyrite to sulfur molecule led to a decrease in bond energy in sulfur ring by 62.6 kJ/mol. Sulfur-sulfur bond becomes loose and lengthens from 210.5 pm to 219.6 pm (Figure 2). The reaction is exothermic ($\Delta H$=-237.4 kJ/mol) with the formation of a strong covalent bond. The energy of iron-sulfur bond is 237.76 kJ/mol, the length of the bond is 216.6 pm. In this case, when iron interacts with sulfur it changes from two-coordinate state to three-coordinate one. It can be assumed that iron polysulfides with various degrees of condensation, probably, of cross-linked structure, are formed. They will encourage the formation of a compact homogeneous material with high physical chemical and performance properties. Therefore, the quantum chemical calculations showed that pyrite would encourage the loosening of bonds in sulfur rings, opening of sulfur rings, and formation of iron polysulfides with strong covalent bonds. These factors ensure that a material is strong and chemically stable.

Mechanical and performance properties of the resulting samples were tested: compressive strength (37-40 MPa), water absorption (maximum 1 percent), density (2.57-2.70 g/cm$^3$), resistance to aggressive environments (0.85-0.97).

## 3. Conclusions

Therefore, sulfur composite materials were made from petrochemical by-products using metal containing modifier (pyrite with high strength properties). These materials can be used as road building materials (paving stones, blocks) to equip areas subjected to aggressive environments in premises or inner facilities in machine building plants.

## References

[1] Korolev E V, Proshin A P, Erofeev V T, Khrulev V M 2003 Building sulfur materials Penza PGUAS Publ. p 372
[2] Akhmetova R T, Yusupova A A, Baraeva L R, Sabakhova G I, Pervushina V A, Khatsrinov A I, Medvedeva G A and Akhmetova A Yu Binder RF patent No 2015 2555166
[3] Proshin A P, Korolev E V and Kalinkin E G 2005 Structure and properties of modified sulfur binder Stroitelnye Materialy vol 7 pp 6-9
[4] Karnaukhov Yu P, Sharova V V and Podvolskaya E N 1998 Binders based on nonutilizable ash and slag mixture and liquid silica glass Stroitelnye Materialy vol 5 pp 12-13
[5] Paturoev V V 1987 Polymer concretes (Moscow: Stroyizdat Publ.) p 286
[6] Yusupova A A, Khatsrinov A I and Akhmetova R T 2018 Activating effect of aluminum chloride in the preparation of sulfur concrete from sulfur and silica Inorganic Materials vol 54 Issue 8 pp 809–14
[7] Akhmetova R T and Gerasimov V V 2002 Sulfur composite materials for thermal power sector Izvestya VUZov. Problemy energetiki vol 1-2 pp 41-3

[8] Fedyaev V L, Galimov E R, Galimova N Ya, Gimranov I R and Siraev A R 2017 Dynamics of coalescence and spreading of liquid polymeric particles during coating formation IOP Conf. Series: Journal of Physics: Conf. Series vol 789 number 012006

IOP Conf. Series: Materials Science and Engineering **915** (2020) 012059 doi:10.1088/1757-899X/915/1/012059

# The laser heat-treatment process control by the non-stationary convective heat exchange modeling

**V V Zvezdin, R R Saubanov, R R Rakhimov, R M Khisamutdinov, A V Boldyrev, D L Karelin**

Kazan Federal University, Naberezhnye Chelny Institute, Naberezhnye Chelny, Russia

saubanov@mail.ru, rafisih88@mail.ru

**Abstract.** Various technological processes allow to obtain high quality indicators, such as laser cutting, welding, hardening, etc. The article presents the results of modeling for laser cutting of thick-walled structures. The creation of computational tools for express analysis in a mathematical setting and a quantitative assessment of temperature fields in spatially variable and non-stationary conditions of laser exposure are of undoubted practical interest to minimize experimental costs.

1. Introduction.Nowadays, the laser technological complexes(LTC) with high density of radiation's power are actively being under development and explored for solving current problems of laser influence in the concentrated flows of energy technology. The exploring dynamics of temperature fields in the zone of laser influence and multifactor optical and thermophysical processes is necessary not only for the quantitative description of phase transition "solid-gas-plasma", but also for controlled modification of the properties of structural materials. There are complex physical and chemical processes occur in the zone of influencing the laser radiation (LR) by implementation all of development cycles and engineering optimization of the technological parameters, generating system processes by energy, and the systems of control in real-time. For the relatively well explored metals the results of exploring the 3D temperature fields and conjugated with them consumable masses characteristics of irradiated metals, spectral-energy melting of thresholds, evaporation of ionisation, obtained by various theoretical and experimental groups, significantly differ. The creating extension tools of express analysis in mathematical statement and quantification of temperature fields in space-time non-stationary conditions of the laser influence are of undoubted practical interest for the minimizing experimental expenses [1-4, 6].

2. Theoretical and experimental research

The features of the LR process and the results of the research should be taken into account by modeling.

The laser cutting process modeling of 17Г1С steel was implemented using a multiphase liquid model (MMJ) based on the method of fixing the phase interface under the influence of a concentrated energy beam and determining the distribution and movement of immiscible phases. This modeling approach assumes that the grid resolution is sufficient to resolve the position and shape of the interface between phases [3, 5-10].

To determine the phase distribution, the mass conservation equation of the phase bulk part was used:

$$\frac{\partial}{\partial t}\int_V a_i dV + \oint_A a_i v da = \int_V \left( S_{a_i} - \frac{a_i}{\rho_i}\frac{D\rho_i}{Dt} \right) dV - \int_V \frac{1}{\rho_t}\nabla\left(a_i \rho_i v_{d,i}\right) dV \quad (1)$$

where A is the vector to the area of the heat transfer surface, v is the mixture velocity (averaged by mass), and is the diffusion velocity, and is the user-defined initial member of phase i or the Lagrangian derivative of the phase densities .

The total mass for all phases is determined using the continuity equation:

$$\frac{\partial}{\partial t}\left( \int_V \rho dV \right) + \oint_A \rho v \cdot da = \int_V S dV \quad (2)$$

where $S$ – a mass source element associated with a phase source element as follows:

$$S = \sum_i S_{a_i} \cdot \rho_i$$

The dependence on the volume fractions of the mixture substance constituent phases are taken into account through density.

The phase velocity was calculated using the mass momentum equation for the corresponding volume part of the phase:

$$\frac{\partial}{\partial t}\left( \int_V \rho v dV \right) + \oint_A \rho v \otimes v \cdot da = -\oint_A pI \cdot da + \oint_A T \cdot da + \quad (3)$$

$$+ \int_V \rho g dV + \int_V f_b dV - \sum_i \int_A a_i \rho_i v_{d,i} \otimes v_{d,i} \cdot da$$

where $p$ – pressure, $I$ –the combined tensor ,$T$ - stress tensor ,$f_b$ - vector mass energy.

The calculation of the temperature field when the material is exposed to concentrated radiation due to the thermal conductivity of the material was performed using the energy equation

$$\frac{\partial}{\partial t}\left( \int_V \rho E dV \right) + \oint_A \left[ \rho H v + p + \sum_i a_i \rho_i H_i v_{d,i} \right] \cdot da = \quad (4)$$

$$= -\oint_A q'' \cdot / da + \oint_A T \cdot v da + \int_V f_b v dV + \int_V S_E dV$$

where E -total energy, H-total enthalpy, - the heat flow vector, - the consumer of the energy source.

Mathematical modeling of laser cutting of 17Г1С steel using the STAR-CCM+v13 software package was performed using equation 4. The simulation results are shown in figures 1 and 2

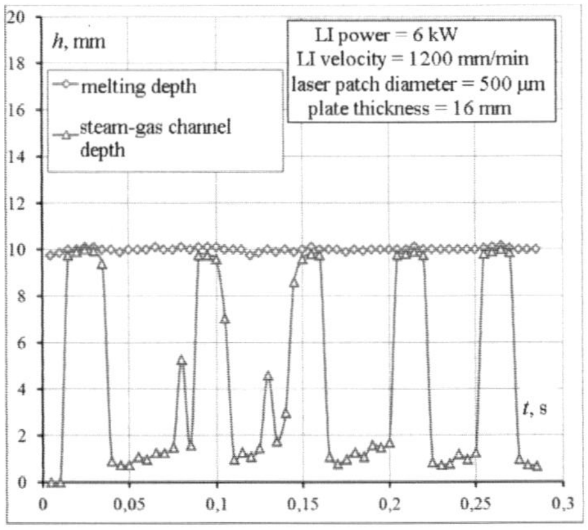


**Figure 1.** Graph of the steam-gas channel depth dependence at the LI power for 6 kW on the exposure time

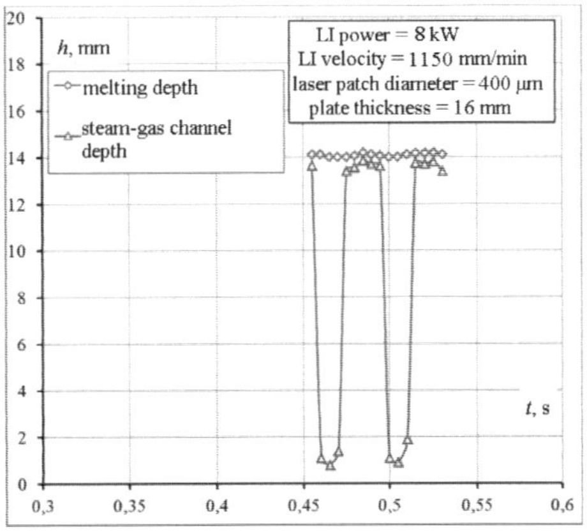


**Figure 2.** Graph of the steam-gas channel depth dependence at the power of LI for 8 kW on the time of exposure

Based on the results (see Fig. 3-6). the presence of unstable pulsation is noted: for 6kW with a frequency of 17-20 Hz; for 8kW with a frequency of 25-29 Hz

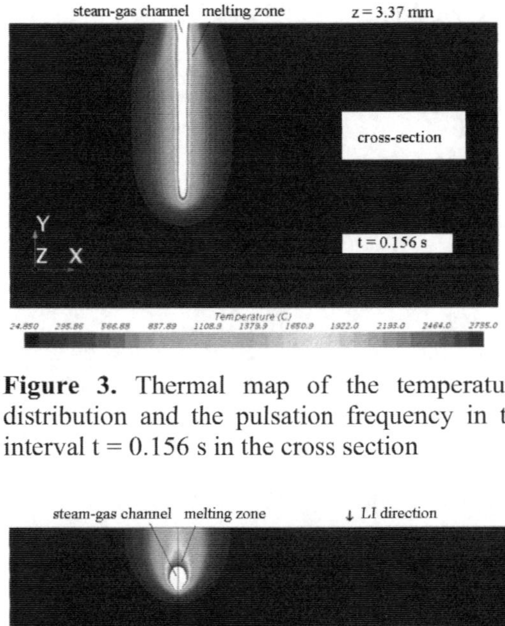


**Figure 3.** Thermal map of the temperature field distribution and the pulsation frequency in the time interval t = 0.156 s in the cross section

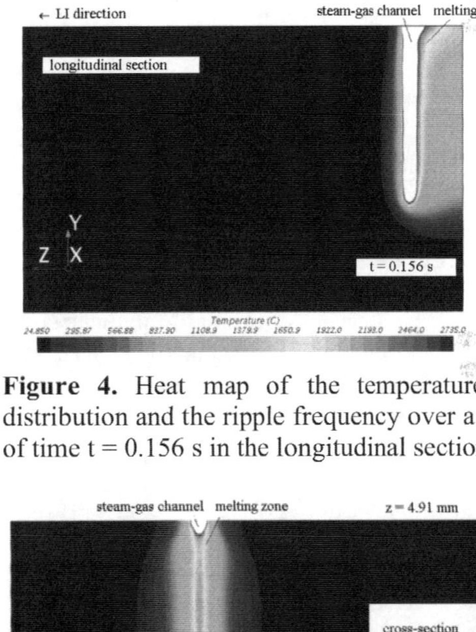


**Figure 4.** Heat map of the temperature field distribution and the ripple frequency over a period of time t = 0.156 s in the longitudinal section

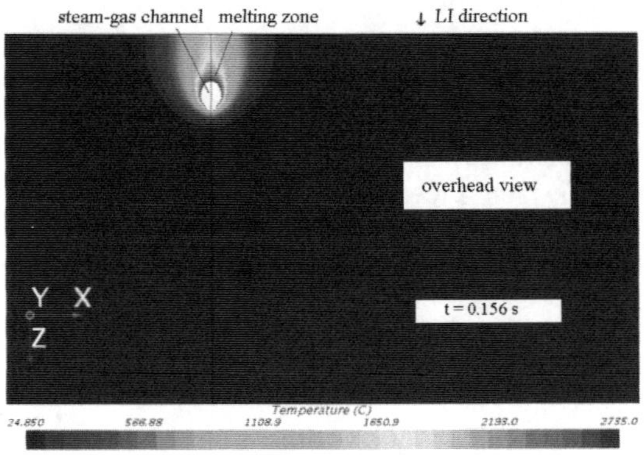


**Figure 5.** Heat map of the temperature field distribution and the ripple frequency over a period of time t = 0.156 s on the front view

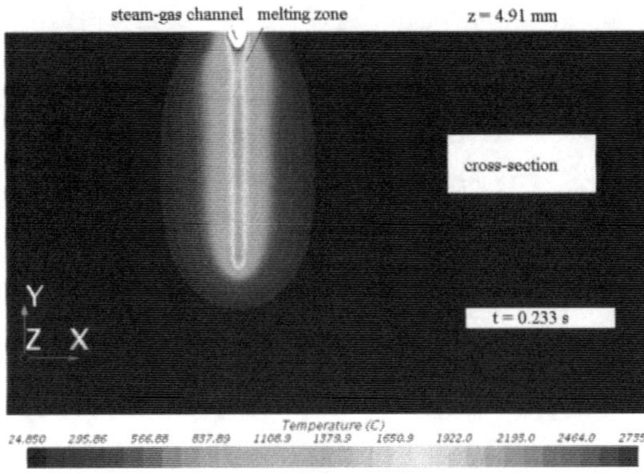


**Figure 6.** Heat map of the temperature field distribution and the ripple frequency over a period of time t = 0.233 s in cross section

IOP Conf. Series: Materials Science and Engineering **915** (2020) 012059 doi:10.1088/1757-899X/915/1/012059

## 3. Conclusion

From experimental studies, it is known that the spread of the liquid phase from the cutting front occurs in all possible directions: up, sideways, and, if the cut is through, down. As a result of double "support" in this part of the cut, a zone of thermal influence of such a depth can develop, the presence of which, as noted at the beginning of the work, is unacceptable in most cases of using laser cutting in manufacturing technologies for those products in which it is either the most effective or the only possible tool.

## References


[1] Kuznetsov I N, Zvezdin V V, Israfilov I H, and Portnov S M 2014 Automatic control system of high precision welding of workpieces in mechanical engineering IOP Conference Series: Materials Science and Engineering Volume 69, Issue 1 Article number 012029.

[2] Valiev R A, Bochkov V E, Bashkirov Sh Sh, Romanov E S and Chistjakov V A 1991 Mössbauer study of surface layers of high-speed steel after laser treatment Hyperfine Interactions 69(1-4) Pages 589-92.

[3] Grigoryants A G, Perestoronin A V, Portnov S M, Zvezdin V V and Israfilov I H 2015 A system for automatic control of precision laser welding in engineering Welding International T. 29, number 10 pp 801-4.

[4] Zvezdin V V, Hisamutdinov, R M, Rakhimov R R, Israfilov I H and Saubanov R R 2018 Technology of overlay laser welding of durable powdery into blade edge of miller IOP Conference Series: Materials Science and Engineering Volume 412, Issue 1 Article number 012083.

[5] Rakhimov R R, Zvezdin V V, Israfilov I H, Nabiullina G I Saubanov R R 2014 Modeling of the laser heat treatment News of the Tula State University (Technical science) number 11-2 pp 476-484.

[6] Boldyrev S V, Boldyrev A V 2014 Numerical experiment optimization to obtain the characteristics of the centrifugal pump steps package IOP Conference Series: Materials Science and Engineering Volume 69, Issue 1 Article number 012005.

[7] Zvezdin V V, Hisamutdinov, R M, Saubanov R R, Rakhimov R R and Sirkin S.S. 2019 Technological complexes of laser welding and hardening tool IOP Conference Series: Materials Science and Engineering Volume 570, Issue 1 Article number 012110.

[8] Rakhimov R R, Saubanov R R, Zvezdin V V, Hisamutdinov, R M. 2019 Surface quality provision of laser welding using plasma cleaning Journal of Physics: Conference Series Volume 1328, Issue 1 Article number 012064.

[9] Zvezdin V V, Rakhimov R R, Saubanov R R, Israfilov I H and Akhtiamov R F 2017 Management of laser welding based on analysis informative signals IOP Conference Series: Materials Science and Engineering Volume 240, Issue 1 Article number 012073.

[10] Zvezdin V V, Klochkova K V, Rakhimov R R, Saubanov R R, Israfilov I H and Pesoshin V A 2018 Technology ability of laser bonding of compacted graphite iron IOP Conference Series: Materials Science and Engineering Volume 412, Issue 1 Article number 012084.

[11] Zvezdin V V, Khisamutdinov R M, Rakhimov R R, Israfilov I H and Akhtiamov R F 2017 Method of control position of laser focus during surfacing teeth of cutters IOP Conference Series: Materials Science and Engineering Volume 240, Issue 1 Article number 012072.

# AUTHOR INDEX


Abdullin, T M ........ 1, 5
Abdullina, A R ........ 208
Abyzbekova, G M ........ 72, 76, 211
Abzhaev, M M ........ 24
Adigamov, N R ........ 192
Agashkin, D A ........ 215
Akhmerov, A V ........ 36
Akhmetov, F M ........ 16
Akhmetova, A Y ........ 246
Akhmetova, R O ........ 249
Akhmetova, R T ........ 246
Akhmetshin, R S ........ 111
Andreev, N K ........ 139
Antonov, A V ........ 11
Astashchenko, V I ........ 171
Aukhadeyev, A E ........ 203
Bagdanurov, F F ........ 16
Balabanov, I P ........ 19, 59
Balabanova, O N ........ 19
Baraeva, L R ........ 246
Baryshnikova, E P ........ 147
Bayniyazova, A T ........ 24
Belousov, A M ........ 27
Bobrishev, A A ........ 249
Bobyshev, A A ........ 246
Boldyrev, A V ........ 27, 253
Boldyrev, S V ........ 27
Butakov, V M ........ 199
Cherepen'kin, I V ........ 135
Chernov, V A ........ 89, 98
Davletshina, G K ........ 131
Degtyaryova, N G ........ 189
Dolgova, A N ........ 36, 40
Dudarovskaya, O G ........ 36
Fandeyev, V P ........ 135, 199
Fashutdinov, A I ........ 59
Filina, O A ........ 139
Gabdrakhmanov, Az T ........ 44, 48
Gabdrakhmanova, T F ........ 44, 48
Gabrakhmanov, A T ........ 215
Gafiatullin, A A ........ 147
Galeev, I A ........ 67
Galeev, R D ........ 52
Galeeva, A I ........ 63, 67, 72, 76, 211
Galiakbarov, AT ........ 48
Galimov, E R ........ 175
Gavariev, R V ........ 186
Gazetdinov, M X ........ 101
Gazetdinov, Sh M ........ 101
Gilman, V N ........ 19, 59
Gilmanshin, I R ........ 1, 5, 52, 63, 67, 72, 76, 211, 235
Gilmanshina, S I ........ 52, 67, 72, 76, 211
Haibullina, A I ........ 40
Hairullin, R A ........ 223
Idiyatullin, R G ........ 203
Ilyushin, O V ........ 203
Israphilov, I H ........ 44
Ivanova, V A ........ 189
Kadyirov, A I ........ 81
Kadyrmetov, A M ........ 31
Kapitonova, G M ........ 231
Karaeva, J V ........ 81
Karasev, I Yu ........ 219
Karasev, I Y ........ 48
Karataev, O R ........ 85
Karataeva, E S ........ 36, 40, 85, 157
Karelin, D L ........ 27, 253
Kashapov, N F ........ 52, 89, 94, 98, 101, 119, 128
Kasyanov, S V ........ 106
Khafizov, A A ........ 111
Khafizov, I I ........ 114, 119, 123, 128, 151
Khamidullina, D A ........ 157
Khamidullina, G R ........ 128
Khatsrinov, A I ........ 246
Khisamutdinov, R M ........ 253
Khizbullin, R N ........ 135
Khizbullin, Rad N ........ 208
Khizbullin, Rob N ........ 208
Khusainov, R M ........ 131
Khusnutdinov, A N ........ 135, 139, 165, 196, 199, 203, 208
Khusnutdinova, E M ........ 81, 128, 135, 139, 144, 165, 168, 196
Khuzyatov, Sh Sh ........ 241
Kisneeva, L N ........ 203
Kolesnikov, M S ........ 171
Kolomeichenko, A V ........ 192
Konakhina, I A ........ 128, 144
Kondrasheva, S G ........ 157
Korchagin, O P ........ 215
Kosmodemyanskaya, S S ........ 63
Kravchenko, I N ........ 192
Kulakov, A T ........ 147
Kuzmina, I A ........ 151, 154
Kuznetsov, I S ........ 192
Lashkov, V A ........ 157
Litvinenko, R S ........ 203
Makarova, O I ........ 98

Malachowski, A F ... 147
Mamonova, A O ... 165
Mandrykin, I A ... 31
Medvedev, V V ... 89, 98
Medvedeva, G A ... 249
Mingazov, R A ... 94
Minnakhmetova, V A ... 72
Moliboga, M N ... 168
Mukha-Metzyanov, I R ... 171
Mukhametzyanova, G F ... 171
Murataev, F I ... 175
Nafikov, M M ... 89, 94, 98, 98, 101
Nazarov, F L ... 147
Nigmatullin, M D ... 89, 98
Nigmatzyanov, A R ... 89, 94, 98, 101
Nizamov, R S ... 223
Novikov, V F ... 85
Nurullin, I G ... 114, 119, 123
Osetinskiy, G V ... 139
Osipov, A L ... 36
Pankratov, D L ... 186, 223
Panov, A G ... 189
Pavlov, P P ... 135, 199, 208
Plahotin, A A ... 31
Plugatyreva, O A ... 196
Puryaev, A S ... 11
Rakhimov, R R ... 253
Romanov, V V ... 31
Romanova, N V ... 215, 219
Rudenko, A S ... 40
Ryzhkova, S A ... 189
Sabitov, L S ... 199, 203, 208
Saduakaskyzy, K ... 63, 211
Sagitova, R N ... 67, 211
Saubanov, R R ... 253
Sayakhov, R I ... 85
Sayfutdinov, R F ... 94
Shaekhova, I F ... 189
Shafigullin, L N ... 16, 44, 215, 219
Shamsutdinov, I R ... 215
Sharifullin, S N ... 24, 31, 192
Shibakov, V G ... 223
Simonova, L A ... 231
Siyetinskaya, A S ... 199
Snarsky, S V ... 147
Snigireva, Yu V ... 36
Snyatkov, E V ... 31
Solovyev, R Y ... 192
Timerbaev, I F ... 235
Valiev, R A ... 241
Valiev, R I ... 111
Valinova, I E ... 219
Velichko, S A ... 192
Yashagina, A V ... 139
Yunak, S A ... 40
Yusupova, A A ... 246, 249
Zdor, V O ... 81
Ziangirova, E R ... 131
Zonina, S V ... 219
Zvezdin, V V ... 253

**Institute of Physics**
Dirac House, Temple Back
Bristol BS1 6BE UK

ISSN: 1757-8981
ISBN 978-1-7138-4758-8

CURRAN ASSOCIATES INC.
proceedings
.com